The RISE *and* REIGN *of the* MAMMALS

ALSO BY STEVE BRUSATTE

The Rise and Fall of the Dinosaurs

The

RISE

and

REIGN

of the

MAMMALS

A New History,
from the Shadow of the Dinosaurs to Us

STEVE BRUSATTE

WITH ILLUSTRATIONS BY
TODD MARSHALL AND SARAH SHELLEY

MARINER BOOKS
New York   Boston

HarperCollins books may be purchased for educational, business, or sales promotional use. For information, please email the Special Markets Department at SPsales@harpercollins.com.

FIRST EDITION

Designed by Bonni Leon-Berman
Illustrations by Todd Marshall

Library of Congress Cataloging-in-Publication Data has been applied for.

ISBN 978-0-06-295151-9 (hardcover)
ISBN 978-0-06-295156-4 (international edition)

22 23 24 25 26   LSC   10 9 8 7 6 5 4 3 2

For Anthony, my favorite little mammal

CONTENTS

TIMELINE OF MAMMALS

PHANEROZOIC															
PALEOZOIC				MESOZOIC			CENOZOIC								
								Paleogene			Neogene		Quaternary		

Carboniferous											
Mississippian	Pennsylvanian	Permian	Triassic	Jurassic	Cretaceous	Paleocene	Eocene	Oligocene	Miocene	Pliocene	Pleistocene	Holocene

359

299

252

201

145

66

56

34 23

5 2.6

.010 (i.e. 10,000 years ago)

First synapsids (stem mammals), as mammal and reptile lineages split from each other

First mammals

Mammals diversify but stay small as dinosaurs rule

Asteroid kills dinosaurs, mammals survive and get larger and diversify

More "modern" placental mammal groups diversify, including true primates

First hominims (human group)

Last Ice Age

First *Homo sapiens*

numerical age

MAMMAL FAMILY TREE

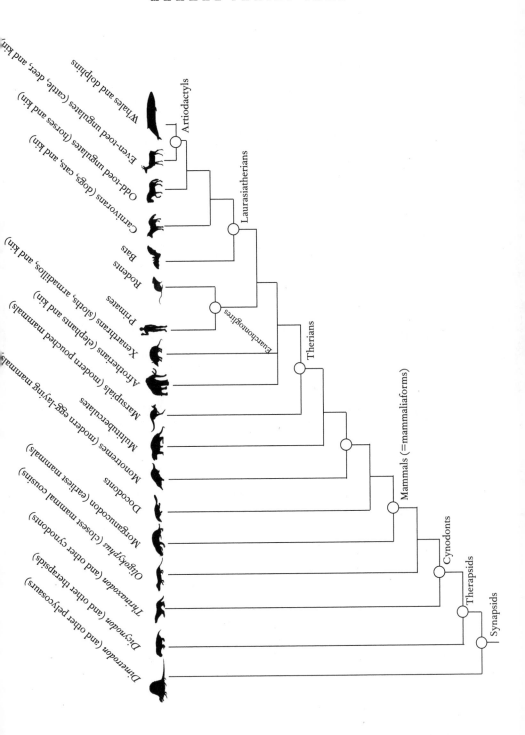

Artiodactyls

Whales and dolphins

Even-toed ungulates (cattle, deer, and kin)

Odd-toed ungulates (horses and kin)

Carnivorans (dogs, cats, and kin)

Bats

Rodents

Primates

Xenarthrans (sloths, armadillos, and kin)

Afrotherians (elephants and kin)

Marsupials (modern pouched mammals)

Multituberculates

Monotremes (modern egg-laying mammals)

Docodonts

Morganucodon (earliest mammals)

Oligokyphus (closest mammal cousins)

Thrinaxodon (and other cynodonts)

Dicynodon (and other therapsids)

Dimetrodon (and other pelycosaurs)

Laurasiatherians

Euarchontoglires

Therians

Mammals (=mammaliaforms)

Cynodonts

Therapsids

Synapsids

MAPS OF EARTH OVER TIME

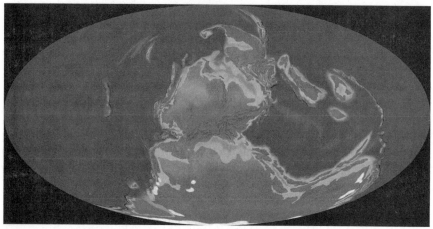

320 million years ago, Carboniferous (Pennsylvanian)

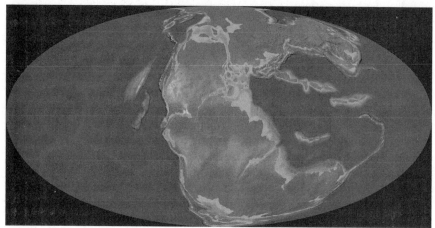

200 million years ago, Triassic-Jurassic Boundary

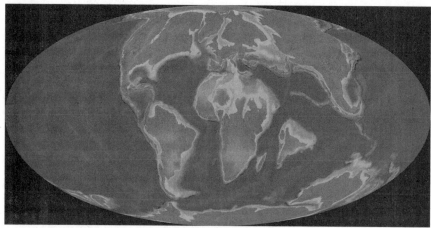

66 million years ago, end of the Cretaceous, time of asteroid impact

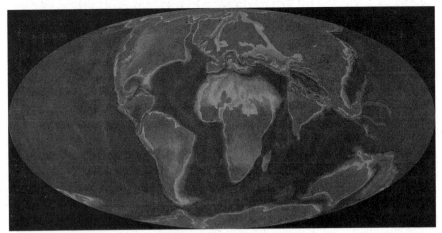

50 million years ago, Eocene

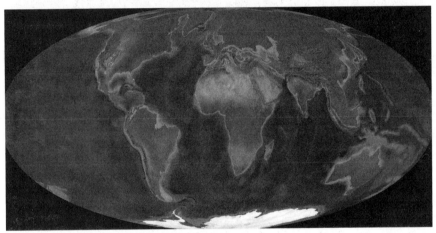

20 million years ago, Miocene

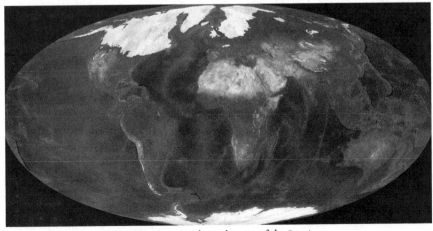

21,000 years ago, last advance of the Ice Age

INTRODUCTION

Our Mammalian Family

FOR THE FIRST TIME IN years, the sun broke through the darkness. There was still a whiff of smoke wafting from the gray clouds, which blanketed the ground in shadow. Down below, the land was wrecked. It was all dirt and mud, a wasteland absent any greenery or color whatsoever. Silence hung in the wind, punctured only by the churn of a river, its currents clogged with sticks and stones and the residue of decay.

The skeleton of a beast lay upon the riverbank. Its flesh and sinew were long gone, its bones a moldy beige. Its jaws were agape in a scream, its teeth busted and scattered in front of its face. Each one the size of a banana, with the sharp edges of a knife, the murder weapons this monster had used to dismember and crush the bones of its prey.

It was, once, a *Tyrannosaurus rex*, the tyrant lizard, the King of the Dinosaurs, the oppressor of a continent. Now its entire species was no more. And little else seemed to be alive.

Then, from somewhere within the behemoth, a soft sound. A clicking chatter, a flutter of footsteps. A tiny nose poked out between a couple of *T. rex* ribs, haltingly, as if afraid to go any farther. Its whiskers trembled, in expectation of danger, but it found none.

Time to come out of hiding. It leapt upward, into the light, and scurried onto the bones.

Clothed in fur, with bulging eyes and a snout full of teeth that looked like mountain peaks and a whiplash tail, this critter couldn't have been more different than a *T. rex*.

It paused for a moment to scratch the hair on its neck, turned its ear to the air, and scampered forward on all fours. Hands and feet planted firmly underneath its body, it moved fast, with purpose. Up the rib, across the backbone, and onto the dinosaur's skull.

There, on the side of the head, where the eye of this *T. rex* once glared at herds of *Triceratops*, the furball stopped. It looked back in the direction of the rib cage, and let out a high-pitched squeak. From the bowels of the beast, out bounded a dozen smaller fur-balls. They raced toward their mother, and latched onto her belly, lapping up a breakfast of milk as they experienced their first minutes aboveground.

As she nursed her babies, the mother stared into the sunlight. The world now belonged to her, and her family. The Age of Dinosaurs was over, put to rest with the fiery destruction of an asteroid and a long, dark, global nuclear winter. Now the Earth was healing. The Age of Mammals had begun.

SOME 66 MILLION years later, more or less, another mammal stood in the same spot, swinging a pickaxe. Sarah Shelley was my first PhD student after I started my job as a paleontologist at the University of Edinburgh in Scotland. We were in New Mexico, on a fossil hunt, searching for the bones and teeth and skeletons that would help us understand how mammals survived the asteroid, outlasted the dinosaurs, and made the world their own, becoming the furry animals that we know, love, and sometimes fear today.

Mammals are the most charismatic and beloved creatures on the planet—with all due respect to reptiles and birds and the other eight-million-plus animal species that are not mammals. Perhaps this is because many mammals are simply cute and fluffy, but in part, I think it's because, on a deeper level, we can relate to them, and see ourselves in them. The cheetahs and gazelles locked in chase on the television screen, as David Attenborough's dulcet tones narrate the drama. The mother otter playing with her pups on the cover of a nature magazine. The elephants and hippos that make every child beg their parents to take them to the zoo, and the endangered pandas and rhinos that pull at our heartstrings when so many other appeals for charity might annoy us. The foxes and squirrels that tolerate our cities, the deer that encroach on our suburbs. Whales with bodies longer than basketball courts, emerging from the abyss to spray blowhole geysers several stories into the sky. Vampire bats that literally drink blood, lions and tigers that make our hair stand on end. Our cuddly pets, of the feline or canine or sometimes more exotic variety. For many of us, our food—beef burgers, pork sausages, lamb chops. And, of course, us. We are mammals, in the same way that a bear or a mouse is.

As a porcupine shaded itself from the New Mexico afternoon in the nook of a cottonwood, and a colony of prairie dogs chirped in the distance, Sarah brandished her pickaxe. Each strike into the rock released a haze of foul, sulfur-smelling dust. Each time she would wait for the dust to clear, to see if anything interesting had loosened from the Earth. For at least an hour, each strike brought only more rock. Until, with one thwack, something with a shape, a different texture, a different color poked out. She knelt down to

Sarah Shelley and me in New Mexico, collecting teeth from mammals that lived soon after the dinosaurs went extinct. (photo by Tom Williamson)

take a look. And then hollered a victory cheer so loud, and so happily profane, that I can't repeat it here.

Sarah had found a fossil, her first major discovery as a student.

I rushed over to see her prize, and she handed me a set of jaws, fused together at the tip. The teeth were coated in gypsum, and as they sparkled in the desert sun, I could see that there were sharp canines near the front and big, grinding molars at the back. Mammal! And not just any mammal, but one of the very species that assumed the crown from the dinosaurs.

We exchanged high fives, and got back to work.

Sarah's jaws belonged to a species called *Pantolambda*, and it was big, about the size of a Shetland pony. It lived just a couple million years after the dinosaur extinction, generations after that little mother peered out from the *T. rex* rib cage, in my fictional but credible story. Already, *Pantolambda* was considerably larger than any mammal that ever saw a *T. rex* or *Brontosaurus*. Some of those meek creatures—none bigger than a badger—endured the asteroid, by virtue of their smallness and adaptability, and suddenly found themselves in a dinosaur-free world. They grew in stature, and migrated, and diversified, and soon began forming complex ecosystems, replacing the dinosaurs—which had ruled the Earth for over 100 million years.

This particular *Pantolambda* lived in a jungle, on the edge of a swamp (hence the nasty odor of its entombing rocks). It was the largest herbivore of this environment. As it waded into the cooling waters after a lunch of leaves and beans, it would have seen or heard an abundance of other mammals. Overhead, kitty-size acrobats negotiated the tree branches with their grasping hands. At the swamp's edge, mutts with gargoyle faces burrowed into the mud, their claws seeking roots and tubers packed with nutrition. In the patchier parts of the forest, daintier ballerinas raced through the meadows on their hoofed toes. All the while, camouflaged in the thickest subtropical weeds of this Paleocene-aged jungle, a terror lurked: the apex predators, built like a stocky dog, with flesh-slicing teeth.

The death of the dinosaurs allowed these mammals—in ancient New Mexico and all around the world—to become ascendant. But mammals have a much deeper history. They—or rather, we—actually originated around the same time as the dinosaurs, over 200 million years ago, when all land was gathered together as one supercontinent, scorched by vast deserts. Those first mammals had an even deeper legacy, tracing back to about 325 million years ago, to a humid realm of coal swamps, when the ancestral mammal lineage split from the reptile line on the great family tree of life. Over these immense stretches of geological time, mammals developed their trademark features: hair, keen senses of smell and hearing, big brains and sharp intelligence, fast growth and warm-blooded metabolism, a distinctive lineup of teeth (canines, incisors, premolars, molars), mammary glands that mothers use to nourish their babies with milk.

From this long and rich evolutionary history came today's mammals. Right now, there are over six thousand mammal species sharing our world, our closest cousins among the millions of species that have ever lived. All modern-day mammals belong to one of three groups: the egg-laying monotremes like the platypus, marsupials like kangaroos and koalas that raise their tiny babies in pouches, and placentals like us, which give birth to well-developed young. These three types of mammals, though, are simply the few survivors of a once-verdant family tree, which has been pruned by time and mass extinctions.

At various points in the past, there have been legions of saber-toothed carnivores (not only the famous tigers, but also marsupials that turned their canines into spears), dire wolves, giant woolly elephants, and deer with ridiculously enormous antlers. There were supersized rhinos that lacked horns but sported long necks to guzzle leaves high in the treetops to sustain their nearly twenty-ton girth—mammals mimicking *Brontosaurus*, setting the record for largest hairy beasts to ever live on land. Many of these fossil mammals are familiar: they are icons of prehistory, stars of animated movies, and exhibits at any reputable natural history museum.

But even more fascinating are some of the extinct mammals that have never quite made it into pop culture stardom. There were once wee mammals that glided over the heads of dinosaurs and others that ate baby dinosaurs for breakfast, armadillos the size of Volkswagens, sloths so tall they could dunk a basketball, and "thunder beasts" with three-foot-long battering-ram horns. There were oddballs called chalicotheres that looked like an unholy horse-gorilla hybrid, which walked on their knuckles and pulled down tree branches with their stretched claws. Before it docked with North America, South America was an island continent for tens of millions of years, and hosted a whole family of wacky hoofed species whose Frankenstein mashup of anatomical features flummoxed Charles Darwin—and whose true relationships to other mammals has only just been revealed by the shocking discovery of ancient DNA. Elephants were once the size of miniature poodles, camels and horses and rhinos once galloped across an American savanna, and whales once had legs and could walk.

Clearly, the history of mammals is far bigger than the mammals we can see today, and it's about so much more than our own human origins and migrations over the last few million years. All of these fantastic mammals I've just mentioned, you'll meet in these pages.

I started my scientific career by studying dinosaurs. Growing up in the midwestern United States, it was *T. rex* that fascinated me most, and I went to college and did a PhD and carved out a niche as a dinosaur specialist. A few years ago, I told the story of dinosaur evolution, from their humble origins to apocalyptic extinction, in my book *The Rise and Fall of the Dinosaurs*. I'll always love dinosaurs, and will continue to study them. But since I moved to Edinburgh and became a professor, I've started to drift. Perhaps it's logical: having studied the dinosaur extinction, I've become obsessed with what happened afterward. I've become obsessed with mammals.

Sometimes people ask me why. Children everywhere dream of growing up and digging up dinosaurs, so why do anything else? And why mammals? My retort is simple: dinosaurs are awesome, but they are not us. The history of mammals is *our* history, and by studying our ancestors, we can understand the deepest nature of ourselves. Why we look the way we do, grow the way we do, raise our babies the way we do, why we have back pain and need expensive dental work if we chip a tooth, why we are able to contemplate the world around us, and affect it so.

And if that's not enough, consider this. Some dinosaurs were huge, as big as Boeing 737 airplanes. The biggest mammals—blue whales and kin—are even larger. Imagine a world where mammals were extinct, and all we had were their fossil bones. No doubt they would be as famous, as iconic, as dinosaurs.

We're learning more about the history of mammals at breathtaking speed. More mammal fossils are being found than ever before, and we can study them with an array of technologies—CAT scanners, high-powered microscopes, computer animation software—to reveal what they were like as living, breathing, moving, feeding, reproducing, evolving animals. We can even get DNA from some mammal fossils—like those strange South American mammals that infatuated Darwin—which like a paternity test tells us how they are related to modern species. The field of mammal paleontology was founded by Victorian men, but is now increasingly diverse, and international. I've been privileged to have mentors who welcomed me—just another dinosaur guy—into their territory of mammal research, and now I find my greatest joy in mentoring the next generation, like Sarah Shelley (whose illustrations grace these pages!), and many other excellent students who will continue to write mammal history with their discoveries.

In this book, I will tell the story of mammal evolution, as we know it now. Roughly the first half of the book covers the early stages of the mammal lineage, from the time they split from the reptiles until the extinction of the dinosaurs. This was when mammals acquired nearly all of their signatures—hair, mammary glands, and so on—and morphed, piece by piece, from an ancestor that looked like a lizard into something we would recognize as a mammal. The book's second part lays out what happened after the dinosaurs died: how mammals seized the opportunity and became dominant, adapted to constantly changing climates, rode along on drifting continents, and developed into the incredible richness of species today: runners, diggers, flyers, swimmers, and big-brained book readers. In telling the tale of mammals, I want to convey how we've pieced together this story using fossil clues, and give you a sense of what it's like to be a paleontologist. I'll introduce you to my mentors, my students, and the people who have inspired me, and whose discoveries have provided the evidence that allows me to chronicle the narrative of mammalhood.

This book does not focus obsessively on humans; there are plenty of others that do. I will discuss human origins: how we emerged from primate antecedents, reared up on two legs, inflated our brains, and colonized the world—after living alongside many other early human species. But I'll do so in one chapter, and give humans about the same attention as horses and whales and elephants. After all, we are but one of many amazing feats of mammalian evolution.

Our story is one that needs to be told, however, because although we're only a single mammalian species and we've only been around for a fraction of mammalian history, we are impacting the planet like no mammal before. Our phenomenal success in building cities and growing crops and connecting the globe with highways and flight routes is having an adverse effect on our closest kin. More than 350 mammals have gone extinct since *Homo sapiens* came out of the forests and swept across the world, and many species are at high risk of extinction today (think tigers, pandas, black rhinos, blue whales). If things continue at the current pace, half of all mammals might succumb to the same fate as woolly mammoths and saber-toothed tigers: dead and gone, with only ghostly fossils to remind us of their majesty.

Mammals are at a crossroads, at the most precarious point in their—our—history since they—we—stared down the asteroid that killed the dinosaurs. And what a history it has been. During our long evolutionary run, there have been times when mammals cowered in the shadows and times when they were dominant. Periods when they flourished and others when they were knocked back, and nearly knocked out completely, by mass extinctions. Eras when they were held in check by dinosaurs and eras when they were the ones doing the checking; times when none were bigger than a mouse and times when they were the biggest things that ever lived on Earth; times they weathered heat spikes and times they faced mile-thick glaciers, during the Ice Age. Times when they could occupy only the lower rungs of the food chain, and times when some of them—us—got conscious and became able to shape the entire Earth, in good and bad.

All of this history laid the foundation for today's world, for us, and our future.

Steve Brusatte
Edinburgh, Scotland
JANUARY 18, 2022

The RISE *and* REIGN *of the* MAMMALS

1

MAMMAL
ANCESTORS

Dimetrodon

SOMETIME AROUND 325 MILLION YEARS ago—give or take a few million years—a group of scaly creatures clung to a tangled raft of ferns and broken logs. They were usually solitary and preferred to lie camouflaged in the dense greenery of the jungle, occasionally emerging to snatch an insect before returning to anonymity. But desperate times had steered them together. Their world was changing, fast. Their swampy paradise, perched at the boundary between water and land, was becoming engulfed by the sea.

The small critters—the largest were barely a foot long—looked about nervously. They had the manner of a gecko or an iguana, in the way that their arms and legs stuck out to the sides and their long, thin tail dragged behind. Some of the smaller ones ambled across the rotting vegetation, holding on with their skinny fingers and toes. The older animals just stared out at the vastness of the sea, their tongues flickering as they bobbed in the waves, water lapping up against them.

A few weeks earlier, all had seemed normal. From their well-hidden dens, they would have peered into a forest dripping with humidity. Greens of every imaginable shade surrounded them. Ferns choked the forest floor, their spores dancing through the sticky air with each welcome gust of wind. Larger seed-bearing shrubs, some of which were distant ancestors of today's evergreen trees, formed the midstory. Whenever it rained—which was much of the time—their marble-size seeds poured down with the showers, covering the ground with a layer of ball bearings that made walking treacherous.

With their tiny eyes, the scaly critters could not have seen the top of the forest, which seemed to stretch infinitely toward the heavens. Two types of trees made up the bulk of the canopy, both of which grew to around a hundred feet (thirty meters) high. One was called *Calamites*, and it looked like an emaciated Christmas tree, with a straight bamboo-esque trunk that spit out intermittent bunches of branches with whorls of needle-shaped leaves. The other was *Lepidodendron*, whose six-foot-thick trunks were naked except for a thicket of branches and leaves only at the very tip-top—a mop of foliage on a giant stalk. They grew remarkably fast, going from spore to sapling to the pinnacle of the canopy in only ten or fifteen years, before dying, being buried and turned to coal, and replaced by another generation.

The scaly critters were one of hundreds of animal species that, at least until today, had called the swamp forest home. These ranged from the mundane to the fantastic. Insects were common, making them a perfect food source. Spiders and scorpions clambered through the leaf litter and up the tree trunks. Primitive amphibians gathered along streams, stocked with fish and patrolled by eurypterids: armor-covered brutes that resembled giant scorpions, some the size of humans, which snapped their prey with nutcracker claws. Back during those calmer times, the streams trickled into a river, which fanned out into a delta that led into the quiet tidal waters of a brackish bay.

Occasionally, a creepy slither pierced the stillness. This was *Arthropleura*, a monstrous millipede more than six and a half feet (two meters) long, slurping up spores and seeds. Sometimes an even more terrifying sound echoed through the swamp: the wingbeats of *Meganeura*, a pigeon-size dragonfly with four enormous, translucent wings that buzzed as it searched for bugs. If it was hungry enough, it might even attack one of the scaly critters—another reason why they preferred to stay hidden.

As the group of scaly critters latched on to their makeshift boat of leaves and twigs, the fear of a *Meganeura* attack seemed quaint. The danger now was much greater. They were encircled by water, and the currents were growing stronger. Far to the south, a massive ice cap was melting, shedding water into the oceans and raising the sea level. All around the world, coastlines were flooding, drowning the mangrove bayous of *Calamites* and *Lepidodendron* and their animal inhabitants. The scaly critters had no way of knowing this. All they could sense—as frothy eddies of dead shrimp and jellyfish rocked up against them—was that their forest was no more.

Then, a flash of lightning. As thunder crashed overhead, a storm wind pushed a wall of water onto the raft, turning it over and breaking it in half. Some of the scaly critters were washed away by the surge, their limp bodies joining the rotting jellyfish and shrimp. Most of them, however, were able to scramble back onto one of the two remnants of the divided raft. As rain pelted the bay and the winds howled, the currents split: one sweeping to the east and the other to the west. The two rafts—and their scaly cargo—headed in opposite directions.

A few days later, as the storm subsided, the rafts washed up on different shores. As the two bands of critters ventured out into their new homes, they were faced with different challenges—different habitats, climates, and predators. Over the course of many generations, both groups became well adapted to their new environments, to the point where each became a new species. Both species then begat other species, and two major lineages were born. One of them developed two windowlike openings behind the eye socket, to provide room for bigger and stronger jaw muscles. The other developed a single, expansive opening.

The first group, with their two skull openings, were the diapsids. They would eventually evolve into lizards, snakes, crocodiles, dinosaurs, birds, and turtles (which closed up their holes). The second group, with their single skull opening, were the synapsids. They would diversify into a dazzling array of species, including—more than a hundred million years in the future—the mammals.

THIS IS A story, and this exact sequence of events probably did not happen. But it is true that around 325 million years ago—during a time in Earth history called the Pennsylvanian Period (also known as the Late Carboniferous Period)—there was an ancestral stock of small, scale-covered critters that lived in lush swamp forests, which were frequently inundated by rising seas. They split apart, with one side of the family tree leading to reptiles and the other toward mammals.

How do we know this? Paleontologists—scientists like me who study ancient life—have two key lines of evidence. It's this evidence that I will marshal throughout this book to tell the story of mammal evolution.

First, there are fossils, and the rocks that entomb them. Fossils are direct evidence of species that used to be alive; they are the clues that paleontologists travel around the world searching for, often braving heat, cold, humidity, rain, lack of money, mosquitoes, war zones, or other obstacles. Many of us fancy ourselves as deep-time detectives, and in this analogy, fossils are the equivalent of hair or fingerprints left at a crime scene. They tell us what lived where, and when, and in some cases fossils can reveal prehistoric dramas of predators slashing prey, victims swept up in floods, survivors enduring grim extinctions. The most familiar fossils are body fossils: actual parts of a once-living organism, like a bone, tooth, shell, or leaf. Others are trace fossils: records of an organism's behavior or something that it left behind, like footprints, burrows, eggs, bite marks, or coprolites (fossilized dung).

We don't find fossils lying on the street or in the soil of our backyards, but rather inside of rocks like sandstone and mudstone. Different rocks formed in different environments, and some rocks can be dated using chemical techniques, which count the amount of radioactive parent and daughter isotopes to calculate an age, based on reference to known rates of radioactive decay from lab experiments. This all provides critical context for understanding when, and in which habitats, the fossilized creatures lived.

The second type of evidence is all around us, and doesn't require any special skill (or luck) to find. It's DNA, which we and all other organisms carry inside of our cells. DNA is the blueprint that makes us what we are, the genetic code that controls what our bodies look like, our physiology and growth, and how we produce future generations. DNA is also an archive; evolutionary history is written into the billions of base pairs that make up our genome. As species change over time, their DNA changes. Genes mutate, move around, and are switched on and off. Stretches of DNA are duplicated or erased. New bits of DNA are inserted. Consequently, as two species diverge from a common ancestor, their DNA will become progressively more different over time as each species goes its own way and adapts to its own changing situations. Thus, you can take the DNA sequences of modern-day species, line them up and compare them, and make a family tree by grouping together species with the most similar DNA. There's also another nifty trick. You can take any two species, count the number of DNA differences, and then knowing something about the rate DNA changes in lab experiments, back-calculate to figure out when those species separated from each other.

I used both types of evidence in creating this story of the flood-ravaged swamp. DNA studies predict that the reptile and mammal lines separated from each other around 325 million years ago. Fossils and rocks tell us what this lost world was like, a landscape that was much different than today.

A map of the Pennsylvanian-aged Earth would have been scarcely recognizable. There were only two big landmasses, one called Gondwana centered on the South Pole and another named Laurasia that hugged the equator, fringed by a series of smaller islands to the east. Over many millions of years, Gondwana drifted northward, at about the rate that our fingernails grow, before colliding with Laurasia. This was the beginning of the birth of Pangea, the supercontinent on which the early stages of mammal and dinosaur evolution would eventually unfold. As the two slabs of crust slammed into each other, they deformed into a long stretch of mountains paralleling the equator, similar in scale to today's Himalayas. Today's modest Appalachian Mountains are a remnant of this once-towering chain.

The tropical and subtropical regions on both sides of the equatorial mountain range were havens for life. These were the coal swamps, so-called because much of the coal that fueled the Industrial Revolution—particularly that mined in Europe and the midwestern and eastern United States—formed in these very swamps. It was made from the dead, buried, and compressed remnants of those gigantic, fast-growing *Lepidodendron* and *Calamites* trees. These were not at all like the palms, magnolias, and oaks that are so common in similar lush environments today. In fact, these ancient trees didn't have flowers, or produce any fruits or nuts. They were close relatives of club mosses and horsetails, primitive plants that persist today as rare parts of the understory, a sad remnant of what they once were. The Pennsylvanian trees—and the supersize dragonflies that buzzed around their branches and millipedes that scurried around their trunks—were able to grow so large because there was much more oxygen in the air then, some 70 percent more than today.

The trees formed vast rain forests, which clutched the shores of the shallow seas that lapped far onto the growing supercontinent, and the many streams, rivers, deltas, and estuaries that fed into them. An aerial survey of these swamps probably would have looked something like the Mississippi River bayous of modern Louisiana: a dense blanket of trees and smaller plants, twisted together, some perched on islands of mud between tangled stream networks, others with spidery roots extending into the water, with all sorts of creatures climbing and jumping and flying about. But no birds, no mosquitoes, no beavers or otters or other fur-covered mammals. All of them would evolve much later, in a very different world, although their ancestors did inhabit the coal swamps.

Why were so many trees getting buried and turned to coal? It's because the swamps were constantly flooded. Sea level was always rising and falling, in a pulsating rhythm. The Pennsylvanian was a glacial world—in fact, it was the last major ice age before the most recent one, when mammoths and saber-toothed tigers reigned (a story we will get to later). The entire planet was not frigid; certainly the coal swamps weren't. But over the South Pole of Gondwana, and then southern Pangea, there was an enormous ice cap. It owed its very existence to the coal swamps: the growth of so many giant trees drew carbon dioxide out of the atmosphere, and with less of this greenhouse gas to insulate the planet, temperatures plummeted. Over tens of millions of years, the ice cap waxed and waned in size, a conductor controlling the global sea level. Ice would melt, seas would rise, swamps would drown, trees would die and get buried. Then the ice would grow, sucking up water from the seas, lowering the sea level, making room for swamps to thrive. Back and forth it went. We know this because Pennsylvanian rocks often form barcode sequences called cyclothems, repeated series of thin layers formed on land and in the water, with coal seams tucked in between.

Fossils from this time are plentiful, especially where I grew up in northern Illinois. They are embedded in the cyclothems, above and below the coal. The best ones are found along the banks of Mazon Creek, a gentle tributary of the Illinois River, and the strip mines to the east. During the Pennsylvanian this was where sea met swamp, where denizens of the rain forests were flushed into the water, sunk to the bottom, and encased in ironstone tombs—

oval, flattened, rust-colored nodules that you can pluck from the creek bed or mine tailings. As a teenager I searched for these nodules, near the tiny Route 66 town of Wilmington, where my mother was raised. I scoured the spoil piles of the long-closed mines, which more than a century earlier had beckoned my Italian great-grandparents to a new life in Middle America. I put the nodules in a bucket, took them home, put them outside in the brutal Chicagoland winter, and let them repeatedly freeze and thaw as the temperature fluctuated. When it looked as if one was starting to crack, I would finish the job with a hammer.

If I was lucky, the nodule would break open and a treasure would be revealed: a fossil on one side, its impression on the other. Each time it was an otherworldly experience, knowing that you were the first person to see this thing—which was once alive!—that had died over 300 million years ago. Many of the cracked nodules had plants inside: fern leaves, pieces of *Calamites* bark, chunks of *Lepidodendron* roots. I was particularly fond of jellyfish—what veteran Mazon Creek fossil hunters dismissively called "blobs"—and always enjoyed catching sight of a shrimp or worm.

What I really wanted—and never had the fortune of finding—was a tetrapod, a land-living animal with bones. I knew from the textbooks I was devouring, after school and on quiet weekend afternoons, that tetrapods had evolved from fishes and crawled onto land about 390 million years ago, before the Pennsylvanian Period. These first tetrapods were amphibians, which still needed to return to the water to lay their eggs. Some primitive amphibian skeletons—remote relatives of frogs and salamanders—had even been found at Mazon Creek.

Sometime during the Pennsylvanian a new group spun off from these amphibians. They were the amniotes, more specialized tetrapods named for their amniotic eggs, whose internal membranes surround the embryo to protect it and prevent it from drying out.

These new eggs unlocked a great new potential: the amniotes no longer were handcuffed to the water but could lay their eggs inland, giving them access to new frontiers. Treetops, burrows, plains, mountains, deserts. Only with the advent of the amniotic egg did tetrapods truly divorce themselves from the seas, and properly conquer the land.

It was from the amniotes that the reptile and mammal lines—the diapsids and synapsids—arose, by splitting from each other like two siblings from their parents. This is not simply an analogy; it is how evolution produces new species, new families, new dynasties. Species are always changing as their environments change—this is Darwin's evolution by natural selection. Sometimes populations of a single species become separated from each other, maybe by a flood, fire, or a rising mountain range. Each population will continue to change through natural selection, and if they are separated long enough, they will change in their own idiosyncratic ways, to become adapted to their different circumstances, so much so that they no longer look the same, or behave the same, or can mate with each other. At this point one species has become two. The two new species might then split again, two becoming four, and so on. Life is always diversifying in this way, branching like a tree that has been growing for over 4 billion years. This is why we use family trees to visualize genealogy—both for extinct species and our own human families—rather than family nets, or road maps, or triangles, or some other type of graphic aid.

The diapsid-synapsid split—which *really would* have begun inconspicuously with one small, scaly ancestral species dividing into two—was one of the keystone moments in vertebrate evolution. And I knew that the diapsids and synapsids—each characterized by their own unique system of skull holes and jaw muscles—were setting off right around the time the Mazon Creek nodules were forming. With each whack of the hammer, I hoped to find a Holy Grail fossil that would help tell this story, but alas, it never came.

Other fossil hunters in other parts of North America, however, were more successful. One important discovery was made in 1956, when a Harvard field team led by the legendary paleontologist Alfred Romer surveyed an abandoned coal mine in Florence, Nova Scotia, near the Atlantic coast. One of his technicians, Arnie Lewis, noticed several fossilized stumps of a tree called *Sigillaria*, a close relative of *Lepidodendron*, whose crown of leaves forked at the top, giving it the impression of a giant brush. The stumps were in life position, as if they had been drowned by rising seas only yesterday, not around 310 million years ago, their true age. Wading through the narrow shafts of the flooded mine, the team was able to collect five of the stumps. When they looked inside, they found quite the surprise: dozens of fossil skeletons! The poor creatures may have sought shelter in the trees as the seas encroached, not realizing they were entering their graves. One particular tree had more than twenty animals inside, including amphibians, diapsids, and synapsids: the trifecta of early land-living tetrapods.

Archaeothyris (illustrated by Todd Marshall)

The synapsids were later described as two new species, *Archaeothyris* and *Echinerpeton*, by a master's student, Robert Reisz, who had just emigrated from Romania to Canada. Now one of the world's leading paleontologists, Reisz cut his teeth on these early synapsids. He chose the name *Archaeothyris*—meaning "ancient window"—to highlight the most important feature of this animal: its large, portholelike opening behind its eye, which housed larger and more powerful jaw-closing muscles than its ancestors. It is this single opening, technically called the lateral temporal fenestra, which defines what it means to be a synapsid. All synapsids—from the coal swamp pioneers to today's bats and shrews and elephants—have this fenestra, or a modified version of it. So do we, and we can feel it every time we close our jaws. Put your hand on your cheekbone, take a big bite, and feel the muscles of your cheek contracting. Those muscles are passing through the remnant of the fenestra, which in modern mammals has more or less merged with the eye socket, but still anchors the temporal muscles that stretch from the side of our head to the top of our lower jaw,

Diapsid **Synapsids**

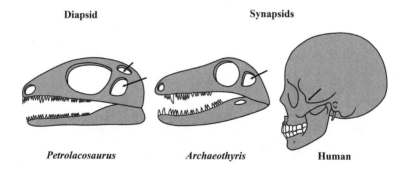

Petrolacosaurus   *Archaeothyris*   Human

The two main skull types of land-living vertebrates: diapsids with two openings for jaw muscles behind the eye and synapsids—including humans—with a single opening. Arrows denote the jaw openings. (illustrated by Sarah Shelley)

powering our bites. This single opening developed early in synapsid history, right after they split from the diapsids, which went on to evolve two such openings behind their eyes.

If you saw it scampering through the coal swamps, *Archaeothyris* would have looked unexceptional. It was about one and a half feet (fifty centimeters) long from snout to tail, with a small head perched on a long, slender body. Its limbs are not well known, but the preserved bones leave no doubt that the arms and legs sprawled out sideways, akin to a lizard or crocodile. Clearly, it was not built for speed. On closer inspection, however, it was exceptional in other ways. Not only were those bigger jaw muscles hiding inside its skull, but its snout bore a series of curved, pointed teeth. One of the front ones was noticeably larger than the others, such that it looked like a minicanine. Amphibians, lizards, and crocodiles do not have canines. All these animals have uniform teeth, which basically look the same across the entire jaw. Mammals, though, have a much more varied dentition, split into incisors, canines, premolars, and molars—a division of labor that allows us to grab, bite, and crush all at the same time. The full mammalian dentition would assemble later, over many evolutionary steps, but the little canines of *Archaeothyris* are the whispers of a dental revolution.

Taken together, the big jaw muscles, sharp teeth, and canines of *Archaeothyris* were an arsenal of weapons for feeding on big insects, and maybe other tetrapods, like *Echinerpeton*. This second Nova Scotian synapsid could have easily curled up and fit in between the pages of this book. But its scrappy fossils do show one peculiar feature, which gives it its name: "spiny reptile." The spines on the neck and back vertebrae—the individual bones that make up the backbone—expand upward as elongate tabs. When lined up together they would have formed a small sail along the back, which might have been used for display, or as a solar panel to heat up the body on cool days, or a fan to shed heat on warm days, or something else entirely.

There is another, much more famous extinct animal with an even bigger sail on its back: *Dimetrodon*, which lived during the next interval of time after the Pennsylvanian, the Permian Period. All too often *Dimetrodon* is mistaken for a dinosaur, sharing space with *T. rex* on dinosaur posters, jostling with *Brontosaurus* and *Stegosaurus* in dinosaur toy sets. But it is not a dinosaur; it is a synapsid. More specifically, it is a type of primitive synapsid called a pelycosaur.

The pelycosaurs were the first big evolutionary wave of the synapsid lineage; they were the first to diversify and spread around the growing supercontinent of Pangea, and the first ones to start developing some of the trademark features that, some 300+ million years later, are still the things that make mammals stand out from amphibians, reptiles, and birds. Features like the temporal muscle opening, and the canine teeth. Features that we've already seen, in *Archaeothyris* and *Echinerpeton*. That's because these two Nova Scotia species are the very oldest pelycosaurs, the founders of the first great dynasty on the journey toward *Dimetrodon*, and ultimately mammals.

AS THE PENNSYLVANIAN Period drew to a close, there were pelycosaur synapsids living across the equatorial regions of Pangea, on both sides of the still-rising mountain range. Some ate insects, others preyed on small tetrapods and fish, and a few started to experiment with a new food type that thus far had been ignored: leaves and stems. They were diversifying, but they remained minor components of their ecosystems, which were overrun with amphibians that could easily reproduce, and thus prosper, in the damp coal forests.

Then, between about 303 and 307 million years ago, the world dramatically changed during a spasm called the Carboniferous Rainforest Collapse. The climate became drier, temperatures swung cold and hot, and the ice caps melted, eventually disappearing for good in the ensuing Permian Period. The coal swamps were devastated, as the soaring *Calamites*, *Lepidodendron*, and *Sigillaria* trees found it harder to grow in the more arid conditions. They were replaced by conifers, cycads, and other seed-bearing plants, which were more drought resistant. The ever-wet rain forests gave way to more seasonal, semiarid drylands in the tropics, and other parts of Pangea became parched deserts. This is reflected in the rock record by the sudden shift from coals and cyclothems to "red beds," full of rusty iron formed in dry climates.

These changes had a striking impact on biodiversity. Plants were hit especially hard. Not only was there a shift from Pennsylvanian coal swamp vegetation to the more dry-adapted seed plants, but there was an extinction event. Many of the Pennsylvanian species disappeared, some leaving no descendants or close relatives, others only their smaller, much less impressive cousins. All told, about half of the Pennsylvanian plant families were extinguished. This is one of only two mass extinctions recognized in the plant fossil record. The other occurred at the end of the Permian, a story we'll get to shortly. That means that the Carboniferous Rainforest Collapse was a greater botanical catastrophe than the end-Cretaceous asteroid impact that killed the dinosaurs.

What about the animals living in the coal forests? A study by a young researcher, Emma Dunne, tells the tale. Emma, who grew up in Ireland and went to England for her PhD, is a leader among the new generation of paleontologists. She collects fossils, like the legions of bone-hunters before, but also specializes in big data and advanced statistical methods. It's always tempting to spin snappy stories based on a couple of new fossils, but to really understand the patterns and processes of evolution, Emma's generation thinks like a stock market analyst or investment banker: collect tons of data, use statistical models to take into account uncertainty, and explicitly test hypotheses against each other using numbers rather than intuition.

In this spirit, Emma constructed a database of over one thousand tetrapod fossils of Carboniferous and Permian age, noting which groups they belonged to and where they were found. She used statistical tools to smooth out sampling biases that, inevitably, can plague paleontological studies that are so reliant on chance fossil discoveries from those few lucky places that preserve bones or teeth for hundreds of millions of years. Finally, she built statistical models to test how the overall diversity and the distribution of these species—amphibians, diapsids, and synapsids among them—changed as the rain forests collapsed.

The results were unsettling. There was a massive drop in diversity as the Carboniferous transitioned to the Permian, as many coal-forest tetrapods went extinct. This likely didn't happen all at once, but over several million years, as the drylands gradually replaced the coal forests in the tropics, in a march from west to east. This change of habitat—more of a transition, it seems, than a collapse—produced more open landscapes, which favored migration. Tetrapods that could tolerate the drier climates now could move much more widely. These were not the amphibians that had long dominated the Pennsylvanian wetland world, because their reproductive strategies tied them to the water. The diapsids and synapsids, however, now found themselves with a superpower that was perfect for this new reality: their amniotic eggs, with membranes to nourish, protect, and keep their embryos moist. They moved freely across the land, establishing connections between regions that were previously isolated, and in doing so, evolving new species, new body types, larger size, new diets, new behaviors.

As the coal swamps became open drylands, and the Permian Period unfolded, Earth became a pelycosaur planet. Nothing epitomizes the new age of pelycosaur dominance like *Dimetrodon*, that sail-backed icon best known from dozens of skeletons from Texas. There's a reason *Dimetrodon* is often mistaken for a dinosaur: its body looks, for lack of a better term, reptilian. It was big and bulky, with a long tail, and sharp teeth, and it couldn't move very fast on its stubby, sprawling limbs. Even its brain was small and tube-shaped, like those of dinosaurs and unlike the larger brains of mammals, which have hugely expanded, spaghetti-textured cerebrums that impart greater intelligence and heightened senses. In these features, *Dimetrodon* probably wasn't much modified from those small, scaly ancestral critters that split into synapsids and diapsids in the Pennsylvanian.

In other characteristics, however, *Dimetrodon* was very different from its ancestors. Nowhere is this more apparent than in the mouth, where the teeth look nothing like the uniform series of blades or pegs seen in most amphibians and diapsids. The front of the snout sported large, rounded incisor teeth, followed by a big canine, and ending with a set of smaller, sharply curved postcanine teeth along the cheek. This was another step in the evolution of the classic mammalian dentition, following the development of canines in early pelycosaurs like the tree-stump-hiding *Archaeothyris*. In concert with the dental changes were alterations to the jaw muscles, which got larger and attached to a stronger and deeper lower jaw, which promoted an even stronger bite. Changes were also happening in the backbone, where the individual vertebrae now connected in a way that limited the awkward, side-to-side undulation so characteristic of reptiles and amphibians.

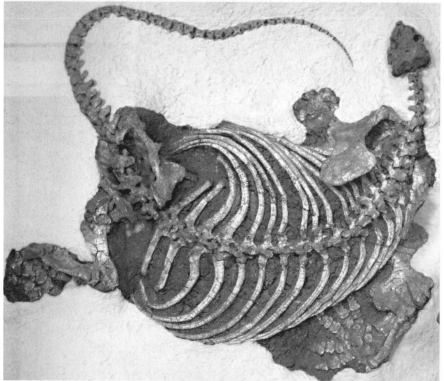

Pelycosaurs, primitive synapsid forebears of mammals: the sail-backed
Dimetrodon (top) and a pot-bellied, plant-eating caseid (bottom).
(photos by H. Zell and Ryan Somma, respectively)

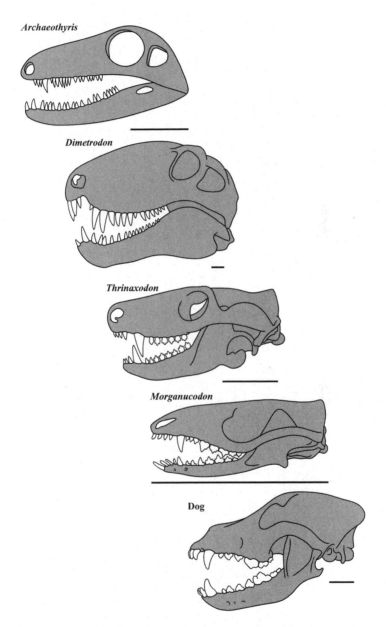

The evolution of skulls and teeth during synapsid history, illustrating how the teeth become more complex and split into incisors, canines, premolars, and molars in mammals. Scale = 3 cm. (illustrated by Sarah Shelley)

Dimetrodon, therefore, has a mixture of primitive and advanced characteristics. It is something of a Frankenstein creature that combines ancient reptilian-style traits and more derived hallmarks of mammals. This is exactly what you would expect, given its position on the family tree. In older textbooks, you might see *Dimetrodon* and similar animals referred to as "mammal-like reptiles," which although evocative, is an outdated term. This is because despite its appearances, *Dimetrodon* was not a reptile nor did it evolve from actual reptiles—because reptiles themselves stemmed from the diapsid group. Its "reptilian" features are simply primitive traits that it had yet to shed. Instead, in the parlance of scientific classification, it and the other pelycosaurs are "stem mammals": extinct species on the evolutionary lineage toward the modern mammals, more closely related to mammals than to any other group that's still alive today. It was on this stem lineage where the body plan of mammals was constructed, piece by piece, over millions of years of evolutionary time. Along the mammal stem lineage, creatures that started off looking like a reptile—even though they were not reptiles!—morphed into small, furry, big-brained, warm-blooded mammals.

And do you know what that means? *Dimetrodon* is more closely related to you and me than it is to *T. rex* or *Brontosaurus*.

During the early Permian heyday of *Dimetrodon*—from about 299 to 273 million years ago—mammals were a still unrealized concept, which evolution had yet to assemble. True, *Dimetrodon* and kin were evolving features that today we recognize as mammal trademarks, but they were not evolving them *to become* mammals. Natural selection doesn't plan for the future, it only works in the present, to adapt organisms to their immediate circumstances. In the grand scheme of Earth history, these are usually trivial things: shifts in local weather or topography, the arrival of predators into a new pocket of forest, the sudden availability of a new food type. For *Dimetrodon* and the other pelycosaurs, diet was probably driving much of their evolution, and thus the development of these initial mammalian characteristics.

Dimetrodon was not an animal to be messed with. It was the top predator in its ecosystems, verdant lowland forests dotted with ponds and crossed by rivers. The coal swamps were long gone, but these ecosystems still had a swampy, aquatic component to them. At around 15 feet (4.5 meters) long, and weighing up to 550 pounds (250 kilograms), *Dimetrodon* ate whatever it wanted. On the menu were other land-living tetrapods, including both synapsids and diapsids, plus the amphibians that moved along the stream banks, and freshwater sharks that swam in the rivers. The sail-backed menace would stalk the evergreen groves and shorelines, using its newly evolved incisor teeth to grab its victims, its scythe-shaped canines to deliver a killing bite, and its cheek teeth to break apart muscle and tendon before swallowing. If the prey tried to escape during any point, then *snap!*, those huge jaw muscles would come clenching down. In behaving this way, *Dimetrodon* was one of the first truly big, powerful apex predators that ever lived on land: a founder of the niche that so many of its distant mammalian inheritors—like lions and saber-toothed tigers—would eventually fill.

If *Dimetrodon* was feeling particularly adventurous, or hungry, it might attack another pelycosaur species, a doppelgänger called *Edaphosaurus*. This similarly sailed creature was a bit smaller than *Dimetrodon* in overall length, but slightly heavier, with a plump belly and a little head. Once *Edaphosaurus* opened its mouth, however, you could immediately tell it was not only a different species than *Dimetrodon*, but had an entirely different diet. Instead of incisors and canines, *Edaphosaurus* had a more standardized row of sharp teeth, shaped like triangles. It also had a second battery of peculiar, flatter teeth on the palate and the inner surfaces of the lower jaws. This one-two punch was perfect for eating plants: the cheek teeth of the upper and lower jaws worked together, somewhat like a pair of garden shears, to cut leaves and stems, and the internal teeth did the crushing and grinding.

Eating plants might not seem very special; it is such a common way for animals to make a living today. During the Permian, however, it was a hot new trend. *Edaphosaurus* was one of the very first tetrapods to specialize in vegetation. Its Pennsylvanian-aged ancestors began testing this diet before the Rainforest Collapse, but it was in the more arid and seasonal world afterward, rich with seed-bearing plants, that it became a normal lifestyle. In fact, different groups of pelycosaurs independently evolved a taste for greens, a sign that this diet was going from fad to mainstream. One such group, the caseids, were perhaps the weirdest synapsids of all. With their tiny heads and barrel-chested bodies, they looked more like *Star Wars* characters than functional animals produced by evolution. But they were very real, and very good at eating plants, and some of them became the heftiest synapsids of their time, like the half-ton *Cotylorhynchus*. It needed such a fat, wide gut to digest all the leaves and twigs it was gobbling. Together, the *Edaphosaurus* group and the caseids inaugurated the big base-of-the-food-chain herbivore niche, which so many mammals—from horses and kangaroos to deer and elephants—would later occupy.

The flesh-gouging *Dimetrodon*, plant-vacuuming *Edaphosaurus*, and pudgy caseids were but a few of the many pelycosaurs that flourished in the early Permian. For a few tens of millions of years, the world—particularly the tropics, which were more humid and less seasonal than other parts of Pangea—was theirs. But then, when they seemed to be at their peak, the pelycosaurs declined. The reasons are not so clear, but probably related to the culmination of the warming and drying trend that began with the Carboniferous Rainforest Collapse, and the final disappearance of the ice cap on the South Pole. As the early Permian ticked over to the middle Permian, about 273 million years ago, pelycosaurs living in the tropics cratered in diversity, as these areas became more arid. Once again, it was not a sudden cataclysm, but a drawn-out death march over millions of years. There were big changes in the higher-latitude temperate regions as well, with a near complete turnover of species. In both the tropics and temperate zones, a new type of synapsid appeared, and quickly diversified into a wealth of new species, among them carnivores and herbivores, shadow-dwellers and giants.

These were the therapsids. They evolved from *Dimetrodon*-like pelycosaurs, and then developed many advanced features relating to faster growth, higher metabolism, keener senses, more efficient locomotion, and stronger biting. They were the next major step on the line to mammals.

THE KAROO OF South Africa is a beautiful, but tough, place. Its endless blue skies give an air of tranquillity, but the cloudless vistas mean little rain. This is classic desert, roasting in the day, shivering in the night, heavy with dry air that barely rustles the aloes and other heat-adapted shrubs that pop up from the sand and rocks. The first European intruders made many attempts to settle it, to no avail. Of course, Indigenous people were able to live there, not that the Dutch and British paid much attention. The locals were safe for a while, until the colonists built roads and railways, and imported windmills to pump water from deep in the ground. Soon the Karoo became farm country, the heart of South Africa's lamb and wool industry.

Building the roads was tricky. Not only did work crews have to endure the cruel climate, they had to blast through quite a bit of rock. Rock is everywhere in the Karoo; it's carved into mountains and valleys and litters the desert floor. Layer after layer of rock—mostly sandstones and mudstones formed in ancient rivers and lakes and dunefields—stacked like a giant wedding cake, some six miles (ten kilometers) thick. Back during the Carboniferous and Permian times, and continuing into the subsequent Triassic and Jurassic periods, the Karoo was a spacious basin—teeming with plants and animals—which accumulated mud and sand as rivers drained mountains on its rim. It was a hungry basin, never quite getting its fill, because as the rivers emptied their loads, fault movements made sure the basin floor kept falling. By the time this standoff ended, the Karoo held a 100-million-plus-year record of Earth history, a sequence of rocks recording the Carboniferous Rainforest Collapse, Permian aridification of the land, transition from glacial icehouse to greenhouse, and the assembly of the supercontinent Pangea.

Good engineers were needed to carve roads through these rocks, and one of the best was Andrew Geddes Bain. Born in the Scottish Highlands, Bain moved to South Africa as a teenager, when his uncle, a colonel, was stationed in the Cape Colony, then part of the British Empire. After trying out many careers—saddler, writer, army captain, farmer—he was commissioned by the military to build roads in the Karoo. With each mile of road that he laid, the more familiar with the rocks he became. Eventually, he added geologist to his impressive list of careers, when he made the first detailed geological map of South Africa. Bain also started collecting curiosities found inside the rocks, including Permian-aged skulls of tusked, dog-size animals that resembled nothing in South Africa's modern savanna fauna. He found the first of these in 1838, while working near Fort Beaufort, a small village founded as a religious mission and later turned into a military encampment. There was no local museum to exhibit the fossils, so he sent some to London, and kept them coming when the Geological Society started paying him.

In the British capital, Bain's fossils found their way to Richard

Owen, anatomist and naturalist extraordinaire. Then in his early forties, Owen was a titan of Victorian Britain's scientific establishment. Just a few years earlier, he had coined the word "dinosaur" to describe the skeletons of ancient giants turning up in southern England. A few years later, he would be made natural history director at the British Museum, and toward the twilight of his life, he helped establish the Natural History Museum, in the posh South Kensington district of London. He was a favorite of the royal family; he served as tutor to some of Victoria and Albert's children, which along with his scientific work earned him a knighthood. If there was a medal or prize for scientific achievements in the Victorian realm, you could bet that Owen won it at some point in his long career. Which speaks to his genius, given that he was an acerbic, paranoid, two-faced egomaniac, who always relished a fight and cultivated far more foes than friends.

In 1845, Owen published a description of some of Bain's fossils, one of which he named *Dicynodon*. It was a puzzling animal, with a reptilian-like head capped by a beak, but also a growling set of canine tusks—hence its name, which means "two-dog tooth." Another species, described in a subsequent publication, he called *Galesaurus*: the "weasel lizard." To Owen, the name reflected what he saw in the fossils: an unusual mix of lizardlike and mammal-like features. He was particularly enamored by the teeth in many of Bain's skulls, which were partitioned into the familiar incisors, canines, and cheek teeth of mammals. But otherwise, these animals seemed like reptiles in their build and proportions—so much so that Owen incorrectly classified some of them as dinosaurs.

Owen encouraged Bain to collect more fossils, which he continued to study and name as they arrived from the distant colony. He even recruited the teenage Prince Alfred—fourth child of Victoria and Albert, and second in line to the throne—to gather additional specimens on a visit to South Africa in 1860. The prince obliged,

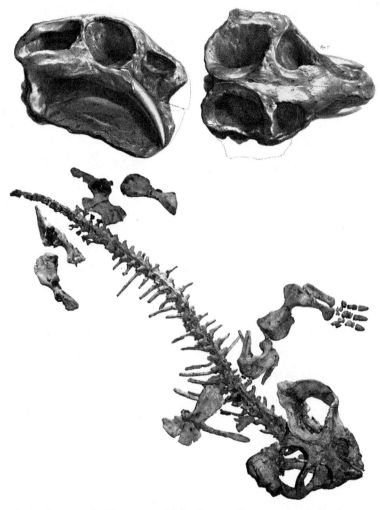

Dicynodonts, primitive synapsid forebears of mammals: skull of *Dicynodon* from Richard Owen's 1845 monograph (top) and skeleton (bottom). (photo by Christian Kammerer)

bringing back two skulls of *Dicynodon*. Even with the growing inventory of Karoo fossils, Owen didn't know what to make of them. They seemed similar to mammals in so many ways, which he conceded in his essays and lectures, and later in his milestone 1876 catalog of South African fossils. Even so, Owen couldn't quite bring himself to consider them mammal ancestors, the links in an evolutionary chain between primitive reptilelike animals and modern mammals. That's because he was embroiled in a feud with Charles Darwin about evolution itself. A staunch social conservative and fiery defender of the status quo, Owen didn't accept Darwin's theory of evolution by natural selection, penning a caustic review of *The Origin of Species* that stands as one of the greatest failed put-downs in the history of science. It's not that Owen didn't think species were able to change, it's just that he thought Darwin's ideas about mechanism were all wrong. That, and he plainly held a personal grudge.

No surprise, then, that Darwin's most steadfast defender—Thomas Henry Huxley, naturalist and coiner of the term "agnostic" to describe his religious views—couldn't bring himself to acknowledge the mammalian features of Owen's Karoo reptiles, or the possibility that mammals evolved from them. Instead, Huxley made an argument that, in hindsight, seems farcical: mammals evolved from salamander-type amphibians. As the years ticked by, Owen and Huxley continued to bicker. When they both passed away in the 1890s, the debate hadn't yet been resolved—although most evidence was breaking in Owen's favor. Among these clues was a certain newly discovered sail-backed "reptile": *Dimetrodon*. In describing it and other fossils from North America, paleontologist Edward Drinker Cope—remember the name, as we'll meet him later in a more adventurous setting—advocated a link between the "reptilian" pelycosaurs, Owen's Karoo fossils, and today's mammals.

Owen and Cope were finally vindicated a few decades later, by another Scotsman who followed Bain's path and emigrated to South Africa. Robert Broom was born in Paisley, famous for its weaving industry, and trained as a medical doctor in Glasgow. For a few years, he was an obstetrician in the Glasgow Maternity

Hospital, but he grew scared of contracting tuberculosis, so he moved abroad, first to Australia and then South Africa. It wasn't just fear that forced him to move, however; it was also obsession. Obsession with figuring out the origin of mammals. Broom was a keen naturalist as a child and took university courses in comparative anatomy, which hooked him on paleontology. After his time in Australia, where he studied the continent's quirky marsupial fauna, he specifically moved to South Africa so he could collect and study "mammal-like reptiles" from the Karoo. For several decades, he practiced medicine, frequently moving jobs between provincial towns, while pursuing fossils as a hobby, and occasionally serving as town mayor.

During the first half of the twentieth century, Broom was one of the most prominent scientists in South Africa. He wrote more than four hundred articles on Karoo fossils and described over three hundred species of "mammal-like reptiles." Before his arrival, most of the Karoo fossils had been studied in a haphazard way, even by Owen, who frankly gave them low priority compared to his dinosaurs, dissections of modern mammals, Victorian societal commitments, and his quarrel with Darwin. Broom, on the other hand, made these fossils his life's mission, and he approached the task like a neurotic comic book collector. He systematically surveyed the Karoo desert, befriending farmers and road builders and training them to recognize fossil skeletons. In a remarkable testament to his legacy, the progenies of one such construction worker (Croonie Kitching's son James) and one such farmer (Sidney Rubidge's grandson Bruce) left their family trades and became two of South Africa's most eminent academic paleontologists. Today, Rubidge and his colleagues at the University of the Witwatersrand pay it forward by working closely with local communities and Indigenous students.

Broom's great achievement was proving, once and for all,

Cope's hypothesized link between pelycosaurs, the Permian-aged Karoo fossils, and mammals. In 1905, Broom invented the term "therapsid" to categorize many of the Karoo "mammal-like reptiles," and argued forcefully that mammals evolved from this group. Then, in 1909 and 1910, he visited the United States, where he studied fossils of *Dimetrodon* and other pelycosaurs. He saw unmistakable similarities between pelycosaurs and therapsids, and in a landmark monograph, he made the case that they were closely related. In doing so, he put the two pieces together and advocated a pelycosaur-therapsid-mammal connection. He recognized therapsids as more advanced than pelycosaurs, particularly in their well-developed and more upright limbs, which allowed them to stand up straighter, with their bellies farther from the ground. In this sense, they were progressively more mammalian. Thus, the pelycosaurs came first, and then the therapsids were the next step on the line to mammals.

What were these therapsids like? There were hundreds of species, and they came in a bewildering variety. We've already met Owen's *Dicynodon*, the namesake for the most diverse Permian therapsid subgroup: the dicynodonts. When Owen named *Dicynodon*, it became something of a celebrity, taking its place alongside the newly discovered dinosaurs *Iguanodon* and *Megalosaurus* in London's famed 1854 Crystal Palace exhibition. The two *Dicynodon* sculptures in the exhibit—which can still be seen today, albeit slightly decayed—helped introduce the Victorian masses to the prehistoric world. This fame also had a downside: as the first representative of the dicynodont group discovered, *Dicynodon* became a taxonomic dumping ground for legions of new fossils. Some 168 new species were assigned to *Dicynodon* over the next century and a half, causing Broom to lament it as "the most troublesome genus we have to deal with," which caused him "utter confusion."

Only in 2011 was this mess resolved, by another paleontologist as obsessive and detail-oriented as Broom: Christian Kammerer. I first met Christian when he was a PhD student, and I an undergraduate, at the University of Chicago. Christian was one of the Big Men on campus; everyone knew who he was and had stories about his prowess. At normal universities this would mean athletic achievements or fraternity stunts, but the University of Chicago— where, as the slogan has it, fun goes to die—is no such place. Christian earned his reputation through Chicago's legendary scavenger hunt, a four-day nerd fest of collecting weird, wonderful, and impossible items, like homemade nuclear reactors (extra points if functional). This collector mindset served Christian well when it came to untangling the Gordian knot called *Dicynodon*.

Years of careful research culminated in Christian's monographic review of *Dicynodon*, which he wrote while sitting across from me in our office at the American Museum of Natural History, the next stop on our strangely entwined academic journey. In it, he recognized only two valid *Dicynodon* species. All the other stuff that was once called *Dicynodon* belonged to a bevy of distinct dicynodont lineages, made up of species of varying shapes and sizes, which formed a bushy family tree. For years, the wastebasket of *Dicynodon* had been concealing incredible diversity.

Dicynodonts were the most numerous synapsids, and often the most numerous vertebrates period, in most middle-to-late Permian land ecosystems, all over the world. They were plant-eaters, and probably lived in large social groups. Most dicynodonts lacked most of their teeth—save their walruslike canines—and replaced them with a beak for cropping leaves and stems, which they then pulverized with a powerful backward-angled bite. With their short legs, blubbery bellies, and ludicrously small tails, they would have cut an unmistakable profile as they were stripping leaves from branches or using their tusks and stubby arms to dig up tubers.

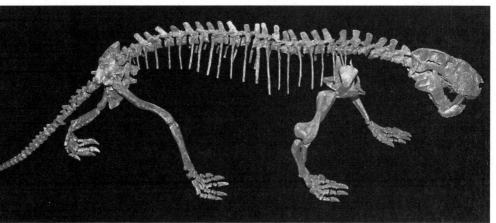

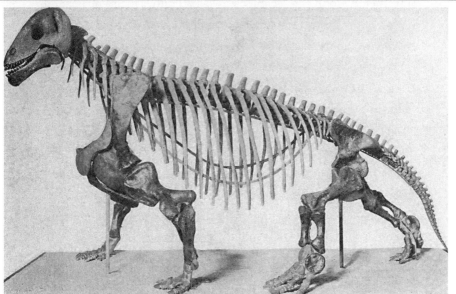

Therapsids, primitive synapsid forebears of mammals: a saber-toothed gorgonopsian (top) and the head-butting dinocephalian *Moschops* (bottom). (photos by H. Zell and AMNH Library, respectively)

For a brief period of time, dicynodonts shared the herbivore niche with another therapsid subgroup, the dinocephalians. These "terrible headed" brutes were named for their ugly craniums, which were big, bulky, and often covered in gnarly knobs, bumps, or horns.

The bones of the skull were exceedingly thick and dense. One species, *Moschops*, had skull bones nearly five inches (twelve centimeters) thick, which it probably used to head-butt rivals in fights over mates or territory. It's a terrifying scene to conjure: two *Moschops* alpha males—their grotesque heads, stocky builds, and arching backs giving them the appearance of characters out of *Where the Wild Things Are*—bashing each other's brains while the rest of the herd gathered ringside. It's even more terrifying when you consider that, while some were herbivores, other dinocephalians were fierce carnivores. One of them, *Anteosaurus*, grew up to sixteen feet (five meters) long and weighed about half a ton, the same general size as a polar bear. It was among the largest synapsid predators that ever lived, before the advent of modern mammals.

There was another type of therapsid predator, a little bit smaller, but probably more ferocious. These were the gorgonopsians, the terrors of the middle and late Permian. They ranged in size from a small dog to monsters like *Inostrancevia*, which was 11.5 feet (3.5 meters) long, weighed around 650 pounds (300 kilograms), and had a 2-foot-long head. Their most vicious weapons were their enlarged canines, of the saber-tooth style. Their jaws could open obscenely wide, clearing enough space for the canines to puncture the hides and windpipes of their prey. But unlike saber-toothed tigers—which are proper mammals like us—gorgonopsians replaced their teeth throughout life, so they could attack their victims with the confidence that a broken canine could simply be regrown if anything went wrong. Rounding out their armory were sharp nipping incisors, for manipulating struggling prey, and expanded jaw muscles that bulged with anticipation, sticking out farther backward and to the sides compared to other therapsids. One thing they weren't, though, was very intelligent, as they retained the unremarkable, tube-shaped brains of their pelycosaur ancestors.

These and other therapsids became triumphant. The middle-late

Permian world was theirs—in the same way that dinosaurs and mammals would later claim dominion. Together, various therapsids formed complex ecosystems, which for the first time in Earth history were completely detached from the water. Pelycosaurs, you might recall, were a waypoint on this journey in the early Permian, but they still lived near lakes and streams, and were parts of food chains that included sharks and other fish. The Karoo therapsids, however, comprised a community that wasn't too different—in general ecological structure—than the modern African savanna fauna. Herds of shrub-eating dicynodonts formed the base of the food pyramid, outnumbering the carnivorous gorgonopsians tenfold. The land plants were the primary producers, the plant-eating therapsids the primary consumers, and the meat-eating therapsids the apex predators. Not too different from the grasses-wildebeest-lion trifecta in the savanna today.

All this therapsid diversity—from dicynodonts to dinocephalians, from gorgonopsians to the many other subgroups I haven't even mentioned—evolved from a common ancestor, which was a midsize, meat-eating pelycosaur, probably weighing somewhere between 100 and 220 pounds (ca. 50–100 kilograms), that lived no later than the early-middle Permian boundary. It seems as if this ancestor, and the early therapsids that sprang from it, came from the temperate regions—distant from the humid tropics, where it was always summer.

These early therapsids started to do something unusual: they began to raise their metabolism and develop better control of their body temperatures. It's not exactly clear why. Perhaps their higher latitude homes forced them to deal with more seasonal climates, and by finely tuning their internal furnaces, these therapsids could better survive both cold spells and heat waves. Or maybe it was driven by hunger. The pelycosaur ancestors of therapsids, with their sprawling limbs and slow-moving ways, were probably "sit-and-wait" predators, which loitered around most of the time, occasionally bursting out to capture prey. Some therapsids, however, became foragers, ranging across wide areas and seeking out prey on foot. This method of hunting would have required more energy, and thus, perhaps, higher metabolisms. Debate continues. No doubt, though, that the physiology of therapsids was changing in the Permian. Regardless of the reason, these animals were taking the critical first steps in the development of one of the most mammalian of abilities: our warm-blooded—or in scientific parlance, endothermic—metabolism.

There are a few lines of evidence that therapsids—although not yet fully warm-blooded—were growing faster and were more metabolically active than their pelycosaur ancestors. The best clues come from making thinner-than-salami slices of bone, mounting them on slides, and looking at the texture under the microscope. Different types of textures are indicative of different rates of growth, and some bones even have yearly growth lines inside— like the rings of a tree trunk—that tell you how old the animal was when it died. The South African paleontologist Anusuya Chinsamy-Turan is one of the pioneers of this field, called bone histology.

Anusuya grew up in Pretoria in the apartheid era. She dreamed of becoming a science teacher, but higher education options were limited for young women of her background. Instead of abandoning her aspiration, she told what she calls a "white lie" when applying to the University of the Witwatersrand, which required nonwhite students to provide a persuasive reason for wanting to attend their largely white university. She said she wanted to study paleoanthropology—a field that Wits was uniquely famous for, by virtue of the rich fossil hominin record of South Africa. That meant she had to take paleontology classes, and much to her surprise, her fib turned into her passion. She went on to do a PhD,

Anusuya Chinsamy-Turan studying microscopic images of bone in her lab. (photo courtesy of Anusuya Chinsamy-Turan)

become a world expert in deciphering the texture of fossil bone to determine growth rates, and in 2005 was named South African Woman of the Year in honor of her scientific contributions.

Anusuya and her colleagues Sanghamitra Ray and Jennifer Botha cut up a lot of therapsid bones, particularly the ribs and limb bones of dicynodonts and gorgonopsians. They found that these bones had a predominant type of texture, called fibrolamellar bone, haphazardly arranged into an entwined pattern. The chaotic arrangement is the result of rapid growth: the bone was deposited so quickly that collagen and minerals were laid down in a random pattern. This differs from the more regular lamellar bone of slower-growing animals, which forms ordered layers of mineral crystals. The presence of widespread fibrolamellar bone meant that these therapsids were growing briskly, at least during parts of the year. Because these bones also had growth lines, growth must have stopped at times, probably during the winter or dry season.

Thus, these therapsids had elevated growth rates compared to typical "reptilian" species, and some control over body temperature, but they probably weren't able to maintain constant high temperatures like fully warm-blooded mammals.

There's another clue that therapsids were raising their metabolisms and better controlling their body heat.

Hair.

Therapsids, it appears, invented hair. Coprolites—fossilized dung—with therapsid bones inside also include tangled masses of hairlike structures. These are controversial, but if they are hair, odds are they belonged to the therapsids. Regardless, there is a stronger line of evidence for hair: many therapsid fossils are pocked with pits and grooves on their facial bones, similar to the network of canals that bring blood and nerves to the whiskers of mammals today. That's not to say that therapsids were furry, with hair all over their bodies. Maybe they were, but more likely, they had a mangier coat of hair, or maybe their hair was restricted to small regions, like around the head and neck. The point is: it seems as if hair originated in the therapsids.

Hair is among the most quintessential of mammalian novelties. It is a fundamental component of our fleshy and glandular skin, so different from the scaly integument of our tetrapod ancestors, maintained in reptiles today. Hair likely began as a sensory aid (like whiskers), a display structure, or as part of a gland-based waterproofing system, and was later repurposed as a body coating to retain heat. Once an animal has a lot of hair on its body, it's a telltale sign that it is producing at least some body heat internally, and doing its best to keep it from escaping. Making heat is expensive. If you're going to run your furnace at full blast, you'll want to close the windows, too, lest your gas bill will become unsustainable. For mammals, hair is that closed window.

The increased growth rates and higher metabolism of therapsids were profound evolutionary acquisitions and were related to a series of other changes to therapsid anatomy and biology. The limbs moved farther underneath the body, resulting in a more upright posture—as Broom recognized when comparing therapsids to their pelycosaur ancestors. Dicynodonts had erect hind limbs but sprawling forelimbs—evident not only from the shape of the shoulder and pelvic joints, but also from fossilized trackways of narrow footprints trailing more sprawling handprints. The more advanced gorgonopsians, however, had straighter forelimbs and hind limbs. The limbs were also becoming more flexible. Therapsids lost the awkward screw-shaped shoulder joint that constrained the arms of pelycosaurs to a slow-moving sprawling gait, thus freeing the forelimbs to do all sorts of new things—running, digging, and climbing.

These changes were happening in concert with one another, and in many cases, it's hard to unravel what was driving what. The renowned early mammal expert Tom Kemp calls this "correlated progression": many anatomical, functional, and behavioral aspects of therapsids were changing in unison, and in doing so these animals were evolving—step by step—features that characterize today's mammals. Put another way, they were progressively becoming more mammal-like as the Permian played on.

By the late Permian, this long march of correlated progression had produced a new type of therapsid, which was smaller, with even more upright limbs, faster growth, and higher metabolism than its dicynodont and gorgonopsian antecedents. Its teeth, jaw muscles, brains, and sensory systems were changing too. These creatures—Owen's "weasel lizard" *Galesaurus* among them—were the cynodonts, and they were the next big step on the line to mammals.

2

MAKING *a* MAMMAL

Thrinaxodon

AS THUNDER BOOMED IN THE distance, and rain streamed down, an animal stuck its head out of its burrow. It twitched its nose and set its whiskers to the wind. Time to leave, fast.

When this weasel-size creature—a *Thrinaxodon*—dug its burrow a few months earlier, the land was parched. It hadn't rained for months. The river was nearly dry, and the ferns and club mosses that had once gripped its banks were reduced to wilted husks. Dust devils swept across the valley, burying herds of potbellied herbivores on a desperate search for the last edible leaves and roots. Some of them stuck out from the sand dunes, dead and mummified, their sharp tusks giving the dystopian scene an even more sinister air.

By then it was clear that there was no food left. No sight of insects, no sniff of tasty amphibians, and only dry, desiccated flesh to scavenge from the corpses. So the furry creature had no choice: burrow, hunker down for a while, and conserve precious energy until conditions improved.

Now, as if a switch had been flipped, the valley was being pelted by monsoonal rains. The river had burst its banks and water was entering the burrow, slowly filling the bulbous chamber in which the *Thrinaxodon* had been resting. Outside, green shoots were starting to poke through the mud, which the engorged river had spread across the dunes, covering up the creepy mummies. The slate, it seemed, had been wiped clean. Life was returning, the dry months a distant memory. But in this bipolar world, the rains wouldn't last long, so the *Thrinaxodon* had to take advantage.

It first needed to eat, to kick-start its metabolism. *Thrinaxodon* was a greedy feeder. Its voracious appetite fueled its fast growth and gave it the energy reserves necessary for hunkering down in its den and keeping its body temperature at a high, steady level during the months of torpor. Fasting had made it even hungrier than usual, and it relished the thought of piercing its sharp, multicusped teeth into the exoskeletons of insects, or the slimy skin of small amphibians that were congregating near the river.

Then, with its belly full, it could move on to the next important task: finding a mate. This *Thrinaxodon* had been born less than a year earlier, toward the end of the last rainy season. For a few weeks it clung close to its mother and siblings, gorging on insects and learning the topography of the river valley, before setting off on its own, finding a nice patch of muddy floodplain to dig in, and curling up in its den when it could no longer handle the heat. Now that the rains had returned, this would probably be its only chance to mate—one shot in a short, strange life of birth, inertia, and a final frenzy of gluttony and reproduction.

At least there were a lot of potential mates, though, as *Thrinaxodon* burrows peppered the flatlands on both sides of the river. Their internal bedrooms opened to the surface as small holes, in a pockmarked pattern that looked like the craters on the moon. All around it, our *Thrinaxodon* could see many of its compatriots sticking their heads out of their burrows, twitching their noses, rain streaming down their hairy faces, their whiskers straining to sense what was happening. They were all considering the same question: Stay or go?

Our *Thrinaxodon* made its choice. It wiggled out of its burrow, scrunching its limbs underneath its body so it could fit through. It ambled onto the mud flat, all soupy and mucky, and spread its arms and legs wider to brace itself. As it watched its burrow fill with water, the *Thrinaxodon* scurried off into an uncertain future. Food and a mate might await, or they might not. Either way, it would all be over soon.

This particular *Thrinaxodon* didn't know it, but it was living in interesting times. True, it didn't have the brainpower to reason out its place in the history of life, of evolution, of Earth. But then again, people living during interesting times usually don't realize it, either, too caught up in the moment, focused on the next meal, or their families, or a myriad of other things. You often don't realize you are living through upheaval until things have settled, and hindsight sets in. These *Thrinaxodons*, as it turned out, were living through the greatest cataclysm in the history of Earth—a short period of a few tens to hundreds of thousands of years, which began with a catastrophic extinction, saw the start of a halting recovery, and helped forge mammals from their ancestral therapsid stock.

THRINAXODON IS A cynodont, which lived about 251 million years ago, during the very earliest Triassic Period. Cynodonts were part of the mammal "stem lineage." They were members of the therapsid group, alongside the tusked dicynodonts (the mummies in the above story), head-butting dinocephalians, and saber-toothed gorgonopsians. Therapsids evolved from pelycosaurs, which stemmed from that "scaly critter" that split into the synapsid and diapsid lines during coal forest times, which in turn can trace its ancestry back to those tetrapods that evolved from fishes, crawled onto the land, and developed amniotic eggs.

All this we learned in the last chapter. But life goes back much, much further: fishes evolved from the first vertebrates, minnowy swimmers that started reinforcing their bodies with bone during the whirlwind of evolutionary change called the Cambrian Explosion, between about 540 and 520 million years ago. This was the same time that so many of today's most familiar groups of ocean-living species invented their own skeletons and began to prosper—mollusks like mussels and clams, echinoderms like sea urchins and starfish, arthropods like shrimp and crabs. Before that, these animals had soft-bodied ancestors that lived during the Ediacaran times, beginning around 600 million years ago, which left ghostly impressions of their blobby bodies in sandstone. These were the first animals, and they evolved from bacteria that were able to clump themselves together into larger, more complex, multicellular forms. This happened around two billion years ago, which was some two billion years after the first single-celled bacteria evolved, which was only about half a billion years after Earth formed from a cloud of gas and dust.

Life has been a four-billion-year evolutionary spectacle, which, of course, is still running today. During this entire time, the closest life has ever come to being extinguished—completely wiped out, leaving Earth a barren planet—was at the transition between the Permian and Triassic periods, between 252 and 251 million years ago. This was not long before *Thrinaxodon* was hiding in its burrows—during the tortured recovery phase of this catastrophe—in what is now the Karoo of South Africa.

The end-Permian was the mother of all mass extinctions, and it claimed something like 90 percent of all species, maybe more. Unlike with many other mass deaths in the fossil record, there is no murder mystery. Volcanoes did it—so-called megavolcanoes, fed by a hot spot of magma deep within the earth's mantle, which parked itself underneath what is now Siberia but was then on the northern fringes of the Pangean supercontinent. These eruptions were different from any that humans have ever witnessed, thankfully. Their size and scale were preposterous. For several hundred thousand years, lava spewed from giant cracks in the ground—a sprawling network of volcanic vents, each one up to several miles long, bleeding lava like the earth had been slashed with a supersized machete. There were bursts of fire and quieter periods, and by the end of it all, several million square miles of North Pangea were crusted over with basalt, the hardened remnant of the lavas. Today, even after more than 250 million years of erosion, this basalt covers about a million square miles—about the same land area as Western Europe.

These volcanoes ruptured the peace of a therapsid world, a time when these early mammal antecedents paraded across the arc of Pangea. There were so many species, of astounding shapes and sizes, with their tusks and beaks and head-smashing domes and stabbing canines, eating so many things and filling so many niches, as snarling predators to plant-connoisseurs. From the vantage point of the latest Permian, right before the first volcanoes bubbled, it might seem like therapsids would continue their dominion, but it was not to be.

During the late Permian, many therapsids lived in what is now Russia, not too far from the volcanoes. Gorgonopsians sunk their sabers into dicynodonts, and cynodonts were there, too, lurking in the seed fern forests. They would have been the immediate victims of the eruptions, and many of them were probably literally engulfed in lava, as in a kitschy disaster film. But they were not the only casualties, because these volcanoes were much deadlier than they appeared. Bubbling up with the lava came silent killers, noxious gases like carbon dioxide and methane that seeped into the atmosphere and spread around the world. These are greenhouse gases, which trap heat in the atmosphere by absorbing radiation and beaming it back to Earth, and they caused runaway global warming. Temperatures increased by about 9 to 14 degrees Fahrenheit (5–8 degrees Celsius) within a few tens of thousands of years—similar to what is happening today, although actually at a *slower* pace than modern warming (a fact that should give everyone reading this pause). Still, it was more than enough to acidify the oceans and starve them of oxygen, causing widespread death of shelled invertebrates and other sea life.

On land it wasn't any better, and the best record of what happened—what died, what lived, how quickly things recovered—comes from the Karoo. As the volcanoes erupted and the atmosphere warmed, the Karoo climate became hotter and drier in the earliest Triassic. The seasons became more pronounced, as did temperature fluctuations during the day. In effect, the Karoo became a desert, not unlike what it is today, with one big exception: it was occasionally pummeled by monsoons, which tore across Pangea. The diverse Permian forests, dominated by the seed fern *Glossopteris* and evergreen gymnosperms, collapsed, as plants endured their second and final mass extinction—the only one after the Carboniferous Rainforest Collapse some 50 million years earlier. They were replaced by ferns and club mosses—much smaller relatives of the coal swamp *Lepidodendron* trees—which grew fast from spores rather than seeds, allowing them to better cope with the intense seasonality and fluctuations in rainfall. As the vegetation changed, the wide, meandering river systems of the Permian gave way to faster-moving braided waters in the Triassic. Without roots of large trees to stabilize their banks, these rivers scoured the land during the rainy season, but trickled to nothing during the arid months.

This environmental cascade had disastrous effects on the animals living in the Karoo, particularly the therapsids. Before the extinction, there was a thriving community with several plant-eating dicynodonts at the base of the food chain, preyed on by smaller carnivores called biarmosuchians—another type of therapsid, with gaudy bumps and hornlets on their heads—and the apex hunter gorgonopsians. Rare cynodonts like *Charassognathus*—the oldest-known member of the group, the size of a squirrel—ate insects, sharing the small-bodied vertebrate niche with a bounty of reptiles and amphibians, some of which also ate bugs, others fish. But, as the climate changed, the forests withered, and about 70–90 percent of surface vegetation disappeared. This caused the entire ecosystem to fall like a house of cards. Food webs simplified—with only a few herbivores and carnivores—in the earliest Triassic, and for about five million years afterward suffered boom-bust cycles before finally stabilizing as the volcanoes stopped erupting and temperatures normalized.

If you were a therapsid, you had three possible fates. The first was extinction—that's what happened to the gorgonopsians, which would never terrorize Triassic prey with their saber canines and wide-jawed bites. The second was survival but degeneration—that's what happened to the dicynodonts, which made it through the destruction and re-diversified, but never quite replicated their Permian successes before wasting away and being put out of their misery by the next mass extinction, at the end of the Triassic. The third was survival and domination—that's the path the cynodonts took, persevering through the volcanoes, global warming, aridification, monsoons, forest collapse, ecosystem implosion, and the 5-million-year slog of recovery all the stronger for it. They would continue to diversify during the remaining 50 million years of the Triassic, spinning out a great variety of species—big ones, small ones, meat-eaters, plant-eaters. One of these cynodont lines led to mammals, picking up more "mammalian" features along the way.

Why were the cynodonts—along with some of their dicynodont cousins—able to survive? I acutely remember when I learned the answer. It was at the annual meeting of the Society of Vertebrate Paleontology, when it was held in Los Angeles in 2013. I had recently finished my PhD and started my new faculty job in Edinburgh, and I was presenting my doctoral research in the Alfred Sherwood Romer Prize session—named for the legendary Harvard paleontologist who led the expeditions to Nova Scotia that discovered the tree stumps full of early synapsids that we learned about in the last chapter. The Romer Prize is the premier award for graduate students in my field, and I was hoping to wow the judges with my work on the origin of birds from dinosaurs. Alas, I did not win the prize, but it couldn't have gone to a more deserving colleague: Adam Huttenlocker. He stood up a few slots after my

talk and captured the audience by explaining the mystery of cynodont survival at the end-Permian. By the time he sat down, I was
resigned to my fate. Once again, mammals (or in this case, proto-
mammals) would best the dinosaurs.

Adam described a peculiar evolutionary phenomenon that happened during and after the extinction: the Lilliput effect. Named
after the fictional island in *Gulliver's Travels*, inhabited by tiny
humans, the Lilliput effect is a decrease in body size in animals
that survive a mass extinction and prosper afterward. It doesn't
always happen, but it did in the case of cynodonts and their close
relatives, and this was a big part of their endurance. Adam marshaled an enormous database of Karoo therapsid fossils and found
a pronounced decrease in both maximum and average body size
in the earliest Triassic therapsids compared to their latest Permian
precursors. The difference was due to the heightened extinction of
larger-bodied species as the volcanoes erupted and the temperature
boiled. Being big, it seems, was a hindrance during this unstable
time. Cynodonts, which were smaller than most other therapsids,
thus had a better chance of making it through the chaos.

Why was small size beneficial? For one, smaller animals could
hide more easily and wait out bad weather, temperature swings,
and dust storms in their burrows. And burrow they did. The
Karoo rocks above the extinction layer—mudstones formed on
a river's floodplain, interspersed with mummies buried by windblown dust—are full of fossil burrows, some of which have
skeletons inside. Skeletons of *Thrinaxodon* itself, the hero of our
chapter-opening story. The most remarkable of these burrow
graves contains a *Thrinaxodon* lying side by side with a small,
injured amphibian. The little salamander relative had suffered
trauma to its ribs but was healing, while it rested alongside the
curled-up, sleeping *Thrinaxodon*. As the burrow was a tight space,
and *Thrinaxodon* a sharp-toothed predator, it seems odd that

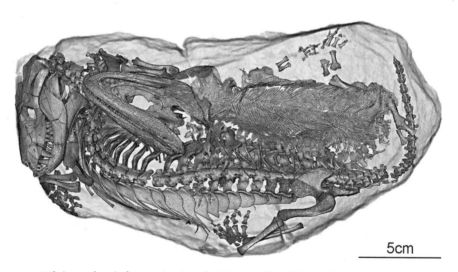

5cm

Thrinaxodon skeletons (top) and CT scan of a *Thrinaxodon* fossilized in a burrow next to an amphibian (bottom). (images by Christian Kammerer and from Fernandez et al., 2013, *PLoS ONE*, respectively)

another creature could just recuperate there unnoticed. The only satisfying explanation is that the *Thrinaxodon* was estivating: lying dormant, perhaps for several weeks or months, to conserve energy and ride out the dry season.

Second, small size is linked to many other aspects of growth and metabolism. Jennifer Botha, working with Adam and a team of colleagues, laid these out in an important 2016 study. The earliest Triassic cynodonts—like *Thrinaxodon*—grew fast, started reproducing at a young age, and had very short lives, probably of no more than a couple of years. How do we know this? Bone histology—examining thin slices of bone under the microscope— holds the key. Permian therapsids, for the most part, have many growth lines in their bones, meaning that it took them many years to reach adult size. The earliest Triassic cynodonts, however, have fewer growth marks, and in fact, *Thrinaxodon* usually doesn't have any. They must have grown at frenzied speed, perhaps becoming mature, reproducing, and dying all within a single year. In effect, they were breeding young to compensate for dying at an earlier age. While no single *Thrinaxodon* individual would live to see old age, this growth strategy was better for perpetuating the species. By growing fast and reproducing earlier, they would have a better chance of successfully making it through a mating season, ensuring that they passed on their genes to the next generation in this harsh, mercurial world.

Thrinaxodon and other cynodonts, it seems, had a strong hand of cards that won them reprieve from the extinction. Skeletons of *Thrinaxodon* start appearing by the dozens about a hundred feet (thirty meters) above the extinction horizon, meaning that they proliferated across the Karoo Basin within a few tens of thousands of years. They may have evolved from Permian ancestors that were rare components of the pre-extinction Karoo ecosystems, or more likely, they immigrated from the intertropical regions of Pangea, where harsher Permian climates had preadapted them for dealing with drought. There are so many *Thrinaxodon* skeletons in the earliest Triassic rocks that it, and a similarly abundant dicynodont called *Lystrosaurus*, are considered "disaster species." They were particularly well suited to the nasty conditions of the postextinction hothouse, which most other species could not handle. They seemed to *like* these conditions, to bloom in them. *Thrinaxodon* and *Lystrosaurus*, thus, were the subway rats and cockroaches of the Early Triassic.

The comparison to modern pests, however, does *Thrinaxodon* a disservice. It really was a champion, one of the few plucky animals that made it through the dark night of prehistory's worst massacre, ensuring that the mammal line wasn't snuffed out long before mammals had a chance to evolve.

Thrinaxodon was an unlikely hero. Not even two feet long, and probably bearing whiskers and covered with at least a patchy coat of hair, *Thrinaxodon* would have spent a lot of time hidden from view, nestled in its dens. It would have to come out to feed, though, and it had a taste for insects and small game. Like its therapsid ancestors, it had a set of incisors, canines, and cheek teeth. Where it differed—and what gave it its name, which means "trident tooth"—is that the cheek teeth had the shape of a clip-art mountain range, with one central peak flanked by a smaller peak on each side. These three sharp prongs—called cusps—were perfect for puncturing insect exoskeletons and tearing through flesh. These cusped teeth were constantly replaced throughout *Thrinaxodon*'s short life, like in a reptile or amphibian.

In many aspects, however, *Thrinaxodon* was remarkably more mammalian than its therapsid forebears. It walked even more upright, belly farther above the ground, in a semisprawling stance. This would have allowed it to run faster, perform more athletically, and fit more comfortably in its burrows. Its vertebral column was not uniform but was divided into sections that bore ribs and others that did not, giving it more flexibility and enabling it to curl up when estivating. Its jaw-closing muscles were massive, anchored to a sheet of bone sticking up from the skull roof, called the sagittal crest. As a *Thrinaxodon* rapidly grew from hatchling to adult, the crest expanded in size, endowing an ever-more-powerful bite. Exceptional fossils preserve several *Thrinaxodons* together, showing that they were social animals that assembled into groups. In some cases, adults are fossilized with smaller, younger individuals—evidence of parental care.

There's one last thing that's important about *Thrinaxodon*. It didn't stick to the Karoo Basin. Its fossils have also been found in Antarctica, a sign that it was so well adapted to the foul Early Triassic world that it was dispersing around Pangea. The earth was now one completely unified landmass, surrounded by a single vast ocean, and as the volcanoes cooled and ecosystems healed from the extinction, the cynodonts were poised to become major players on this new stage.

LIKE MANY WRITERS, Walter Kühne did some of his best work while in prison. How he—a paleontologist with a fondness for microscopic teeth—got there is quite the story.

Kühne was born in Berlin in 1911, the son of a drawing teacher. While studying paleontology at Friedrich-Wilhelm Universität in Berlin, and later at the University of Halle, he became known for two things: his interest in medieval church bells, about which he wrote an effusive article in a travel magazine, and his communist sympathies. The latter was of much more concern to the Nazi authorities, who were just beginning their rule of terror. The young Kühne was sent to prison for nine months—his first experience behind bars—and then forced to immigrate to Great Britain in 1938.

How was a poor political refugee to support himself and his young wife in a foreign land? By collecting fossils, naturally. Kühne had heard about a handful of Triassic-aged mammal teeth discovered in a cave near Holwell, a village of a few hundred people, in the mid-nineteenth century. It should have been a monumental discovery but was largely ignored. Few paleontologists, it appeared, wanted to spend months picking through bushels of rubble in the hope of finding minuscule teeth. Still, he was told by a curator at the British Museum not to bother. "All English fossil deposits are well known," the curator apparently said. "It would be absurd to dream of making major new discoveries."

Kühne was not to be deterred. He was both desperate for money and obsessed with mammals. He also had a few tricks up his sleeve. He had developed a sharp eye for fossils as a student, and he possessed an even more important quality: patience. He went to Holwell and gladly collected, washed, and screened more than two tons of cave-fill clay—a job made easier by the assistance of his wife, Charlotte, whose affection (or lack thereof) for teeny teeth has gone unrecorded by history. The Kühnes' diligence paid off: they found two premolars. Walter promptly strutted into the University of Cambridge and showed them to the paleontologist Rex Parrington, who was so impressed that he put Walter on the payroll. From then on, each mammal tooth would be worth £5.

Walter Kühne's illustration of *Oligokyphus* from his 1956 monograph. (image modified from Kühne, 1956)

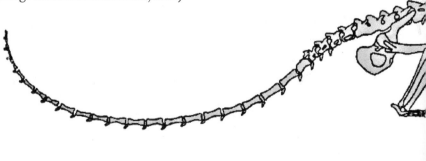

Brimming with confidence, Walter and Charlotte expanded their search to other caves and fissures in southern Britain. Soon after, in August 1939, they found new fossils in the Mendip Hills, a bucolic landscape south of Bristol, in the Somerset countryside. They collected dozens of isolated teeth and bones belonging to a very mammal-like cynodont called *Oligokyphus*, which had been named a few decades earlier from a mere couple of teeth from Germany. Then they moved on, seekers on the prowl for the next great discovery. In September, hammer and geological map in hand, Walter made his way to the Atlantic coast, where he began scrutinizing some limestone cliffs. It is unclear whether he realized that his erstwhile fatherland had just invaded Poland.

The British soldiers patrolling the coast certainly knew that a world war had begun. They found it peculiar that a German was wandering the English seaside with a fistful of maps, so they arrested him. That's how Walter eventually found himself incarcerated a second time, in an internment camp on the Isle of Man, a speck in the Irish Sea between Great Britain and Ireland. From 1941 until 1944, the camp would be his home.

If there was one bright spot, it was that, by this time, Walter had earned the respect of the English scientific establishment. Not long after their own city of London was blitzed, the curators and scientists at the British Museum—who had once discouraged the

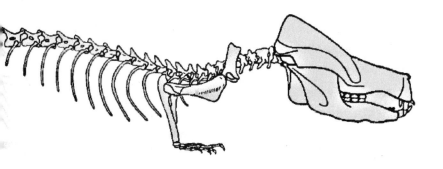

enthusiastic German from collecting British fossils—organized further trips to the Mendip caves, which netted scores of additional bones and teeth. They shipped some two thousand of these fossils to the internment camp.

Walter had, in his words, "considerable time at my disposal." So he laid the fossils out, put the bones together, and assembled much of a skeleton of *Oligokyphus*. Meticulously describing the fossils kept him occupied, and by the time he was released from detention during the waning days of the war, he had started writing up his findings, which would culminate in his 1956 monograph on *Oligokyphus*. It still stands as one of the benchmark works on those cynodonts that were on the cusp of becoming mammals.

Oligokyphus—the size and shape of a runt-of-the litter miniature dachshund—was not a mammal, nor was it a direct mammal ancestor. Instead, think of it as a very near relative—a first cousin. It was a member of a subgroup of advanced plant-eating cynodonts called tritylodontids, which sit next to mammals on the family tree. Like any close cousins, tritylodontids and the earliest mammals were extremely similar, in their bodies and behaviors. For example, both had limbs that were placed directly underneath the body, so they walked fully upright, unlike other cynodonts that still sprawled a little bit. Both tritylodontids and mammals were part of a pulse of cynodont diversification during the Late Triassic, beginning around 220 million years ago. This was a good 30 million years after *Thrinaxodon* weathered the end-Permian mass extinction and shepherded the mammal line through its most vulnerable moment—as of yet. During those 30 million years a lot had changed: the mammal "stem lineage" continued to accumulate "mammalian" features—like the straight-up limb posture—as the cynodonts navigated a maze of fierce weather and even fiercer competition.

The biggest change was, well, small. Diminutive size had already helped cynodonts at the end of the Permian, so they kept a good thing going. Over the course of the Triassic, the mammal line got progressively tinier. What began as weasel-size species like *Thrinaxodon* had, by the Late Triassic, turned into a variety of mostly rat or mouse-size vermin. There were some exceptions to this rule, as side branches on the family tree occasionally turned out bigger species, like *Oligokyphus* and its tritylodontid brethren, which needed bigger guts for digesting plants. But by and large, Triassic cynodont evolution was a march of miniaturization.

Why were the cynodonts shrinking in size? For one, they were not alone in the brave new world of the Triassic. Other animals survived the end-Permian calamity, too, and they all jostled for space as Pangea was on the mend. Out of this evolutionary crucible came not only mammals, but also many of the most familiar animals that continue to live with mammals today: turtles, lizards, and crocodiles. Plus, something even more terrifying was fanning across the supercontinent, expanding and diversifying from a humble, cat-size ancestor that withstood the volcanoes.

Dinosaurs.

The founding dinosaurs were battling their crocodile cousins for supremacy, and both were getting bigger and bigger. By the end of the Triassic there were thirty-foot-long (nine-meter-long), ten-ton long-necked dinosaurs like *Lessemsaurus*—primitive kin of the colossal sauropods like *Brontosaurus*—and a variety of knife-toothed carnivores that preyed on them. As the dinosaurs went big, the mammal ancestors went small. This was the beginning of a recurring story line: the fates of mammals and dinosaurs, intertwined.

In concert with their decreasing body size, many cynodonts on the mammal line may have become nocturnal. Crawling out in the dead of night to eat and socialize was a good tactic for avoiding the jaws, or the crushing footstomps, of dinosaurs, which were probably daytime creatures. Moving into a nighttime niche wouldn't have been too difficult, as it seems like many earlier synapsids on the mammal stem lineage—Permian pelycosaurs and therapsids—had first experimented with this lifestyle. However, the darkness had its consequences. Mammal ancestors essentially gave up on keen vision, going all in for scent, touch, and hearing. Most mammals cannot see in color, which is why most mammals have drab brown, tan, or gray fur. Why dress yourself in flamboyant hues—like many day-living, sharp-eyed birds and reptiles do—if your mates or rivals cannot see them? This may seem strange to us; after all, we can see in color! But we are highly unusual among mammals, one of the few modern species that can sense color, along with some of our closest primate relatives. When a matador flaunts a red cloth at a bull, the bull sees it as black.

As if avoiding dinosaurs wasn't reason enough, small size may have also conferred other benefits to the mammal ancestors. Pangea was a unified landmass, but it was not a safe place to call home. It was hot, there were no ice caps on the poles, and much of the interior was endless emptiness. Violent air currents streamed across the equator, powering intense weather systems called megamonsoons, which as their hyperbolic name implies, were supersize versions of today's tropical storms. Although theoretically a proto-mammal could walk from pole to pole, this would have been a fool's journey. The megamonsoons helped divide Pangea into climatic provinces, with different amounts of precipitation and winds, and different temperatures. The equatorial regions were a broiling and humid hellscape, bordered on both sides by impassable deserts. The midlatitudes, however, were slightly cooler and much wetter than the deserts, and it was here where many of the rapidly evolving Pangean animals lived. Small size may have been a survival strategy in this dangerous world. The smaller you were, the easier it was to hide, to burrow, to sleep away the megamonsoonal rains and the carnage they wrought.

Whatever the reasons cynodonts got small, it profoundly transformed their entire biology, and their evolutionary trajectory. As they shrunk in size, they also changed many aspects of their growth, metabolism, diet, and feeding styles. They already had elevated body temperatures and metabolisms inherited from their therapsid ancestors, but now they developed full-on warm-bloodedness. They already had strong jaw muscles and bites, a legacy of their even deeper pelycosaur ancestors, but now they innovated new ways of eating lots of food fast while continuously breathing at this same time—eating on the go.

This was a continuation of the "correlated progression" of evolution that Tom Kemp described in Permian therapsids. Many things were changing in harmony, and it's difficult to untangle what was driving what. Perhaps small size necessitated higher body temperatures to buffer against sudden climate shifts or demanded more efficient ways of gathering and processing smaller parcels of food. Maybe warm-bloodedness mandated that these cynodonts ate bigger meals to fuel themselves, or possibly it was the other way around: changes to the jaws and muscles came first, allowing them to eat more, and thus providing more energy for warm-blooded physiology to develop. We don't really know the answer. What we do know, though, is that small size, warm-blooded metabolism, and stronger and more efficient bites developed together as part of a package deal.

How do we know this was happening? The astoundingly rich fossil record of Triassic cynodonts—linking *Thrinaxodon* to *Oligokyphus* and mammals—is our guide.

First, let's consider physiology and metabolism. We touched upon this subject in the last chapter, but it deserves greater explanation. The term warm-blooded is convenient shorthand for a whole range of sophisticated body temperature control mechanisms. It is not the case that warm-blooded animals have hot

blood, and cold-blooded animals cool blood. In fact, stick a ther-
mometer into an average warm-blooded mammal and an average
cold-blooded lizard, and you will probably get a similar reading.
Or you might even get a higher reading for the lizard, particularly
if it's a sunny day. That's because cold-blooded animals rely on
their environment to heat their bodies, meaning they are at the
mercy of changes in the weather, changes in the seasons, even
temperature swings from night to day, or between sun and shade.
Warm-blooded animals—technically called endotherms—have
broken free of this handicap. They produce their own heat—
often by packing more energy-producing mitochondria into their
cells—and thus can maintain a body temperature warmer than
their environment. We experience this every time we go out into
the winter frost and don't literally freeze.

In effect, warm-blooded animals have their own internal furnace, which is always on and is always running hot. This enables
higher metabolism, faster growth, more energetic lifestyles, greater
stamina, and more athletic behaviors. Mammals can sustain running speeds some eight times faster than lizards, for example, and
they can also forage for food across wider areas. These superpow-
ers come at a cost, though. Warm-blooded animals have higher
resting metabolic rates, meaning they burn more calories than
cold-blooded animals while at rest. And of course, they burn *a
lot* more calories when they are active—when they are running,
jumping, chasing prey, fleeing predators, scampering up trees,
digging burrows, and doing the many other things that a warm-
blooded metabolism makes easier. So they have to eat a lot more
calories than cold-blooded animals of similar body size, and they
have to breathe in a lot more oxygen.

Animals aren't necessarily warm-blooded or cold-blooded,
however. There are intermediates. Today's mammals and birds are
fully warm-blooded, which means three things. They *control* their

body temperatures internally, and have *high* body temperatures and *constant* body temperatures, regardless of whatever their external environment throws at them. This type of system did not evolve at once, but over time—through transitional stages in which their ancestors became increasingly better at both making heat and controlling the temperatures of their furnaces. For mammals, this process began with therapsids in the Permian, as they dealt with the seasonal climates in their higher-latitude homes. Their finer control of body temperature and faster growth were probably central to their survival—particularly of the *Thrinaxodon*-type cynodonts—at the end-Permian. Then, during the Triassic, these cynodonts continued to move toward total warm-bloodedness.

There is a wealth of evidence that cynodonts were becoming fully warm-blooded as the Triassic unfolded. Fibrolamellar bone—the chaotically arranged texture indicative of rapid growth—became increasingly prevalent on the line to mammals. The bone cells and blood vessel channels within the bones got smaller across cynodont evolution, a signal that their red blood cells were getting smaller, too—another mammalian feature, which allows these cells to take up more oxygen, faster. And one clever study measured the oxygen composition of various Permian and Triassic species by crushing up their bones and teeth. The ratio of the two most stable types of oxygen—the lighter and more common 16O and the heavier and rarer 18O, which differ in the number of neutrons—depends on the temperature at which the bones and teeth grew. In effect, this oxygen ratio is a paleo-thermometer, and its signal is clear: the Triassic cynodonts had higher and more constant body temperatures than the animals they were living alongside, including most other therapsids.

How did these cynodonts pay the heating bill? The same way any warm-blooded animal would: by taking in lots of oxygen, and lots of calories. Along the line to mammals, cynodonts acquired many anatomical features that allowed them to increase their consumption of both.

Critically, cynodonts freed themselves from a thorny problem called Carrier's constraint, which plagues amphibians and reptiles that wiggle their bodies sideways, left to right, during their sprawling walks. Such lateral flexion means that one lung is always expanded while the other is compressed, making it difficult for these animals to move and breathe at the same time, greatly limiting their speed and agility. Cynodonts developed both a more upright stance, as we've seen, but also bony brakes on their vertebrae, which prevented the spine from moving too much side to side. These skeletal modifications completely changed the way they walked: their limbs now moved back and forth rather than side to side, and their backbone bent up and down (in the manner

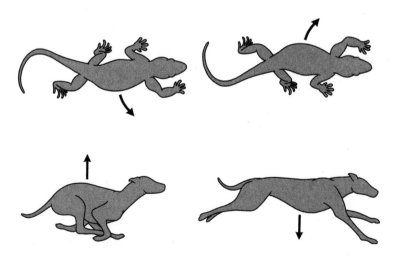

Differences in locomotion between reptiles that move side to side (top) and mammals that move up and down (bottom). Arrows show direction of movement. (illustrated by Sarah Shelley)

of a bounding gazelle) rather than left to right (like a slithering snake). They could now breathe comfortably as they moved.

The vertebral column changed in another way: whereas the backbones of amphibians and reptiles are all pretty much the same, cynodonts divided their spine into different regions, each with specialized functions. Along the torso, the vertebrae abruptly lost their ribs, where the thoracic region transitioned to the lumbar region. This is a telltale sign of a diaphragm—the powerful muscle that draws air into the lungs.

Changes were happening in the skull, too. Cynodonts developed a secondary palate—a hard roof of the mouth—which divides the mouth from the nasal passage. Air now had its own dedicated pathway to the lungs, and these cynodonts could eat and breathe at the same time. Within the new nasal passage were eccentric new structures, called turbinates: convoluted, curled scrolls of cartilage or bone that protrude in the middle of the airflow. They might seem like a hindrance, but in fact, they are key to keeping body temperature constant when breathing in lots of air. Some turbinates are covered with blood vessels, which warm and moisten the inhaled air before it hits the lungs. We have them— inside our skull, in between our nostrils and the back of our throat. They ensure that even our coldest, driest breaths on the most frigid of winter days are warmed and humidified to rain-forest-like conditions within a split second. They also work in reverse, recouping precious water as we breathe out.

All these seemingly nuanced anatomical changes added up to something important: lungs that could take in much more oxygen. At the same time, cynodonts were also increasing their calorie intake, through new ways of biting.

The first step, it appears, was dividing the jaw-closing muscles into a set of separate strands, which allowed for more powerful and more complex jaw motions. These divided muscles—the signature temporalis, masseter, and pterygoideus system of mammals—also got larger, so they needed more space and stronger anchors. The bones of the upper skull simplified and solidified into what essentially became a single structure—the reason we have a "skull bone" and not dozens of separately, loosely articulated individual head bones like amphibians or reptiles. The lateral temporal fenestra—the hole that evolved back in the coal swamp times that defines the synapsid lineage—merged with the eye socket, creating a single, master chamber for the muscles.

The most extraordinary changes, however, were occurring in the lower jaw. The ancestors of mammals had complicated mandibles made of numerous bones, with a variety of names. There's the dentary, which was the primary tooth-bearing bone, and the articular, which formed the mouth-closing joint with the upper skull, along with many others: the angular, surangular, prearticular, splenial, coronoid. But we—and all mammals—have only one: the dentary. It's all because of what was happening in Triassic cynodonts. As their jaw muscles got larger, they shifted and exclusively attached to the dentary. It made perfect sense: because the dentary was the only bone that had teeth, rearranging the muscles this way allowed stronger bites, and better precision in optimizing peak bite forces in different parts of the tooth row during the biting cycle.

However, this necessitated a complete redesign of the bones themselves. The dentary became bigger, and much deeper. It developed a huge coronoid process—a flange for muscle attachment—that stuck far above the tooth row. Meanwhile, now that they were no longer anchoring muscles, all the other bones of the lower jaw began to atrophy. They got smaller and moved backward, to the point where they were only loosely attached to the dentary. Some of them even made contact with the stapes bone of the ear, and now that they were retired from their jaw-snapping functions, started to help with relaying sound to the ear.

There was one major problem, though: the joint with the upper skull was still located on the articular bone, which was reducing to obsolescence. The bone of the upper skull that it connected to, the quadrate, was also dwindling. Some cynodonts developed a brace on the surangular bone of the lower jaw to strengthen the contact with the upper skull, but because the surangular was also shrinking, it didn't do much good. No engineer would design a jaw-closing system this way, and the only reason it worked was because these cynodonts were getting so small that they didn't have to endure the fierce bite stresses of their larger ancestors.

Somehow, the cynodonts needed to find a solution. They did— and it is what defines mammals.

CLASSIFICATION IS A human exercise. Nature doesn't put labels on things, people do. A mouse is a mammal, a snake is a reptile, a *T. rex* is a dinosaur.

Mammals, reptiles, dinosaurs. Each of these is a group on the family tree of life, but how are they defined?

It's easy to define a mammal if you're only looking at the modern world: mice, elephants, humans, bats, kangaroos, and thousands of other species share a unique set of features: we all have hair, warm-blooded metabolism, big brains, teeth differentiated into incisors and canines and premolars and molars, we feed our young with milk, and so on. But as we've seen, these "mammalian" attributes evolved piecemeal, over 100 million years of evolution, as coal swamp pelycosaurs gave rise to therapsids, therapsids survived the end-Permian extinction as small cynodonts, and cynodonts shrank further during the Triassic. Where along this evolutionary lineage should we put the dividing line between nonmammal and mammal?

Faced with a bounty of new Permian and Triassic synapsid fossils in the mid-twentieth century, paleontologists collectively came to a decision. Mammals would be all those animals that evolved from the first creature to develop a key innovation: a new jaw-closing joint between the dentary bone of the lower jaw and the squamosal bone of the upper skull. This novel jaw hinge was the simple, but elegant, solution to the problem of the ever-shrinking bones at the back of the jaw. Walter Kühne's *Oligokyphus* did not have a dentary-squamosal connection, so it is not a mammal. But one branch over on the family tree, species like *Morganucodon*—another of Kühne's discoveries, from a cave in Wales—did. So these are mammals, by convention. As are all animals—those living then in the Triassic, those evolving later, including us—that can trace their genealogy back to this ancestor with its new jaw joint.

This might sound a little unsatisfying, a little subjective. In a sense it is. But these paleontologists didn't just pick a random feature to pin the mammal name on. They chose one of the trademarks of all modern mammals, which neatly sets us apart from amphibians, reptiles, and birds. The development of the new, stronger dentary-squamosal joint was also a major evolutionary turning point, and as we shall see, it triggered a domino chain of changes to mammalian feeding, intelligence, and reproduction. These were the finishing touches in the making of mammals after a long process of evolutionary assembly, the last Lego pieces to complete the castle.

At this point, I should pause for a moment and acknowledge that the above definition is not used by most contemporary paleontologists. There has been a move over the last couple of decades to define mammals in a different way, using what is called a crown group definition. This approach takes all mammals that still survive today—the 6,000-plus species of monotremes, marsupials, and placentals—and traces back to their most recent common ancestor on the family tree. This ancestor is considered the dividing line, the off-ramp leading from nonmammals to mammals. The "crown" definition has its advantages, particularly its simplicity. But it also has its disadvantages: many hundreds of fossil species that looked, behaved, grew, metabolized, fed, nursed, and groomed their hair like today's mammals evolved before the common ancestor of the modern mammals split off from the family tree, making all of these fossil animals nonmammals. Like *Morganucodon*, for example.

Here I'll be honest: in my scientific writings, I use the crown group definition. In a research paper, I wouldn't call *Morganucodon* a "mammal," but a "basal mammaliaform," or a "nonmammalian mammaliaform." As you can see, the terminology quickly becomes unwieldy. To save us from tripping over all the names of these near-mammals right outside of the crown group, for simplicity I will just call everything with the dentary-squamosal jaw joint a mammal and hope my colleagues forgive me.

All definitions aside, what is most important is that during the Late Triassic, cynodonts like *Morganucodon* figured out an answer to their jaw dilemma. Their dentary became so large and deep that, probably inevitably, it made contact with the upper skull. A ball-and-groove joint was formed, and hence a new jaw hinge. What had been a very weak—and ever-weakening—articulation between the skull and jaw was strengthened. No longer was the only connection point between the withering articular and quadrate, but now there was a second fulcrum between the dentary and squamosal. For a while, both joints coexisted, the dentary-squamosal hinge on the outside of the quadrate-articular one. Eventually, the quadrate and articular became so small that they could no longer

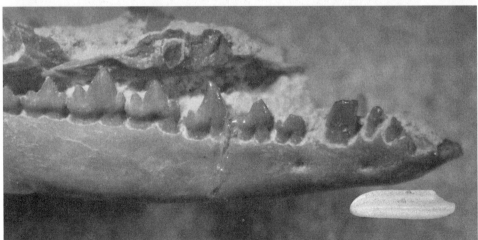

Morganucodon, one of the first mammals. Skull and head reconstruction based on CAT scans (top) and fossilized lower jaw with a grain of rice for scale (bottom). (images by Stephan Lautenschlager and Pamela Gill, respectively)

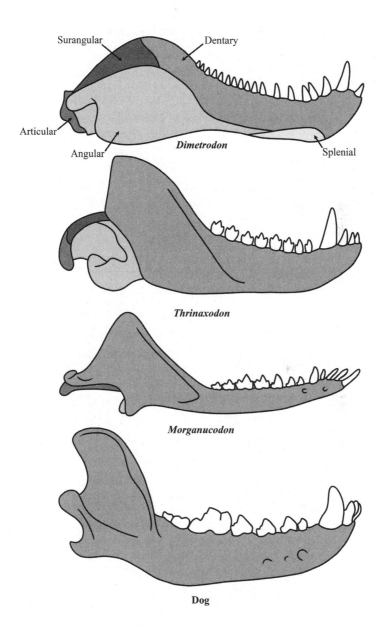

The reduction and simplification of synapsid jaws over time, culminating in the single lower jaw bone (dentary) of mammals. (illustrated by Sarah Shelley)

play any role in mouth opening, so they shriveled even further, and eventually detached from the rest of the jaw. Instead of disappearing, though, they took on a surprising new function—which we'll get to next chapter.

The dentary-squamosal joint was a game-changer. Suddenly the jaws, which had been so tenuously hitched to the skull, were firmly fastened. This new joint—operated by the jaw muscles that had gotten bigger earlier in cynodont evolution—was now capable of generating much stronger bites. It could also deliver much more controlled bites, orchestrated by the divided temporalis, masseter, and pterygoideus muscles, the three strings of a puppeteer. Thus, these jaws could accurately focus the strongest bite forces on particular teeth at particular times. This unlocked a completely new way of eating, one we take for granted, but which is extremely rare among animals: chewing. By chewing their food into mush, these early mammals could do most of the processing in the mouth—essentially beginning digestion before the food hit the stomach. It was yet another way to take in more calories, more efficiently.

The ancestors of mammals—the various pelycosaurs, therapsids, and cynodonts we've met so far—mostly had very simple bites. They were like lizards, or meat-eating dinosaurs: they snapped their jaws shut and that was that. An up-and-down chop. It was a grip-and-rip way of eating; the food was torn and swallowed, without much processing in the mouth. Some advanced cynodonts like *Oligokyphus* were an exception, as they found ways to move their lower jaws backward while they closed their mouths, in order to grind plants. But this was rare.

The Triassic mammals, though, became food-processing machines. Their jaws could move up and down, back and forth, and side to side. They developed a chewing motion, consisting of three actions in sequence, like the coordinated steps of a swimmer's breaststroke. First the lower jaw would move upward and inward to approach the upper jaw (the power stroke) as the mouth closed, then downward as the mouth opened (recovery stroke), and then slightly outward (the preparatory stroke) before going through the process again and again. All this was enabled by the unique ability of the lower jaw to roll outward and inward as the jaws opened and closed. Give your jaws a good wiggle, feel how they can move in all directions, and you're experiencing these new chewing motions pioneered by our distant Triassic ancestors.

Complex and coordinated jaw motions were only part of the equation. In order to chew, the upper and lower teeth of these mammals needed to come together. This is called occlusion, and it differs from the condition in dinosaurs or reptiles, in which the teeth of the upper jaw usually bite over the teeth of the lower jaw, or the two slot together in a zigzag array, but without really touching, when the jaw is shut. Mammals are drastically different, as we can tell when we close our mouths. While our very front teeth do make a small overbite, the teeth along our cheeks come together firmly, the upper ones fitting into the lower ones. Our teeth occlude so strongly that, when we bite down completely, our jaws are effectively locked. It is this broad contact and tight fit between the upper and lower teeth that provide the surface area necessary for chewing. This developed in Triassic mammals like *Morganucodon*, as they took the last step in crafting the classic mammalian dentition: dividing their cheek teeth into premolars and molars.

Occlusion, however, introduces a problem. The upper and lower teeth need to match each other precisely, the cusps in one fitting into the valleys in the other. Otherwise they won't slot together, and then, chewing would be inefficient at best, and impossible at worst. This is what would happen if animals that replace their teeth throughout their lives—like the ancestors of mammals— tried to chew. Even if they had upper and lower teeth that fit together perfectly, if an upper one fell out, then the opposing lower one wouldn't have a partner to occlude with. The upper one would of course grow back, but that would take time, and it would keep changing shape as it grew, never quite fitting with the lower tooth until it was full size. Clearly, this won't work for a chewer, so once again, mammals came up with an ingenious solution. They stopped replacing their teeth throughout their lives, trading the unlimited tooth generations of their ancestors for a mere two sets of teeth that have to last for their entire lives: a series of baby teeth during the formative years, and then a permanent array of adult teeth. This is called diphyodonty. Next time you chip (or lose) a tooth and need expensive dental work, you can curse your Triassic ancestors.

How do we know all this was happening in the first mammals? Fossils again tell the story. For starters, you can simply see that the upper and lower teeth fit together—the premolars and molars essentially interlocking—in fossil skulls of *Morganucodon* and other early mammals. Furthermore, the tri-peaked molar teeth of *Morganucodon* and other Triassic mammals have wear facets: flattened surfaces, marked by striations, formed from the repeated contact of the opposing tooth. In wearing their teeth, these mammals were striking a delicate balance. Too much wear and the teeth would grind down to stubs—a catastrophe for an animal that can no longer grow new teeth at will. But just the right amount of wear sculpted flat panels with sharp edges, which worked together like scissor blades when the jaws came together, adding a cutting component to the chewing.

Fossils of these earliest steady-jawed, chewing, tooth-occluding mammals are also noteworthy in another way. They have huge cavities in their skulls, which held brains that were much larger than their pelycosaur, therapsid, and cynodont ancestors. Most of the size increase was at the front, turning the tube-shaped ancestral brain into the more globular shape of modern mammals, with bulbous cerebral hemispheres. The brain of *Morganucodon* has

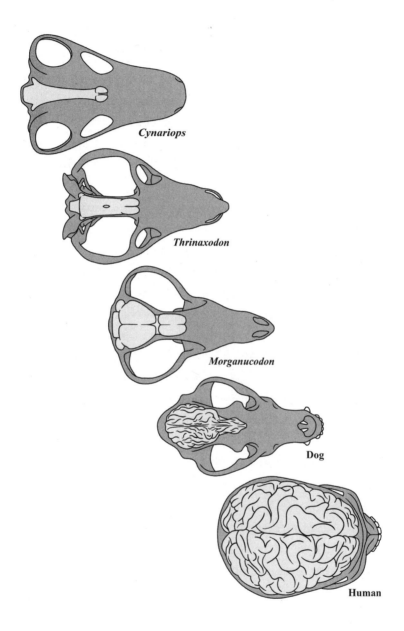

The enlargement of brain size in synapsids over time, culminating in the large brains of mammals, with a convoluted texture and enlarged neocortex of the cerebrum. Scale = 3 cm. (illustrated by Sarah Shelley)

been digitally reconstructed based on x-ray CAT scans, and aside from a larger and more rotund shape, it exhibits two key features of mammals. The olfactory bulbs—which orchestrate the sense of smell—are massive. And the top of the cerebrum boasts a new structure: a six-layered mass of nervous tissue called the neocortex, one of the most sublime mammalian inventions. Neurologists herald the neocortex as key to mammalian sensory integration, learning, memory, and intelligence. The first Triassic mammals, therefore, were becoming brainier as they were becoming better chewers. Maybe greater food intake helped them grow larger and more complex brains, or perhaps the expansion of the brain forced changes in the position of the jaw muscles, which enabled stronger bites and sophisticated jaw motions. Either way, these monumental changes to both feeding and intelligence were happening in concert in the first Triassic mammals.

Morganucodon is the best known of this first wave of mammals—an exemplar of the basal mammalian stock that was blossoming in the Triassic and persisted into the Jurassic. It was named in 1949 by Kühne—who, in a remarkable turnaround, had become a lecturer at the University of London after being freed from the internment camp—based on a single broken tooth, barely one millimeter long, picked from a twenty-pound sack of Welsh cave rubble. Further work in South Wales, carried on by the husband-and-wife team of Kenneth and Doris Kermack after Kühne returned to Germany in 1952 to assume a professorship in Berlin, produced hundreds of additional bones and teeth.

Meanwhile, in China, a team of paleontologist-priests at the Fu Jen Catholic University in Beijing were on their own quest for early mammals. One of them, Father Edgar Oehler, went out west to Yunnan Province, where he found a 1-inch-long (2.6-centimeter) skull, complete with lower jaws. He sent it back to Beijing, to his brother of the cloth, Father Harold Rigney. Rigney was studying the fossil when the Communist secret police—recently established by Mao's new government—knocked on his door, arrested him, and threw him in jail for four years. Rigney would not be as lucky as Kühne, and there would be no paleontology behind prison walls for him, as the Communists were serious about purging China's religious institutions. Through the maneuverings of backdoor diplomacy, Rigney was eventually freed from prison and returned to the United States, where he was reunited with the skull, which had been secretly whisked away from the Communists. He put it aside for a few years while he wrote his autobiography—gracefully titled *Four Years in a Red Hell*—and then described it in a famous 1963 paper, announcing it as a new species of *Morganucodon*. Along with the growing record of bones and teeth from Wales, this gorgeous fossil would make *Morganucodon* the standard-bearer for early mammals.

There were other early mammals, too: a whole variety of species living in the Triassic and Jurassic. Many of these are British and, like *Morganucodon*, were found in the English and Welsh caves. These scroungers may have been living underground, but more likely, they were washed below during floods or fell into a minefield of cracks and fissures, their bones accumulating over time into prehistoric ossuaries. One of them was found by Kühne, but he didn't realize it at first. These were the two teeth he discovered soon after his exile to Great Britain, which he sold to Rex Parrington in Cambridge, and which Parrington named as a new species: *Eozostrodon*. Another was found in the Welsh caves and named *Kuehneotherium* by Diane Kermack, in honor of Kühne.

And then there were those that Kühne had nothing to do with. In South Africa, the amateur archaeologist Ione Rudner—a self-made woman who, after studying at the University of Cape Town, became a research assistant to the director of the South African Museum—discovered a remarkable skull and skeleton, which was called *Megaʒostrodon*. On the opposite end of the globe, the Harvard professor and former US marine Farish Jenkins—who carried guns in the field to ward off polar bears and ran his camp like a military barracks—discovered the jaws and teeth of *Morganucodon*-like mammals in Greenland. There have been more recent discoveries, too, like the superlative *Hadrocodium* from China. With a head that could fit on your fingernail and a body that could perch on a paper clip, it is about the same size as the smallest mammal that lives today, the one-and-a-half- to two-gram bumblebee bat from Thailand.

The first mammals were prospering. They were diversifying at a dizzying pace, taking advantage of their big brains and their new license to chew. They were spreading around Pangea and quickly became a global phenomenon. Mammals were one of the first major groups of animals to disperse widely across the supercontinent, no longer held in check by the megamonsoons and deserts. Most of them were probably insect-eaters, their small size and strong, precise bites perfectly suited for the agile, bug-snatching niche. Not all of them were eating the same insects, however. Pam Gill and colleagues compared the tooth wear of *Morganucodon* and *Kuehneotherium* and realized that these two species, which coexisted in the earliest Jurassic of Wales, were specializing in different foods. *Morganucodon* ate insects with harder shells, like beetles, whereas *Kuehneotherium* preferred softer prey like butterflies. This was the first whisper of a trend that would repeat itself throughout mammalian evolution: when a new wave of mammals came, it usually started with tiny insect-eaters diversifying in the shadows.

WHEN HE WASN'T collecting mammal jaws with a rifle strapped to his back, Farish Jenkins was a debonair Ivy League professor, who delivered lectures in crisp suits. He was a consummate per-

Farish Jenkins's revolutionary image of the early mammal
Megaʒostrodon. (image modified from Jenkins & Parrington, 1976)

former, amusing generations of students with his dry wit and his
use of props, most notoriously a Captain Ahab prosthetic leg that
he sported to demonstrate different styles of human locomotion.
More than anything, though, his former students recall his meticulous drawings of skulls and skeletons, his primary teaching
tool in those days before PowerPoint. He would show up many
hours before class, sometimes in the dark of the morning—when
the nocturnal mammals of Harvard Yard were on the scurry for
insects—to chalk his masterpieces on the blackboard.

His most famous drawing is of *Megaʒostrodon*. It first appeared,
perfected by a professional artist, in his 1976 description of the
skeleton that Ione Rudner had found a decade prior. It has become

one of those classic images in science, surfacing often in textbooks and, yes, the PowerPoint lectures that most of us now prefer. Like that iconic image of Che Guevara, it signifies something greater than just a single individual. It is a portrait of a revolution, of the first mammals moving across Pangea, starting a dynasty.

The wee *Megazostrodon*, about the size of a shrew or mouse, is startled. It stands at high alert at the base of a tree, its limbs proudly erect beneath its lithe body. A moment of indecision. Its feet point upward, as if it might scamper up the tree and disappear in the branches. But its hands point outward, giving it the choice of running, across the ground. It senses something—a danger perhaps, or maybe a tasty bug, because its jaws are slacked open, bearing a full set of incisors, canines, premolars, and molars. The teeth are tiny, but they are sharp, and the ones from the upper and lower jaw look like they could snap together at a millisecond's notice. The socket for the eye is small, so it must be relying more on other senses to gauge its surroundings. Whatever called it to attention, it must have heard or smelled.

This skeleton is something we would recognize today, if we saw it as a fleshed-out animal. It is clearly, unmistakably, a mammal. It was covered in fur, walked upright in a back-and-forth scuttle, and had the full complement of mammalian teeth, which it used to chew, and which it replaced only once in its entire life. It was agile, able to climb trees and move along the ground, and it was warm-blooded or becoming so, able to keep a comfortable body temperature when it hunted insects in the cool of the night. Inside its head was a big brain, which endowed it with intelligence and keen smell, and it could hear well too.

If you were walking through a forest, or running to catch a subway, and this little critter raced in front of you, I'm pretty sure you would just think it was a mouse.

Mammals had arrived, the inheritors of so many remarkable adaptations that their pelycosaur, therapsid, and cynodont ancestors had accumulated over more than 100 million years of evolution. These new mammals were spreading around the world and poised to take over Pangea—to reclaim what their therapsid forebears had lost in the fury of end-Permian volcanism.

Or were they?

The supercontinent was starting to break apart, and the dinosaurs were getting larger and fiercer. The new mammals—despite all their evolutionary innovations—had limited options.

They would have to become really good at living incognito.

3

MAMMALS *and* DINOSAURS

Vilevolodon

WILLIAM BUCKLAND KNEW HOW TO work an audience.

At the University of Oxford, during the first decades of the nineteenth century, he was the quirky professor, whose lectures on geology and anatomy were must-see entertainment. Dressed in full academic regalia, he buzzed up and down the aisles, shouting questions at his students while passing around severed animal parts.

His home was a hoarder's den of bones, stuffed taxidermy animals, seashells, and other curios, and for a time he kept his own personal zoo. During dinner parties, he would trot out his pet bear—clothed, like its master, in academic robes—while he fed his guests mystery meat from across the British Empire. Mice on toast was a staple, as were panther and porpoise. Occasionally his friends might get lucky and only have to stomach ostrich or crocodile, which Buckland surely considered boring.

His life goal, you see, was to eat his way through the animal kingdom. Anything was fair game—even humans, if you believe the legend of what transpired when the Archbishop of York showed him a pickled heart in a silver casket, which reputedly belonged to Louis XVI. "I have eaten many strange things," Buckland is alleged to have said, "but have never eaten the heart of a king before." He grabbed it and gobbled it down, while his stunned audience stood aghast. It must have been quite the show.

Tonight, however, would be his biggest performance yet.

On an early winter evening in 1824, Buckland rose to address the Geological Society of London. He had just been made president of the members-only club, a who's who of naturalists, theologians, and aristocratic rock collectors of pre-Victorian society. Ever the consummate showman, Buckland wanted to make his inaugural address memorable, and he had an ace to play. For years there had been rumors that he acquired some giant fossil bones, found by workers quarrying limestone slabs in the English countryside, near the quaint village of Stonesfield. The flat, platy layers were perfect for roofing tiles, but occasionally they had bones and teeth embedded inside—if the whispers were true. Now, after nearly a decade of study, Buckland was ready to make an announcement.

With a dramatic flourish, he told the audience that there were indeed bones among the roof stones, and they were big—much larger than the bones of any animals living in modern-day England. They were reptilian in shape and proportion, as if they belonged to a giant lizardlike beast, a creature that seemed more out of a dragon myth than anything the audience members had ever seen or experienced. This was something radically new, and Buckland thought up a perfect name to describe it: *Megalosaurus*.

The enraptured audience didn't yet grasp it, but Buckland had just unveiled the very first dinosaur.

That evening was a seminal moment in scientific history, sparking humankind's enduring fascination with dinosaurs. It has been recounted in countless stories, but what these retellings usually fail to mention is that Buckland made another announcement that day, of another discovery much smaller in physical size but equally revolutionary. For amid the big bones, the limestone slabs entombed another type of fossil, which Buckland—in an uncharacteristic understatement—considered "most remarkable."

Two tiny jaws, barely an inch long if that, bearing a series of cusped teeth.

They were the unmistakable jaws of mammals, about the size of the mandibles of mice or shrews. The teeth were, to Buckland's eyes, eerily similar to those of an opossum, an animal that he probably knew well from his dinner party feasts. Because they were found alongside the bones of *Megalosaurus*, these jaws were evidence of giant saurians and primitive mammals once living side by side. They were the first sign that mammals had a much deeper history than anybody had fathomed.

For a while, it was the minuscule jaws—and not Buckland's

dinosaurs—that were the subject of fierce controversy, during the formative years when paleontology evolved from a gentleman's pursuit into a scientific discipline. Many of the key figures of early- to mid-nineteenth-century paleontology pontificated about them—such as Richard Owen, the irascible anatomist who would later describe the first "mammal-like reptiles" from the Karoo and coin the name "dinosaur" to classify Buckland's *Megalosaurus* and other giant reptiles emerging from across Victorian England. Buckland, Owen, and others argued for decades about the identification of the jaws, a battle in the larger war about whether species evolved over time.

Ultimately, however, the dinosaurs won the popularity contest. *T. rex*, *Triceratops*, and *Brontosaurus* became household names, whereas Buckland's wee jaws—belonging to species called *Phascolotherium* and *Amphitherium*—were relegated to the lexicon of scientists. It was the cold, hard reality of discovery. Over the course of the nineteenth century, big dinosaur bones continued to turn up in England and elsewhere in Europe, and then later colossal skeletons were pried from the badlands of the American West by roughnecks bankrolled by fame-hungry industrialists like Andrew Carnegie. Meanwhile, mammal fossils from this time interval— the Jurassic and Cretaceous periods—were rare and uninspiring, limited mostly to individual teeth or scraps of jaw. There were no complete skeletons, and even if there were, their mousey stature surely would have never inspired the public in the same way as an airplane-size *Diplodocus*.

No wonder, then, that Jurassic and Cretaceous mammals acquired a lackluster reputation. They were said to be dull, unspecialized generalists that all looked and behaved like mice, confined to eke out a meager existence in the long shadow of the dinosaurs. Barely role characters, and more like extras, in a dinosaur drama.

And no wonder that most young people who become enthralled by fossils—me included—are drawn to dinosaurs rather than Jurassic mammals. My teenage obsession became a career, which is how I found myself in the northeastern fringes of China a few years ago, hopping between museums to study the feathered dinosaurs of Liaoning. These gorgeous fossils—skeletons sheathed in downy coats and boasting feathery wings—were formed by volcanic eruptions during the Jurassic and Cretaceous. Bad luck for the dinosaurs, which were killed and buried by ash and sludge, going about their everyday business, Pompeii-style. Good luck for paleontologists, as the fossils are preserved in incredible detail, down to the pigment-bearing melanosomes that gave the feathers color. My mission on this particular trip was to take tiny samples—as delicate as a flake of dandruff—of the feathers to bring back to my lab, where my students and I could view them under high-powered microscopes, identify the melanosomes, and figure out what colors these dinosaurs once were.

Several days into the trip, after a morning of scraping feathers in the dark corridors of the Beipiao Pterosaur Museum, I needed a break. My friend Junchang Lü—one of China's leading dinosaur hunters, who sadly passed away in 2018—exchanged a glance with the museum director. After a few hushed words of Mandarin, Junchang motioned for me to follow. "We have something secret to show you," he said. "And it's not a dinosaur!"

We left the museum, hopped into a car, and snaked through the narrow streets of Beipiao city, clogged with bicycles and noodle vendors. A quick turn took us into a hidden alleyway, which opened into a small courtyard. The car came to a sudden stop and I was instructed to get out. The museum director pointed to a doorway, covered by a metal grate, on the ground floor of what looked to be an apartment block, densely populated, and showing its age. No way such a wealthy businessman—who used his own money

Junchang Lü (center) and his team showing me the mystery mammal fossil in Beipiao, China. (photo by Steve Brusatte)

to build the museum and fill it with fossils—lived in such a humble place, I thought, as he unlocked the door.

As light filled the entryway, a strange scene came into focus. The room was what I imagine Buckland's Oxford home to have looked like, two centuries ago. Dust clung to every surface. A clutter of boxes and wooden crates littered the floor, and stacks of newspapers formed precarious towers on the tables and work-benches. Hammers, chisels, and brushes were strewn about, along with bottles of glue and small plastic bags. And then there were fossils, lots of fossils, implanted in limestone slabs about an inch thick, which probably would make nice roofing tiles. This was no ordinary apartment. It was a workshop—a makeshift laboratory where fossils, purchased from local farmers, were cleaned up before being displayed in the museum.

One of the director's assistants ducked into a side room and came out holding two rock slabs that fit together like puzzle pieces.

He pushed aside a mess of newspapers and set the slabs on a table and then fetched a lamp. As he illuminated the treasure, he asked me to come forward.

On the gray limestone surface, speckled with tiny fossil shells, was a brown smear, about the size of an apple. I looked closer. The brown stuff was hair, and there was a spinal column running down the middle.

It was a fossil mammal! A mammal that lived alongside the feathered dinosaurs during the Jurassic Period, some 160 million years ago. In one sense, this mammal fit the stereotype: it was small and could easily have been swatted away or stepped on by one of the winged dinosaurs. But in another, much more important sense it broke the mold for what a Jurassic or Cretaceous mammal was supposed to be. There was a flap of skin, sticking out from both sides of the spine, stretching between the arms and legs. Just like the wing membrane of a flying squirrel.

Far from being a nondescript mouse-mimic, this mammal could glide between the trees. Clearly, the mammals living with dinosaurs were much more interesting than paleontologists—going back to Buckland—long thought.

BEFORE MAMMALS COULD start gliding over the heads of dinosaurs in the Jurassic and Cretaceous periods, they had to make it out of the Triassic. As did the dinosaurs. This was no easy feat.

As the first mammals like *Morganucodon* were evolving from their cynodont ancestors in the Triassic—developing their new jaw joint, full set of mammalian teeth, chewing abilities, tiny body sizes, enlarged brains, and warm-blooded metabolism—the earth was changing too. Pressure from deep below began to stretch Pangea, one pull coming from the east and one from the west. It was slow and imperceptible to the mammals living on the surface, for many millions of years, until suddenly it became a catastrophe. About 201 million years ago, at the end of the Triassic, the supercontinent started to crack apart, unzipping down its middle. North America separated from Europe; South America from Africa. Our modern continents were born from the Pangean divorce, and today the Atlantic Ocean marks the dividing line. But before water rushed in to fill the gaps between the newly diverging landmasses, the earth hemorrhaged lava.

For about 600,000 years, megavolcanoes erupted along the future Atlantic seaboard. There were four violent pulses of activity, which engulfed the edges of the new continents in fire. Added together, some of the lava flows and magma vents—like those that can be seen today as cliffs of basalt rock in the vicinity of New York City or the deserts of Morocco—were up to three thousand feet thick, more than twice the height of the Empire State Building. But like what happened some 50 million years earlier, at the end of the Permian, the real terror was not lava or ash, but the gases that rode up the volcanic vents from the deep earth, into the atmosphere—greenhouse gases like carbon dioxide and methane, which catalyzed a spurt of global warming. Just like at the end of the Permian, the temperature spike caused oceans to acidify, starved shallow waters of oxygen, and triggered ecosystem collapse on land and sea. Once again, there was a mass extinction. This time, at least 30 percent of species died, but probably many, many more.

There were several notable victims of this extinction, including the very last tusked dicynodonts—close cousins of mammals, holdovers from the Permian diversification of therapsids that had staggered into the Triassic and re-diversified into plodding, pot-bellied plant-eaters, some like the Polish *Lisowicia* the size of elephants. Many amphibians perished, as did almost all the crocodile cousins that were competing with dinosaurs in the Triassic.

Two of the great survivors, however, were the mammals and

the dinosaurs. Why dinosaurs endured remains an open question, and one of the biggest mysteries debated by dinosaur researchers. Maybe it was because they were freed from competition with their deceased croc competitors, or perhaps these dinosaurs were endowed with feathers that shielded them from temperature fluctuations or faster growth rates that allowed them to mature from hatchling to adult quickly. Maybe they were simply lucky. When it comes to mammals, though, it's easy to see how they were primed to weather the meltdown. They were holding a perfect hand of cards: small size, fast growth, keen senses and intelligence, and an ability to hide in trees or burrows. The same way rats have no issues handling the dark confines and toxic fumes of subway tunnels, the *Morganucodon*-type mammals kept on trucking through the global warming.

As the volcanoes switched off, and the new continents continued to inch farther apart, the earth recovered, as it always does. The Triassic gave way to the Jurassic, and dinosaurs and mammals had a new and much emptier world to explore. The dinosaurs responded by getting bigger, much bigger: by the middle of the Jurassic there were long-necked dinosaurs larger than five elephants put together, which would have literally shook the ground as they walked. The dinosaurs diversified, too, their family tree blooming into all sorts of new groups: jeep-size meat-eating theropods with fantastic crests on their heads, plate-backed stegosaurs that ate low-hanging plants, tanklike ankylosaurs whose bodies were covered with armor and spikes, feisty raptors with eviscerating claws and feathery coats, and by the Middle Jurassic, pigeon-size creatures that flapped their wings to power themselves through the air—the very first birds.

The mammals, on the other hand, stayed small. With so many dinosaurs about, they probably had to. But like the dinosaurs, they did diversify—into many new species, with a variety of diets, behaviors, and ways of moving. They became experts at filling those hidden niches underground, in the undergrowth, the darkness, the treetops, the shadows. Wherever the giant dinosaurs couldn't reach, you could bet that mammals were there, and prospering.

The mammal family tree grew exponentially during this time, budding into a tangle of so-called dead end branches on the trunk of the tree, sandwiched between the earliest Triassic mammals like *Morganucodon* and the modern-day species at the tip. That phrasing, however, is unfair. The only reason these groups—like the docodonts and haramiyidans, of which we will learn more in a moment—are considered "dead ends" is because they don't survive today. This is the privilege of hindsight. During the Jurassic and Cretaceous, these early mammal groups evolved at hypercharged rates, and experimented with many of the same feeding and locomotion styles as modern mammals. In their time and place, these mammals were anything but obsolete.

MY INTRODUCTION TO Jurassic and Cretaceous mammals came right around the time their dreary stereotype was being shattered.

It was the spring of 1999, my first year of high school, and dinosaurs were my obsession. My parents didn't quite understand it, but they didn't raise any objection when I asked if, during the Easter holiday, we could visit the Carnegie Museum of Natural History in Pittsburgh—the temple Andrew Carnegie built to showcase the gigantic dinosaurs his mercenaries collected out west. Not content with seeing the exhibits, I wanted a behind-the-scenes tour of the fossil collections, and I knew just who to ask.

A couple weeks earlier, one of the Carnegie curators—Zhe-Xi Luo—was in the news. Luo and his Chinese colleagues had described a stunning new Cretaceous-aged mammal, a skeleton only a few inches long, with a skinny tail, mobile limbs, and three-cusped teeth that it used to eat insects. They called it *Jeholodens*, after an ancient name for the part of China where it was found: Liaoning Province. An artistic rendering of its spotty, furry body leaping up from its fossil skeleton had graced many magazines and newspapers across the country, even, as I recall, my small-town paper in rural Illinois. It wasn't a dinosaur, but I still thought it was pretty cool, so I scoured the Carnegie Museum's website until I found Luo's email, and then sent a shot-in-the-dark request for a private tour.

Luo answered, quickly, and with such generosity for a busy, famous scientist. A short time later, my family was meeting him at the entrance to the museum on a crisp April morning. For well over an hour he took us through the back rooms stacked with fossil bones, and he didn't even mind when I kept asking him questions about dinosaurs. Toward the end of the tour, he introduced us to the artist who had made the painting that had appeared in our newspaper, Mark Klingler. As Mark nodded along, Luo told me to keep my eyes on China. *Jeholodens* was just the start, he said, with the cheeky grin of an insider trader about to cash in on the market. Farmers across Liaoning were finding so many new mammals, alongside feathered dinosaurs. Some, like *Jeholodens*, were Cretaceous, between 130 and 120 million years old, and formed a community called the Jehol Biota. Others, as it turned out, were much older: these were Jurassic mammals that formed an earlier community, called the Yanliao Biota, from around 157 million to 166 million years ago.

Over the following two decades, I watched with awe and admiration as Luo's prediction turned out to be true. The Chinese mammal fossils kept on coming—and they still do, as we speak, with several new Liaoning mammals continuing to adorn the pages of leading journals like *Nature* and *Science* every year. Nearly

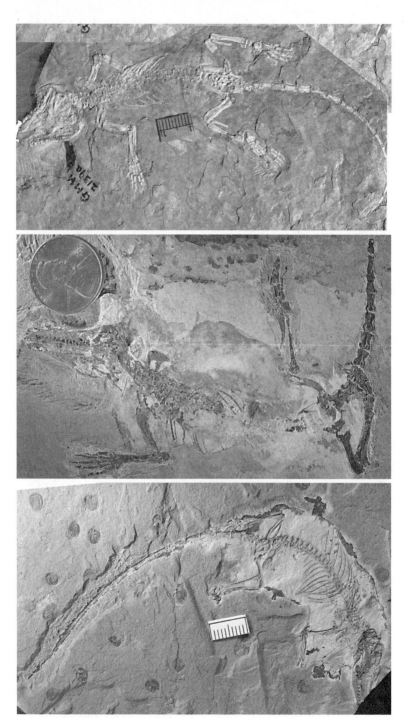

Stunningly preserved mammal fossils from Liaoning, China:
Jeholodens (top), *Agilodocodon* (middle), *Microdocodon* (bottom).
(photos by Zhe-Xi Luo)

every fossil is described by one of two teams that have developed a friendly rivalry, both headed by Chinese-born paleontologists who made their way to the USA. There's Luo's group, now based at my alma mater of the University of Chicago, and a crew headed by Meng Jin, who is a curator at the American Museum of Natural History, where I did my PhD. Although I started out as a dinosaur geek, my research has increasingly moved toward fossil mammals, and it's been an honor to work with both Luo and Meng, two humble world experts who have mentored me on the mammal learning curve.

What makes the Liaoning mammals so special is that they are more than just teeth and jaws—the bane of Buckland, and all mammal paleontologists studying Jurassic and Cretaceous fossils before Luo and Meng's generation. Many of them are complete skeletons, with exquisitely preserved bones and soft tissue, bestowed by the same rapid volcanic burial that turned dinosaur feathers to stone. These skeletons revealed something that teeth and jaws never could: Jurassic and Cretaceous mammals were hugely diverse, in nearly every way imaginable. Their only limitation was size: most were about the size of shrews or mice, as the scrappy record of fossil teeth had correctly foretold, and none that we know of was bigger than a badger. But still, at least one of them—a Cretaceous species called *Repenomamus*—was big enough to do something extraordinary. It was found with the bones of baby dinosaurs in its stomach, turning the age-old story of dinosaur versus mammal on its head. Some dinosaurs, in fact, would have lived in fear of mammals.

Two types of mammals from the Yanliao Biota exemplify the unexpected diversity of the Jurassic. These are the docodonts and the haramiyidans, the first two great radiations of mammals. Neither survived past the Cretaceous, but you could imagine if fate had turned out slightly differently, and a few of them held on today, they would probably be regarded like monotremes, such as

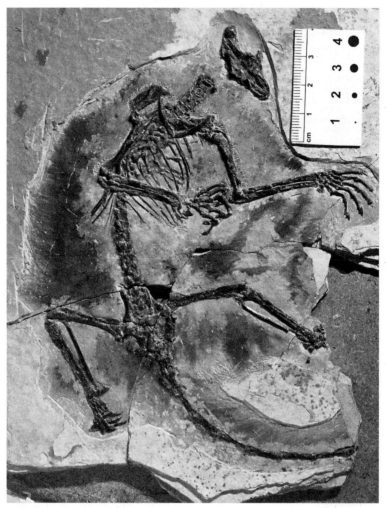

The gliding haramiyidan *Maiopatagium* from the Jurassic of Liaoning, China. (photo by Zhe-Xi Luo)

the platypus. Slightly strange, somewhat primitive vestiges of the early waves of mammal diversification, but bona fide mammals nonetheless, celebrated in cute online videos and must-see zoo attractions. Their branches on the family tree may now be dead, but in their time, both docodonts and haramiyidans were prosperous.

Here is a snapshot of some of the Yanliao docodonts. There's teeny *Microdocodon*, a scrambler who, dare I say it, looks like a pretty standard mouse or shrew, weighing in at less than ten grams— about a third of an ounce. That's as orthodox as these mammals got, however. Another species, *Agilodocodon*, had slender limbs, long fingers, curved hand claws, and highly flexible ankles—all hallmarks of tree-climbers, like modern primates. *Docofossor* had a radically different skeleton, with a massive elbow joint, a reduced number of finger bones, shovel-like claws, and wide hands. These are digging adaptations, characteristic of today's golden moles that burrow and tunnel underground. And then there's *Castorocauda*, which had a long, broad, flat tail and webbed feet, like a beaver. Its molars bore a line of five curved cusps, like in some early whales: a perfect tooth shape for feeding on slippery fish and aquatic invertebrates. *Castorocauda* was a semiaquatic creature, which could paddle in the water and amble along the shoreline.

The haramiyidans were remarkable in different ways. Their molars had multiple, parallel rows of cusps, which formed a series of low strips. The jaws moved in a backward power stroke— called a palinal motion—that worked the cusp rows of the upper molars against the lower ones, providing a broad grinding surface for pulverizing seeds, leaves, stems, and other plant matter. And what plants were they eating? Ones from high up in the canopy. That's because many haramiyidans—like the secret one I saw in the apartment-turned-workshop in Beipiao—were gliders. They were the first mammal aerialists, who could soar from branch to branch, tree to tree.

Several species—*Vilevolodon*, *Maiopatagium*, and *Arboroharamiya* among them—have been found with a trifecta of skin membranes. The main one extends outward from the backbone, stretching between the forelimb and hind limb. The others join the neck and forelimb and the hind limb and tail, respectively. Together they

5 cm

The dinosaur-eating mammal *Repenomamus* from the Cretaceous of Liaoning, China. (photo by Meng Jin)

form a series of airfoils, similar to modern "flying lemurs," the dermopterans (which are not true lemurs, but a nonprimate group of mammals). But there's more: the long fingers and toes of haramiyidans, which are nearly equal in length to one another, and deep ligament grooves on the undersides of their toes suggest that they were capable of upside-down suspension. In other words, they could hang from a tree branch or cave ceiling by their hands and feet, like roosting bats. They may have formed vast, spooky, stinky colonies as bats do today.

Ground-living scurriers. Climbers. Diggers. Swimmers. Gliders. Fish-catchers, plant-munchers, seed-grinders. In the Jurassic, docodonts and haramiyidans were already experimenting with so many of the lifestyles of modern mammals, and filling so many niches across the landscape, from the forests to the lakeshores to underground. Bland generalists, they were not. Surely they had nearly as much ecological variety as the dinosaurs—maybe more!—but just in smaller form.

This realization has led to a radical reinterpretation of the Jurassic and Cretaceous world. The mammals were *better* than the dinosaurs at being small. The tiniest dinosaurs of the time were primitive birds, about the size of pigeons. No stegosaurs, tyrannosaurs, horned dinosaurs, or duck-billed dinosaurs ever shrunk anywhere close to the size of an average docodont or haramiyidan. While it is true that dinosaurs kept mammals from getting big, mammals did the opposite, which was equally impressive: they kept dinosaurs from becoming small.

WHAT WAS HAPPENING in the Jurassic was not just a Chinese story, but a global one. Docodonts, haramiyidans, and many of the other early-diverging mammal groups spread around the world as the continents were fragmenting. Indeed, it may have been the breakup of Pangea that drove their diversification, as the newly separating landmasses became home to their own unique mammal communities. Fossils of these mammals are found in many places around the world, including one of my favorite spots for fossil hunting: the Isle of Skye in Scotland.

I first visited Skye—an island of craggy peaks, misty moors, and wave-battered cliffs—a few months after starting my faculty job in Edinburgh. Once again, it was dinosaurs that drew me there. In the 1980s, a dinosaur footprint fell from a cliff, and in the 1990s a massive limb bone from a long-necked sauropod was found sticking out of a sandstone boulder on the shore. These were the first scanty clues that dinosaur fossils were hiding in the enchanted landscape. Over the past few years, we've found many more: a dance floor of hundreds of sauropod handprints and footprints, tracks left behind by stegosaurs and carnivorous theropods, possible footprints of early duck-billed dinosaurs, knife-blade teeth of a predator similar to Buckland's *Megalosaurus*, and plenty of still-unidentified bones that are now in my lab. As I type, my students are using pneumatic drills and dental tools to remove the bones from their concrete-hard tombs.

My team and I are newcomers to Skye. Other paleontologists came before us, but most of them were—shockingly—not interested in dinosaurs. Jurassic fossils from the Hebrides—the chain of islands that parallels the west coast of Scotland—were first reported by Hugh Miller in the 1850s. Miller was many things: an accountant, a stonemason, a writer, an expert on rocks, an evangelical priest. One summer, he chartered a boat called the *Betsey* and island-hopped across the Hebrides, geologizing and preaching. On a flicker of land south of Skye, called the Isle of Eigg, he came to a beach covered with rusty-red boulders, with smooth black bits protruding from all sides. "They were bones, real bones," he recounted with glee in his travelogue, *The Cruise of the Betsey*, published to great acclaim in Edinburgh after his untimely death in 1856. Most of the bones belonged to ocean-living, noodle-necked reptiles called plesiosaurs, whereas others were from crocodiles and fishes—the animals that lived offshore while dinosaurs thundered across the land.

Over a century later, in the early 1970s, Miller's writings inspired another fossil-hunter—a schoolteacher named Michael Waldman—to bring his students to the Isle of Skye. A bigger island than Eigg, with much more Jurassic rock exposed along its coasts, Skye seemed a more promising place to find fossils. Waldman quickly proved his hunch correct by finding a jaw, barely a centimeter long, full of cusped teeth. It was a mammal jaw, which he named as a new species of docodont: *Borealestes*, the "northern brigand." For the next decade Waldman and his mentor, the Uni-

Moji Ogunkanmi, a member of our team, scrutinizing the Jurassic rocks of Skye, Scotland, for small fossils. (photo by Steve Brusatte)

versity of Bristol mammal expert Robert Savage, returned to Skye and collected many more mammal fossils—particularly the usual: teeth and jaws. But for whatever reason, they never got around to describing them, and at some point in the 1980s their fossils seemed to disappear.

Around the time I was planning my first trip to Skye, I made a reconnaissance visit to the off-site warehouses of the National Museum of Scotland, where most of their fossil collection is stored. I wanted to get an eye for the types of fossils we might find on Skye—shapes, colors, textures—so I did what most paleontologists do in such a situation: rifled through the drawers with no real plan, observing and photographing every fossil that looked remotely interesting. Most of the fossils were what I expected: fragments of bones, vials of teeth, and unidentifiable bags of rubble. And then I opened another drawer—and stopped cold.

There was a block of limestone, about the size of a football, chalky in color. Its surface was pockmarked, from countless tides washing across the wave platform on which it was found. In the center was a jumble of black, shiny bones, in a mass about the size of an apple. I could see vertebrae from the backbone, ribs, and limbs.

It was a skeleton—a mammal!

Here was one of the specimens feared missing, collected by Waldman and Savage, at least fifteen years before the first skeletons were reported from Liaoning, back when Jurassic mammals were thought to be all teeth and jaws. They had a skeleton, but didn't do anything with it. At least it found its way into the National Museum, where it could be kept safe for future researchers.

Almost immediately I wrote a grant application to study the fossil, writing in effusive prose how it was a "crown jewel" of Scottish paleontology and could give critical new insight into the early evolution of mammals. The reviewers didn't share my enthusiasm, and like the first dozen or so grants I wrote as a young faculty member, it was rejected. I needed another tack, so I joined up with Nick Fraser and Stig Walsh—my paleontologist friends at the National Museum of Scotland—to propose a PhD student project focused on the skeleton. Because none of us were specialists on Jurassic mammals, we recruited another colleague who knew these animals better than anybody: Zhe-Xi Luo. For the first time, I would be working with the man who introduced me to fossil mammals as a teenager. We would together supervise the project.

Now all we needed to do was to find a student, and we found a good one: a young woman from the Scottish Highlands named Elsa Panciroli, who had gone back to college to study paleontology after working for a marine conservation charity. "The fact this project takes a Scottish specimen and places it in a wider context really excites me," she wrote in her application, enthused about studying such an important fossil so close to home. Over the next few years she painstakingly studied the fossil, using CAT scans to isolate each individual bone and put them together, digitally, into a fully articulated skeleton. She identified it as *Borealestes*— a body to go with Waldman's jaw. When Elsa presented her research at the 2018 Palaeontological Association conference—the largest gathering of fossil researchers in the UK—she impressed the audience and earned the President's Prize for best talk by a student researcher.

The fossils from Skye and Eigg—mammals, dinosaurs, sea creatures, and much more—paint a vivid picture of life at the time. Ancient Scotland was part of an island, perched in the middle of the still-narrow Atlantic Ocean, which was actively widening as Europe moved away from North America. Raging rivers drained soaring mountain peaks, meandering across the land before emptying, through sprawling deltas, into crystal blue seas. Beaches and lagoons fringed the coasts, and it was here where the mammals and dinosaurs commingled with crocodiles and salamanders, while Hugh Miller's plesiosaurs swam offshore. *Borealestes* may have lived like its Chinese cousin *Castorocauda*: paddling in the subtropical waters of the lagoon, spearing fish, venturing on to land when it wanted to supplement its diet with insects, or to avoid the plesiosaurs.

Dinosaurs would have dominated this world, at least when it came to sheer bulk. The sauropod tracks my team discovered a few years ago are the size of car tires, made by beasts whose necks could stretch a couple stories into the sky. I suppose at least a few dozen *Borealestes* could fit inside a single sauropod footprint. With one footfall, one of these dinosaurs could have wiped out an entire colony of docodonts. But *Borealestes* was one of many species of mammals that lived on this ancient island, not only surviving, but thriving. It was the Age of Dinosaurs, but in the smaller and hidden niches, it was already the Age of Mammals.

GIVEN THE GREAT diversity of docodonts and haramiyidans—the numerous species, their global reach from Chinese forests to Scottish lagoons, their stupefying variety of diets and habitats and lifestyles—it's silly to denigrate them as dead ends in mammal evolution, as I have unfortunately done before, in some of my writings and lectures.

But there's another reason why that phrase is problematic. While it's true that no docodont or haramiyidan survives today, these groups branched off from the trunk of the mammal family tree, and it's that trunk that was the pathway to today's mammals. Along the trunk, Jurassic and Cretaceous species acquired many features that—when added to what pelycosaurs, therapsids, cynodonts, and *Morganucodon*-type mammals had already developed—became components of the blueprint that defines mammals today, including us. We can see many of these new attributes in fossils of docodonts and haramiyidans.

Let's start with hair. As we learned earlier, it was likely that hair first appeared in Permian therapsids—dicynodonts and cynodonts—not as a dense body covering, but as sensory whiskers, display structures, or part of a skin waterproofing system. The evidence, however, is circumstantial: hairlike strands in therapsid coprolites and pits and grooves on the snouts, in the position of whiskers. What is undeniable, however, is that many of the Jurassic and Cretaceous mammal skeletons from Liaoning—including both docodonts and haramiyidans—have whole-body pelts of fur. There's no guesswork: the fur is right there, surrounding the bones, frozen in place by the volcanic preservation just like the dinosaur feathers.

It is, therefore, indisputable that these mammals were fully warm-blooded, as only endothermic animals making their own body heat and keeping a constant temperature would need to cover their entire selves in hair. In fact, this would be detrimental to a cold-blooded animal, as it could cause them to overheat on a sunny day. There is still debate about where exactly on the therapsid-cynodont-mammal line total warm-bloodedness evolved. The scenario I presented last chapter, identifying Triassic cynodonts as the instigators of endothermy, could be proven wrong by future research. Regardless, by the time docodonts and haramiyidans were going about their sundry lifestyles in the Jurassic, mammals must have developed the same type of high-energy, sophisticated metabolism that we now have.

There's another, even more sophisticated feature of mammals that was developing in this part of the family tree. Indeed, it's *the* thing that gives mammals our name, which sets us apart from all other animals: mammary glands. These glands—the largest and most complex in our fleshy skin—are what produce milk, which mammalian mothers use to nourish their young, a process called lactation.

There are many benefits to feeding our babies with milk. It's a remarkably nutritious food source, which neither the mother nor the baby needs to go out and forage, collect, or hunt. The mother can build up her milk supply and control when she feeds her children, which buffers against changes in the weather or seasons that may make food scarce. Bird parents, for example, which gather worms to feed their hatchlings, don't have such luck. If a drought wipes out the worm supply, the babies are in trouble. A ready source of food also permits newborn mammals to grow fast and fosters a bond between mother and child, which is so important for cognitive and social development. I've seen the latter firsthand. As I'm writing this, my five-month-old son is cooing at my wife, but ignoring me.

There are many theories as to how lactation evolved, a mystery that Darwin himself spent considerable time fretting over. There are two leading ideas today. The first is that skin glands began secreting antimicrobial fluids, to help protect the newborns from bacterial infections, and this later developed into a full-on food source. The second is that milk was initially used to keep the tiny eggs of mammals moist, so they wouldn't dry out, but then the hatchlings started to eat it, and natural selection turned it into their nourishment.

Yes, you read that correctly: eggs, and hatchlings.

Although we are used to thinking about mammals as animals that give live birth, this is an advanced ability of therians: the derived group of marsupials and placentals, like us. The most primitive living mammals—monotremes like the platypus and echidna—lay eggs, and it is likely that all the Triassic, Jurassic, and Cretaceous mammals we've been discussing so far did as well.

While we don't have actual fossilized eggs from early mammals (yet!), we know that the very closest relatives of mammals—the tritylodontid cynodonts, the group including Walter Kühne's *Oligokyphus* that we met last chapter—must have been egg-layers. In 2018, Eva Hoffman—another bright PhD student whose research is changing our views of mammal evolution—teamed up with her advisor, Tim Rowe, to describe a fossilized family of the tritylodontid *Kayentatherium*. Huddled together, a mother with at least thirty-eight tiny offspring, maybe more, was buried by a freak flood. There's no way that this cat-size near-mammal would have given live birth to many dozens of babies, so they must have hatched from eggs. And there's also no way that she could have fed them all with milk, at least not exclusively. Perhaps they were out foraging for food as a family when they met their fate.

When, then, did lactation evolve? So far, not even the most divinely preserved Liaoning mammals have been found with fossilized mammary glands. Thankfully, other lines of evidence tell us that a mammal must have been drinking milk as a baby.

The first is diphyodonty: the condition of having only two sets of teeth—baby and adult—which, as we saw in the last chapter, first appeared in *Morganucodon*-like mammals in the Triassic. There is a reason baby teeth are also called "milk teeth": they develop when an infant is still feeding from its mother. They are often poorly formed, with cusps that do not occlude properly when the jaws are brought together. And they do not make up a complete tooth row: molars (and sometimes other teeth) do not have a baby-tooth precursor, so they don't form until after the baby has been weaned off milk. Baby teeth, therefore, are substandard as chewers, crushers, and grinders, but that's not a problem if the child is subsisting on an all-fluid diet.

There's another, more serious issue. As attested by the gummy smiles of most human newborns—or the teething wails of an infant—many mammals are born without teeth, because the jaws need to grow larger and stronger before the first baby teeth can erupt. These youngsters—already so vulnerable—have to survive at least several weeks or months without any hope of even subpar chewing with their baby teeth. Once again, milk provides the solution, as all the baby needs to do is suckle and swallow.

This raises the second type of evidence for lactation: the skeletal structures required for suckling. One is a secondary palate: the hard roof of the mouth, which evolved in Triassic cynodonts. The palate separates the mouth from the airway, ensuring that even the feeblest babies don't choke while guzzling milk. Not only that, but when they feed, babies pull their mother's nipple deep into the mouths with their tongues, usually forcing the nipple against the palate, which releases the milk. A palate on its own, however, is not enough. To provide sufficient force to draw out milk, the infant needs a muscular throat, with complex and highly mobile hyoid bones to cradle the throat cartilage and anchor the muscles. This type of peculiar hyoid system first undoubtedly appears in docodonts, like the diminutive *Microdocodon* from Liaoning.

Putting this evidence together, mothers must have started nursing early in mammal history, probably right around the time the first mammals like *Morganucodon* were scuttling around in the Triassic, and definitely by the time docodonts were flourishing in the Jurassic. It is probably no coincidence that enlarged brains appeared right around the same time. Big brains are metabolically expensive, and a nutritious, sustainable, readily available food source like milk would provide the necessary energy for more neural tissue—particularly the new six-layered neocortex, the superprocessing center that is the seat of mammalian intelligence and sensory integration.

Milk—that most mammalian of substances—not only sustains us when we are young, but it also made us smart. Intelligence, however, is just one component of the neurosensory arsenal of mammals.

AS EARLY AS the sixteenth century, anatomists noticed something unusual in the middle ear cavity of human cadavers, just inside of the eardrum. There were three bones, each about the size of a grain of rice, easily making them the smallest bones in the body.

This was most peculiar. Bones were supposed to be strong and sturdy. They propped up the body, protected vital organs, were a scaffolding for muscles. Even if you couldn't see your own bones, you could sense them. They sculpted the profile of your face, shaped the contours of your waist, hoisted the bulging muscles of your arms, cracked when you made a fist, creaked in old age. And bones were creepy, mystical—decor on tombs, an emblem of death.

The ear bones were none of these things. What were they, and why were they there?

Whatever they were, they weren't unique to us. Later anatomists identified this same three-bone set in other mammals—but only mammals. Always tiny, they became known as "ossicles," as if they weren't deserving of being proper bones. They were given latinized names: malleus, incus, stapes. Most of us, however, learn about them in school under their informal names: the hammer, anvil, and stirrup, so-called because their shapes kind of resemble these objects, in the same way star constellations kind of look like bears and crabs. As anatomists continued to dissect mammal ears, they noticed that the three ossicles were associated with another minuscule bone: the ectotympanic, given the much more intuitive nickname the "ring" because of its hooplike shape.

When counting the bones of the body, it's easy to forget about the hammer, anvil, and stirrup. And nobody ever mentions the ring, because in humans, it is fused to the broad temporal bone, which forms much of the side profile of our head, behind our eyes and cheeks. But the tiny size of these bones is far out of scale to their importance, as scientists and medical doctors would later come to understand.

The ectotympanic ring supports the eardrum—the taut membrane that receives sound waves from the air—like the wooden frame of a tambourine. The hammer, anvil, and stirrup form a chain between the eardrum and inner ear: the eardrum contacts the hammer, which has a mobile joint with the anvil, which touches the stirrup, which strikes the cochlea, the soft part of the inner ear that actually processes sound and sends auditory signals to the brain. The ossicle chain serves three key functions. Like a telephone line, it transmits sound from the eardrum (receptor) to the cochlea (processor). Like a bunch of bullhorns lined together, it amplifies this sound. And like a power plug converter you may bring on vacation to another country, it converts airborne sound waves to waves that flow through the liquid inside the cochlea. It is the motion of these microscopic waves that trigger teeny hairs in the cochlea, whose movements are then transformed to an electrical signal, which is relayed by nerves to the brain, producing what we sense as "sound."

The ring, hammer, anvil, and stirrup enable one of the most advanced neurosensory skills of mammals: our ability to hear a wide range of sounds, particularly high-frequency ones. Birds, reptiles, and amphibians can all hear. They can all take sound waves and convert them to liquid waves in their cochleas. But they can't hear anywhere near as well as mammals, and across such a wide range of frequencies, because they have only a single ear bone to do it all—the stapes, the equivalent to the stirrup.

Where, then, did the other three bones—the hammer, anvil, and ring—come from? Did evolution fashion three entirely new bones to help mammals hear better? That's a reasonable first guess, because often natural selection does create new structures that end up serving specific purposes—like, say, a horn or whiskers.

When anatomists in the early nineteenth century began looking at the developing ears of mammal embryos, however, they realized something astonishing. Armed with little more than a magnifying glass, the German embryologist Karl Reichert observed that the hammer and anvil did not begin forming within the middle ear. In fact, they were nowhere near the ear. Rather, they were positioned at the back end of the jaw, and in early-stage embryos, it was these two bones themselves that actually *made the joint* between the upper skull and lower jaw.

The bones that form the jaw joint in nonmammal vertebrates have a name: the articular and the quadrate. Reichert recognized something simple, yet profound. During early mammal development, the hammer and anvil were essentially identical in size, shape, and position to the articular and quadrate of reptiles. Which could mean only one thing: the hammer and anvil *are* the articular and the quadrate. They are jaw bones.

The anatomist Ernst Gaupp—a fellow German, like so many of the great bone experts of the era—moved Reichert's work forward, and devised a unified theory of mammal ear development, now taught in medical courses everywhere, that bears both of their names: the "Reichert-Gaupp theory." As techniques for studying embryos improved, Gaupp was able to use microscopes to better track fine details of the ear bones as they developed. He confirmed that Reichert was correct in identifying the hammer as the articular and anvil as the quadrate, and he finally solved the mystery of the ectotympanic ring. It, too, began development at the back end of the lower jaw, in a position equivalent to the angular bone of reptiles. Thus, the ectotympanic is the angular. It, also, is a jaw bone.

What this means is extraordinary: none of these mammal ear bones were new inventions. They were jaw bones that evolution repurposed for a new function: hearing.

Modern biologists can study embryonic development in incredible detail, using CAT scans and microscopic slide sections in which different tissues—bone, cartilage, muscle, and so on—are stained different colors. What happens as a mammal grows is now well understood. As development persists, and an early-stage embryo turns into a late-stage embryo and then a newborn, the hammer and anvil change. They both stop growing and turn to solid bone

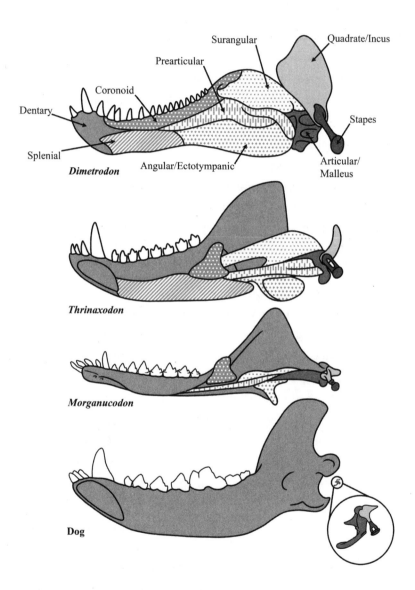

The transformation of jaw bones of mammal ancestors into the tiny ear ossicles of mammals. (illustrated by Sarah Shelley)

earlier than most of the rest of the skull. They move backward and inward, lose all connection with the jaw save for some tenuous ligaments, and become encased in a bony bubble called the tympanic bulla. Meanwhile, the ring bone begins as a strip at the back angle of the jaw, curls up, and moves toward the hammer and anvil, the latter contacting the stirrup. As this is happening, a new joint materializes between the dentary bone of the lower jaw and the squamosal of the upper skull, which becomes the hinge point for suckling, and then later, chewing.

Usually this all happens in utero. In humans, for instance, the hammer and anvil become disconnected from the jaw during the eighth month in the womb—although there are rare clinical cases where the dentary-squamosal joint fails to develop, and human adults close their jaws via their hammer-anvil joint, *in the ear.* Marsupials like the opossum, however, do something ridiculous. Their frail newborns need to start suckling immediately in the pouch, before the dentary and squamosal bones fully form. Thus, they use the joint between the hammer and anvil as their primary jaw joint for the first twenty days of their postbirth lives. During this time, their ears physically power their milk guzzling. As this is happening, a second joint develops between the dentary and squamosal, and for a short while the two joints function side by side, before the connection between the hammer and the lower jaw is severed. The hammer and anvil then become exclusively used for hearing, and the dentary and squamosal exclusively used for jaw closing.

Jaw bones becoming smaller, moving backward, losing their mouth-closing functions; a new, stronger jaw joint developing between the dentary and squamosal. If this sequence sounds familiar, it's because we already learned about it in the last chapter.

Recall that, over the course of cynodont evolution, the quadrate, articular, and many smaller bones at the back of the jaw withered in size while a new, stronger dentary-squamosal jaw joint—the very feature that defines mammals—replaced them. This evolutionary sequence over millions of years of cynodont history impeccably mirrors the developmental sequence within a single growing mammalian embryo today. As biologists say, ontogeny is recapitulating phylogeny. Or put another way, the developing embryo is like a film that captures, in high-speed, the evolutionary journey of jaw bones changing to ear bones.

Fossils also tell this story. There is a transitional sequence—spanning from cynodonts to *Morganucodon*-type early mammals to contemporary mammals—showing how jaw bones were refashioned into ear bones, and shedding light on why this happened.

First, a brief review. Pelycosaurs, therapsids, and the first cynodonts had normal reptilian-style jaws. The dentary held the teeth, and various postdentary bones comprised the back of the jaw, including the articular, which contacted the quadrate bone of the upper skull to form the jaw hinge. As we learned in the last chapter, during cynodont evolution the dentary progressively got bigger and stronger, while the postdentary bones atrophied, until it became necessary to build a new dentary-squamosal jaw hinge. This novel joint first appeared in animals like *Morganucodon*, and those species that have it are called mammals. This is the dividing line between nonmammal and mammal (as I've defined them in this book).

In these first mammals, there were two jaw joints: the new dentary-squamosal one and the ancestral quadrate-articular one. The dentary-squamosal joint was doing most of the jaw closing and was the primary source of the strong bites and precise chewing of these animals. The quadrate and articular joint, however, was still bearing some of the jaw load. At the same time, the quadrate and articular bones had moved backward to the point where the quadrate contacted the stapes bone—that ancestral middle ear bone that transmits sound from eardrum to cochlea in reptiles, amphibians, and birds. Therefore, in *Morganucodon*-type mammals, the quadrate-articular joint was serving dual functions: it was transmitting sound to the ear, but also participating in jaw closing. It was a delicate balance, and one that was not fit to last.

Some fossils show an intermediate condition. The most striking is a footlong critter called *Liaoconodon*, encased by the Liaoning volcanoes and described by Meng Jin and his team. The dentary and squamosal form the only functional jaw joint, and the quadrate, articular, and angular are all small ossicles that have moved backward and inward, into the middle ear cavity. The jaws and ears seem to be separated—but not quite. The articular and angular are linked to the dentary via a thin strip of bone. This probably helped support the delicate ossicles, but because they were still physically tethered to the jaw, it also meant they were still affected by chewing motions.

The next stage in evolution is obvious: the connection between ear and jaw needed to be severed, the bony strip linking them snipped. This is seen in another Liaoning mammal called *Origolestes*, described by Meng, his colleague Fangyuan Mao, and their team. Because these two former jaw bones are now fully detached from the jaw, we can call them by their new names: the hammer and the ring. This small step was revolutionary. Now the jaws could go their own way, and become more efficient at biting and chewing, without worrying about interfering with hearing function. The ears could go their own way, too, and become even better at hearing high-frequency sounds without being disturbed by the jaws.

The full separation of the jaws and ears created the "detached middle ear." The haramiyidans—those wildly diverse Jurassic mammals that glided between branches and may have lived in bat-like colonies—are some of the first to show evidence of a fossilized detached ear. The ear ossicles—now freed from the jaw—nestled within the middle ear cavity and functionally became part of the upper skull, as they were surrounded by the tympanic bulla, their protective bubble. The cochlea of the inner ear also became encased by the petrosal—the "rock" bone, so named for its extreme density. Both bulla and petrosal serve as noise-canceling headphones, allowing mammals to still hear exceptionally well while chewing. Remember this next time you're watching the TV when eating dinner.

There is one final twist to the story. The transition between jaw bones and ear ossicles sounds clean and easy, a gradual transformation as mammals perfected themselves in the Triassic, Jurassic, and Cretaceous. When all the fossils are placed on a family tree, however, and their anatomical features are mapped out, a more complicated tale emerges. The detached ear did not evolve only once, but multiple times: at least three times, maybe four, perhaps five or more. There was a stage in mammalian evolution, along the trunk of the family tree, where the jaw bones had gotten smaller and moved into the ear, but they were still attached to the jaw via a bony strand. This caught-in-between phase wasn't optimal for either chewing or hearing.

Intriguingly, some of the different groups of mammals that independently separated their ears have differently shaped joints between the hammer and anvil. Some, like us, have an interlocking system, where a ball on the anvil fits into a socket on the hammer. Others, like monotremes, have a simple overlapping joint. One idea—still being debated—is that these different ear joints reflect the different jaw joints that they evolved from: jaw joints that articulated in different ways and could move in different planes.

There's only one way this would make sense: these mammal groups developed distinctive ways of chewing first. They constructed their jaws so that the shape of both the quadrate-articular *and* dentary-squamosal joints allowed whatever type of chewing they were doing: back-and-forth jaw sweeping in some, a freer range of side-to-side and up-and-down motions in others, and so on. Then, each group independently cut their jaws off from their ears, probably as a means of streamlining. Having two jaw joints, which needed to work in unison, was at best redundant, and at worst an impediment—like training wheels if you're trying to pedal fast on a bike. Chewing would become more efficient if it was orchestrated by a single, solid, muscular jaw joint: the dentary-squamosal joint. The articular-quadrate joint was liberated to become the malleus-incus (hammer-anvil) joint, to dedicate itself entirely to transmitting sounds from eardrum to cochlea, but it would forever be constrained by the shape of the jaw motions of its ancestors.

Why were mammals so invested in chewing that they would divorce their jaws from their ears, many times over? Because during the Jurassic, and particularly the Cretaceous, there were lots of new foods to eat. Swarms of tasty new insects appeared, and they pollinated a totally new type of bright, beautiful plant, which produced all sorts of delicious flowers, fruits, leaves, roots, and seeds.

The three modern groups of mammals—the placentals, marsupials, and monotremes—can trace their ancestry back to this time, a manic waltz of evolution and ecological change called the Cretaceous Terrestrial Revolution.

4

The
MAMMALIAN
REVOLUTION

Kryptobaatar

AS WE PULLED UP TO the cottage on the outskirts of Warsaw, I was an exhausted mess. Unshaven, greased hair, fingernails blackened with dirt. My forehead was starting to peel with sunburn, and I kept my V-neck flannel open wide, to try to catch some cool air in the back seat of our van, which had no air-conditioning.

It was mid-July, 2010. I was a PhD student and was in Eastern Europe doing fieldwork with the British paleontologist Richard Butler and our Polish friends Grzegorz Niedźwiedzki and Tomasz Sulej. We were back in the capital, after spending a week and a half bouncing around Poland and Lithuania, on the hunt for Triassic-aged dinosaurs.

Fortune was not on our side. During the first half of our trip, in Poland, we didn't find many fossils. Then it was off to Lithuania. A few hours into our drive, the van came to a sputtering halt, the victim of a malfunctioning alternator. What might have been a long exile in the flatlands of northeastern Poland was avoided, thanks to a dodgy mechanic who "found" a replacement part in the engine of another vehicle, whose owners had conveniently gone away for the evening. We arrived in Lithuania and checked into a hotel the following night. Hungry and weary, we lamented the lost day of fossil collecting but consoled ourselves that tomorrow would be better. It wasn't. Rain pelted the clay quarry in which we were working, making it nearly impossible to collect fossils or map the rocks. Grzegorz found a single tooth—that was it, and it wasn't even from a dinosaur.

As we approached the cottage and rang the doorbell, I was in a sour mood. I thought we had it rough. I would soon come to understand that we did not.

The door cracked open, and a small mammal came bounding out. A Pomeranian dog, groomed like a beauty pageant contestant, yelping with a high-pitched screech. It made a beeline for my ankle and clamped on for a moment.

An elderly lady appeared on the porch, and in accented English, apologized for her dog's misbehavior. Small mammals could be feisty, she said, with a coy smile. She was a slight woman, a few inches north of five feet tall, who walked with a hunch. Her hair was gray, her hands wrinkled, her eyes kind. A billowing, striped dress covered her thin frame, fastened by a narrow white belt and big red buttons. In look and manner, she seemed no different from the many babulas—grandmothers—we had seen strolling through the villages on our travels. We were told that she had just turned eighty-five years old.

She motioned for us to come inside, where a table was laid with cakes, sweets, and tea. Her husband, a slightly taller man with gray, parted hair, joined us. After the travails of our fieldwork, it was a welcome bit of relaxation.

Through mouthfuls of cream-filled cakes, we told her about our misadventures. The broken van, the rain and heat, the missed dinner, the scanty fossils. She nodded along, and when we had finished our sob story, she chuckled.

"It was much worse in the Gobi Desert," she said.

You wouldn't know it from glancing at her, or making small talk with her, but this gentle grandmother was one of the world's greatest fossil collectors, a pioneer who once ventured deep into the sand dunes on the trail of dinosaurs and mammals, the captain of one of the first major female-led fossil hunting expeditions.

Her name was Zofia Kielan-Jaworowska, and she was one of my heroes. In truth, I had been looking forward to this meeting more than the fossil collecting. Sitting in Zofia's kitchen, as she recounted tales of hardship and adventure, made the comparatively minor tribulations of our fieldwork melt away.

Zofia was born in 1925, east of Warsaw. She was a teenager when the Nazis invaded. They weren't keen on Poles getting an education, so she had to finish high school and start university by

Richard Butler and me meeting Zofia Kielan-Jaworowska at her home in Poland in 2010 (top) and Zofia in the Gobi Desert of Mongolia in 1970 (bottom). (photos by Tomasz Sulej and Institute of Paleobiology Warsaw, respectively)

taking clandestine courses. While she was secretly studying zoology at the University of Warsaw in 1944, the local resistance forces took their fight to the Germans. The promised Soviet help never arrived, and some two hundred thousand people were killed, and the city mostly destroyed, in what became known as the Warsaw Uprising. During those bleak two months, Zofia put her studies on hold and worked as a medic, tending to the wounded.

The war ended and the university reopened in 1945, although there wasn't much of a physical university left. Courses were held at random places throughout the city, including Professor Roman Kozłowski's apartment, which had survived the destruction. One of the professor's homebound lectures recounted the Central Asiatic Expeditions, fossil-collecting odysseys to Mongolia in the 1920s and early 1930s, led by the charismatic explorer Roy Chapman Andrews, who rumor has it was an inspiration for the character of Indiana Jones. These journeys were the stuff of legend, in which Andrews and his team of Americans dodged bandits and braved sandstorms, using the newly invented automobile to penetrate deep into the desert to wrestle fossil treasures out of the rusty red sandstones. Their discoveries were epic: the first dinosaur nests, the first skeletons of the villainous *Velociraptor*, and nearly a dozen mammal skulls, which at that time were the most complete records of Cretaceous-aged mammals ever discovered.

Zofia was entranced and decided to pursue paleontology as a career. But while she had "audacious dreams" of retracting Andrews's steps, Mongolia was a long way away, and there were fossils closer to home. At Kozłowski's suggestion, she went on to study trilobites—extinct buglike arthropods with hard exoskeletons, which mobbed the oceans hundreds of millions of years before dinosaurs and mammals lived on the land. For nearly fifteen years, she spent her summers collecting these little fossils in central Poland. All the while, she never shook the romance of the Gobi Desert.

By the early 1960s, Zofia had become a world-leading expert on trilobites and other backbone-less creatures. She had also been drafted into management, becoming director of Warsaw's Institute of Paleobiology after Kozłowski's retirement. This put her in a position of influence and allowed her to pitch new research ideas. It was the height of the Cold War, and the Polish authorities were eager for their scientists to make partnerships with comrades in other Communist countries. Countries like Mongolia.

Zofia took her chance. She proposed a joint Polish-Mongolian expedition to the Gobi Desert. It seemed a long shot: not only was she not (yet) a dinosaur or mammal expert, she was a woman, and no woman had led such a large paleontological expedition, anywhere. But her pitch was approved, and suddenly she needed to assemble a team. She recruited several other young women, including three who would become renowned paleontologists in their own right: Halszka Osmólska, Teresa Maryańska, and Magdalena Borsuk-Białynicka. They had little experience working so far away from Poland, and many of them had never even been to a desert. Nevertheless, they persisted, and joined with young Mongolian researchers—Demberlin Dashzeveg, Rinchen Barsbold, and Altangerel Perle, who also became well-respected paleontologists—to make a formidable crew. They undertook eight expeditions from 1963 to 1971.

These were difficult trips. Temperatures could reach more than 100 degrees Fahrenheit (40 degrees Celsius) in the shade, but dip below freezing at night. Often it would take many days to travel from one site to another. Most of the driving was off-road, with little more than a map, compass, and old tire tracks to guide Zofia's team to the fossil-hunting spots. Water was scarce, and much of the logistical planning was concerned with how to get enough water to their campsites, which were usually tens of miles of unforgiving terrain from the closest wells. Their six-wheel-drive trucks, bor-

Zofia Kielan-Jaworowska and team searching for tiny mammal fossils in the Gobi Desert in 1968. (photo courtesy of Institute of Paleobiology Warsaw)

rowed from the Polish state-run automotive factory, were sturdy but cumbersome, and made for an uncomfortable ride. Once, Zofia sat for too long near an open window on a long journey, and the wind ruptured her eardrum, necessitating an emergency evacuation back to Poland for medical treatment. Still, the expedition continued, with Teresa Maryańska deputized as leader . . . until Zofia returned a few weeks later.

During their eight years in the Gobi, Zofia's team found a lot of fossils. Many of these were dinosaurs, which was why I (initially) idolized her. They collected a *Velociraptor* and its victim locked in deadly combat, a long-necked sauropod whose bones weighed twelve tons, and innumerable skeletons of *Tarbosaurus*, *T. rex*'s closest cousin, which I had spent a week studying in Warsaw for my PhD thesis.

Dinosaurs made Zofia and her expeditions famous. But what Zofia really wanted were mammals. She was awed by the tiny skulls collected by Andrews's expeditions and recognized their importance in illuminating the early phases of mammalian evolution during that murky time when they were living underfoot of dinosaurs. She also recognized something else: the Andrews team lusted after the big, glitzy fossils that would make newspaper headlines and blockbuster museum exhibits. The mammals they brought back from the Gobi—although the premier record of such fossils at the time—were afterthoughts, randomly yanked from the desert floor as they kept their eyes peeled for dinosaurs. They had no dedicated strategy to search for fossil mammals, yet they still found several skulls and partial skeletons, which meant one thing.

More mammals must be out there, waiting for the right paleontologist to come along.

Zofia was that paleontologist. She committed to searching for small skulls, jaws, and teeth by literally getting on her hands and knees, nose to the rock, with a magnifying glass. It was backbending, knee-bruising, eye-straining work, and it had none of the glamour of Andrews's macho dinosaur hunts, but it was effective. In 1964, her team discovered nine mammal skulls, nearly as many as Andrews's team found during their decade of exploration. And this was only the start.

In 1970, Zofia took her team to the dark-red rock exposures of the Barun Goyot Formation, which Russian paleontologists—who had explored the Gobi in the years after World War II—had considered barren of fossils. Zofia had a hunch that her magnifying-glass method might prove them wrong. She was quickly proven correct. Within a few hours, Halszka Osmólska had found a gorgeous skull of a mammal, unlike any found previously, by Andrews, the Russians, or Zofia's earlier teams. A new species! After a quick lunch back at camp, Zofia marshaled a larger crew and took them to a site called Khulsan, which soon became—in her words—their own El Dorado. That afternoon alone they found five mammal skulls. They stayed for the next ten days and found another seventeen skulls. In less than two weeks' time, they had accumulated the world's best record of Cretaceous mammals.

After a short winter break in Poland, Zofia and her crew returned to Mongolia in the spring of 1971, brimming with enthusiasm. They went straight to their El Dorado and on that first day, Zofia discovered three more mammal skulls. They expanded their search and found another fossil trove at a place called Hermiin Tsav, where they ran into another group of bone-hunters: a large team of Soviets, who for the last few years had been fanning across the desert, encroaching into prime fossil territory. They had more resources, and better political connections—after all, the Soviet Union was the superpower, and Poland a satellite. And they were adept at playing games, tempting to their side some of the young Mongolian scientists who had been working with Zofia. For a while, the Mongolians tried to have it both ways, dividing their time between the two competing brigades, but at the end of the 1971 season the tension could no longer hold.

Zofia was summoned to meet the president of the Mongolian Academy of Sciences, who broke the bad news. His Mongolian scientists could now work only with the Soviets. The Polish-Mongolian expeditions were terminated. "The news was absolutely crushing for me," Zofia later wrote, confused about the Cold War politics behind the decision. "But I tried to comfort myself (by) remembering that we had a number of fossils from Mongolia which we would be able to describe during the coming years."

Among these fossils were some 180 Cretaceous mammals—the accumulated total of their years of work at Khulsan, Hermiin Tsav, and other sites. By far and away, it was the largest, most complete, most diverse, and most spectacular collection of

Cretaceous mammals ever assembled. Zofia spent the next several decades describing these skulls and skeletons, doing much of the work in her home office.

After we finished our tea and cakes, she took us there. The wall was lined with colored binders—a carefully cataloged library of the great publications on mammal evolution—and clear plastic boxes full of teeth, jaws, and other mammal parts. They were all small: none of these mammals would have been even half the size of her Pomeranian.

Zofia flicked on a lamp and fetched a box from one of the shelves. Her hands shaking ever so slightly, she removed the lid and took out a small plastic tube with a jaw inside, studded with flattened molars covered by a pavement of cusps. She placed it underneath her microscope, braced herself against the table, and slowly hunched over, to take a closer look.

"Maybe a new species," she said, as she invited us to take a glance. It had been nearly forty years since the final Polish-Mongolian expedition, and she still had work to do. She continued that work right up until March 2015, when she passed away, a month shy of her ninetieth birthday, with a house full of Mongolian fossils that remain to be studied.

THE MAMMALS ZOFIA and her team collected come from rocks sculpted into cliffs, crags, gullies, badlands, and other desert landforms across much of southern Mongolia, about two days' drive from Ulaanbaatar. Mostly these rocks are sandstones and mudstones, some of which were deposited in river channels draining lush forests, whereas others are remnants of ancient sand dunes and desert oases that wouldn't look out of place in contemporary Mongolia. All these rocks were formed during the Campanian and Maastrichtian stages at the end of the Cretaceous Period, between about 84 and 66 million years ago. This was a good 80 million years, give or take, after the Jurassic heyday of the docodonts and haramiyidans, those first hugely diverse mammal groups that we learned about last chapter.

A lot had changed during this time. One hundred forty-five million years ago, the Jurassic gave way to the Cretaceous. The transition wasn't triggered by megavolcanoes or another geological calamity, and it wasn't marked by a particularly noticeable mass extinction or ecological collapse. Instead, there were slower changes to the continents, oceans, and climates that gradually added up, ushering in a new Cretaceous world. Sea levels fell, then rose again, over a span of around 10 million years. The blazing hothouse of the Late Jurassic went cold, then arid, and then back to normal in the Early Cretaceous. Meanwhile, the continents remained restless. Old Pangea further cracked apart, the new lands moving ever farther away from one another, at the imperceptible rate of a few centimeters a year. Multiply that by 80 million years, and by the end of the Cretaceous the continents were more or less in their present positions.

The map wasn't quite modern, though. South America was widely separated from North America, but remained tenuously tethered to Antarctica, which nearly touched Australia. India was an island continent off the east coast of Africa, plowing rapidly northward. There were no ice caps, so sea levels were high, and Europe was nothing but a speckle of islands poking out of a tropical sea. Another sea lapped far onto North America, at times stretching from the Gulf of Mexico to the Arctic, bisecting the continent into a mountainous western slice called Laramidia and an eastern portion called Appalachia. The European islands were convenient stepping-stones between North America and Asia, but there was a wide oceanic barrier between these northern lands and the southern continents.

Zofia's Gobi mammals would have lived in the middle of a large Asian landmass that, occasionally, was connected to Laramidia across the Bering Strait, and which had easy access to the European archipelago to the west. The nearly two hundred mammals she pried from the desert constitute a diverse fauna, of many species, belonging to many subgroups. Some are archaic holdovers from the early stages of mammalian history, others are much more derived, and classified on the ancestral lineages leading to today's placentals and marsupials. The vast majority of them, however, are multituberculates.

Multituberculates were the next big wave of mammalian diversification, after the decline of the docodonts and haramiyidans. They may have evolved from haramiyidians, as both groups had similar molar teeth, with long cusp rows, and chewing strokes, in which their jaws moved backward in a grinding motion.

Multituberculates were *the* quintessential Cretaceous mammals, at least on the northern continents. Over one hundred species have been found, and in the Gobi, about 70 percent of all mammal fossils are multituberculates. Zofia described several new species, mostly based on her team's discoveries from Khulsan, Hermiin Tsav, and another splendid site: a ridge of sandstone, cutting a fiery orange figure against the desert sky, that Roy Chapman Andrews called the Flaming Cliffs when his crew found the first fossils there in the 1920s. Among the Gobi multituberculates are *Kryptobaatar*, the most common variety, known from several skulls and skeletons, which Zofia named in 1970. In the same research paper, she also introduced *Sloanbaatar* and *Kamptobaatar*. Later, she christened *Catopsbaatar, Nemegtbaatar, Bulganbaatar, Chulsanbaatar*, and *Nessovbaatar*. You've probably sensed a pattern: "baatar" is Mongolian for hero, the same root that forms the country's capital— Ulaanbaatar, meaning "Red Hero," a relic of the Communist past. There were possibly more potential *something-baatars* lurking in the boxes in Zofia's office when I visited, awaiting identification as yet more new species.

Although Zofia had to stop collecting in 1971, other teams have since found a myriad of new Gobi multituberculates. Almost immediately after Mongolian Communism fell in 1990, Andrews's Central Asiatic Expeditions were reprised by an American Museum of Natural History crew, led by Mike Novacek—whose travelogue pop-science books like *Time Traveler* inspired me as a teenager—and Mark Norell, who would later be my PhD advisor. They have since returned to the Gobi every year for three decades and counting, and eventually eclipsed Zofia's record by amassing several hundred mammal skulls. Many are from an extraordinary site they discovered, called Ukhaa Tolgod: a one-and-a-half-square-mile (four-square-kilometer) Cretaceous dunefield. Here, thousands of mammals, lizards, dinosaurs, and turtles—living on the dunes and in the oases between them—became entombed in sand as dunes collapsed during deluges of sudden desert rain. As would be expected, most of the mammals are multituberculates, including yet another new bataar, *Tombaatar*.

If you were back in the Mongolian Cretaceous, taking in the sun on a dune-top or hunkering in the brush of a desert oasis, hiding from the giant dinosaurs, odds are you would be surrounded by multituberculates. They would have been like rats in a sewer pipe, mice in a derelict building, voles in a field. You probably wouldn't always see them, but you might hear them, and you could be sure that they were there, lurking. Filling the hollows, perched in the shadows, crawling through burrows and the leaf litter.

Jurassic and Cretaceous mammals are still, all too often, unfairly stereotyped as rodent-mimics. When it comes to multituberculates, however, for once this comparison is apt. They had bucktooth incisors, could chew and gnaw, and moved in many of the same ways mice and rats do today. Some had long, straight limbs for scurrying quickly along the ground, others burrowed, and some could bound and hop. Many of them could twist their ankles so that their feet pointed backward, allowing them to gracefully and securely descend headfirst down a tree, like the squirrels outside my window. But multituberculates were not rodents. Whereas rodents are placental mammals like us, multituberculates were a more primitive family, nestled between monotremes and the marsupial plus placental group on the family tree. They evolved all their rodent-like specializations independently from modern rats, mice, and shrews, probably to fill the same general ecological niches. During the Cretaceous, the multituberculates were wildly successful in these mouse-to-guinea-pig-size roles.

What made them so successful? They were champion chewers and developed their own unique feeding style that allowed them to gorge on many types of foods, particularly plants.

Multituberculate teeth have a diversity of highly complex shapes, whose profiles rise up from the jawbone like a city's skyline. The incisors at the front of the snout are procumbent, sticking outward in a goofy bucktoothed grin. Behind them is usually a space, followed by at least one large, thin, serrated premolar that juts upward like the blade of a table saw. The rest of the tooth row is composed of big, flat molars with elongate rows of cusps, which resemble Lego bricks. The name multituberculate, which means "many tubercles," comes from these peculiar molars.

These teeth worked together to mutilate food, like the many tools of a Swiss army knife. The incisors collected and ingested the food, and in many species could gnaw, as in rodents. The premolars crushed and sliced; straight up-and-down motions brought premolars of the lower jaw upward, as pointy scissor blades to cut against their upper counterparts. The most interesting action, however, was with the molars. When the jaws snapped shut, the

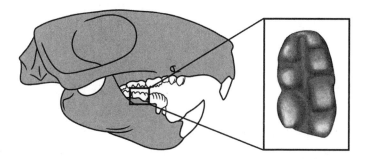

The skull of a Cretaceous multituberculate, with a close-up of the chewing surface of its Lego brick–shaped molar. (illustrated by Sarah Shelley)

lower jaw was constrained to move backward against the upper skull, bringing the lower and upper molars into sliding contact. Any food caught in between was ground down, pulverized by the long rows of upper and lower cusps grating against each other. This backward power stroke was steered by big mouth-closing muscles that attached far forward on the lower jaw—much farther than in most other mammals, including rodents. These long muscles necessitated a long jaw, and hence a long snout, giving multituberculates a signature long-nosed, big-eyed, and probably chubby-cheeked mug shot.

The first multituberculates lived in the Middle-Late Jurassic, when docodonts and haramiyidans were booming. It was during the Cretaceous, however, when multituberculates came into their own. It was then—particularly over the final 20 million years of the Cretaceous—that they became stupendously abundant, the swarms that dominated the small niches, as exemplified by all the skulls and skeletons Zofia collected in the Gobi.

As the Cretaceous unfolded, multituberculates refined their dietary tastes. Their molars got bigger, more ornate, and developed more cusps. A clever study by Greg Wilson Mantilla (who we'll meet properly in the next chapter) and colleagues used methods of cartography—called Geographic Information Systems (GIS)—to show that the landscape of multituberculate molars became increasingly complex during the Late Cretaceous, with more and higher peaks, deeper valleys, and a more convoluted surface texture. From studies of modern mammals, it is known that tooth complexity increases along the dietary spectrum from carnivores to omnivores to herbivores. These Late Cretaceous multituberculates, therefore, were developing ever-more baroque teeth as they became increasingly specialized for eating plants. In doing so, they became more diverse, splitting into hordes of new species, like Zofia's army of bataars. And they got bigger—although they were still, on average, just a couple of pounds (one kilogram) in weight and could fit comfortably in your arms.

Multituberculates spread around the northern continents during this time, chasing new plants to eat. At least a half dozen species, maybe many more, chewed away as *T. rex* and *Triceratops* did battle in the end-Cretaceous ecosystems of Hell Creek, on the Laramidian subcontinent. Many thousands of miles away, an endemic family called the kogaionids populated the European islands, apparently first washing up on a Hispaniola-size sliver called the Hațeg Island, adapting to their new insular environment, and then hopscotching across the mosaic of land specks peeking out from the sea.

In 2009, while fossil hunting along a river with his sons, Mátyás found a dinosaur he thought resembled some of the meat-eaters' fossils Mark had collected in the Gobi. His email to Mark led to a wintry flight to Bucharest, where we identified Mátyás's dinosaur as a new species closely related to the Gobi *Velociraptor*. This led to a longer-term collaboration with annual fieldwork

Fieldwork in Romania: collecting fossils from the Multi-Bed (right), Mátyás Vremir gathering fossils from the river (left). (photos by Akiko Shinya and Steve Brusatte, respectively)

jaunts in the early summer. Our field team came to include Meng Jin, who has described so many of the Liaoning mammals from China, the soft-spoken and dry-witted Romanian paleontologist Zoltán Csiki-Sava, and many others who have filtered in over the years. As it turned out, we've found more mammals than dinosaurs. These mammals, too, have a Mongolian flair, because they are all multituberculates. But they have their own unexpected peculiarities.

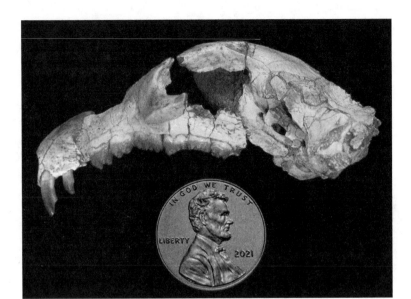

The pea-brained multituberculate *Litovoi*. (photo by Mick Ellison)

The Hațeg Island multituberculates—which number at least five species, with several more soon to be described—all belong to the kogaionid clan, which lived only on the European islands. Their bones and teeth are found at many fossil sites across the rolling Transylvanian hills The most stunning fossil site is a graveyard of skeletons, nicknamed the "Multi-Bed," near the village of Pui. In keeping with the vampiric ambiance, the corpses are found within a dark red layer of mudstone, about a foot and a half (half a meter) thick. This was a soil horizon on a floodplain during the Late Cretaceous, which the multituberculates might have burrowed into, perhaps as a safe haven from the raptors, but then became their tomb. Maybe they were buried alive, by the rising waters of a river.

While I'm speculating about the Cretaceous crime scene, there's no doubt that floodwaters are a problem at the Multi-Bed today. To our indescribable frustration, the bone layer has been engulfed by the white-water rapids of the Bărbat River, meaning that the best

fossils are exposed on the riverbed itself. This makes it impracti-
cal to collect the skeletons without damaging them, and no matter
how many fossils we gather there, it never gets easier seeing flakes
of white bone float down the river, the pieces that got away. Every
day, every hour, perhaps every minute, bones are ripped up from
the river bottom and carried away by the currents.

Our team does its best, and thankfully, Mátyás was an incredibly skilled collector. He had a superhuman eye for bones, sharper
than any paleontologist I've ever worked with. Sometimes he'd use
goggles to peer through the water, but he preferred his own makeshift invention of a cut-up plastic soda bottle, which he wielded
like a spyglass. Once he spotted a white shape, he'd dart his hands
into the water, with the speed and precision of a spear fisherman,
dig his nails into the rock, and scoop up as much as he could. He
had very steady hands, which was remarkable given how much
caffeine and nicotine he ingested. While his method of necessity
can't rescue entire skeletons, one time he came tantalizingly close.

On a June day in 2014, our fieldwork team split into two. I was
tasked with removing a couple nests of dinosaur eggs, using hammers and chisels. Mátyás, Mark, and a few others grew tired of the
tedium and decamped to Pui, to enjoy a fine summer afternoon of
wading in the river, looking for tiny bones in the Multi-Bed. When
we reunited back at the guesthouse that evening, I was sore and
drained, but Mátyás was buoyant. He cracked open a jumbo-size
bottle of Romanian beer, made a toast, and then showed us what
he had found. It was a skull, bearing multicusped molars, plus the
limbs, vertebral column, and ribs of a mammal, which probably
weighed 5 or 6 ounces (140–170 grams) and could easily curl up in
his hands—which is why he was able to grab so much of it out of
the river. "Best Cretaceous mammal specimen from Europe??!!" I
wrote in my field notebook.

Indeed, it was easily the best preserved, most complete

kogaionid ever discovered from the European islands, and one of the finest multituberculates from anywhere outside the Gobi. In 2018, we named it as a new species, *Litovoi*, after a thirteenth-century Romanian ruler of the same rank as Vlad the Impaler, the real Dracula.

There was one more surprise: when Mark took the bones to New York for further study, he did what many paleontologists now routinely do when we find a nice fossil: he stuck it in a CAT scanner, to see details of the internal anatomy. The images on the computer screen floored us: the brain cavity was tiny. Astoundingly tiny. When we measured the volume of the brain, *Litovoi* had one of the smallest brains of any mammal ever recorded, with a brain-to-body-size ratio more similar to the primitive cynodont ancestors of mammals than to other, non-kogaionid multituberculates.

We think we know why. Recall that *Litovoi* lived on an island. Islands are arduous places: there is often little space and few resources, at least compared to the mainland. Many modern mammals that find themselves washed up on islands are known to change aspects of their biology and behavior to adapt to the constraints of their new home. One of the most common changes involves the brain: it often gets smaller, probably to save energy, because big brains are expensive to maintain. *Litovoi*, it seems, was capable of a very advanced survival trick of modern mammals, way back in the Cretaceous.

This is another striking example of how multituberculates were thriving in the Late Cretaceous. They were hugely adaptable, able to fine-tune their teeth, their diets, even their brains as their environments changed, and they even burrowed and nested together in social groups, as shown by a 2021 discovery of a fossilized colony. While they remained small, unable to beef up in a world still stirring with dinosaurs, the Cretaceous underworld was theirs.

MULTITUBERCULATES, FOR ALL their talents and success, were one of many animal groups rapidly changing during the Cretaceous. Conducting this evolutionary overture were plants—but not just any plants.

A special, new type of foliage was in control of the music, *the* driving force of Cretaceous evolution. These were the angiosperms, more commonly known as the flowering plants. Like a maestro furiously sweeping her baton, the angiosperms steered the bugs, dinosaurs, mammals, and other animals in unanticipated new directions, leading an evolutionary movement that reshaped the earth. As a result, today angiosperms are dominant; they are, by far, the most abundant plant type in nearly every terrestrial landscape on the planet. They decorate our gardens, their wood frames our homes, they're romantic gifts for our loved ones, and they provide so much of our food, from fruits and most vegetables to grains like wheat and corn, which are domesticated grasses (a very specialized angiosperm).

I use the analogy of music, but other paleontologists prefer the metaphor of rebellion. They refer to a span of the mid-Cretaceous to the Late Cretaceous, from about 125 to 80 million years ago, as the Cretaceous Terrestrial Revolution. It was a time of diversification and upheaval, a changing of the guard from primeval communities to a more modern world, of forests alive with colorful flowers, fragrant fruits, buzzing insects, chirping birds, and most importantly for our story, a multitude of new mammals, including the immediate ancestors of today's placentals and marsupials.

This insurrection was mounted by some of the meekest revolutionaries imaginable: small herbs and shrubs, living in the shade of evergreen forests and on the margins of lakes. By developing fruits, flowers, and more efficient ways of growing, these unassuming angiosperms overtook the base of the food pyramid, changing the structure of the entire ecosystem. In the spirit of the analogy, they were the peasants who overthrew the king and created a new system of government. And they did it through stealth insurgency. They didn't invade from a foreign land or seize control after a disaster. They were there all along, biding their time in the exile of the shadows, an insurrection from within.

It's a classic underdog story. Angiosperms seemingly originated long ago, perhaps in the Late Triassic or Early Jurassic, as indicated by the huge amount of variation in modern flowering plant DNA that could only accrue over such a lengthy time span. Yet they do not appear as fossils until the Early Cretaceous, about 140 to 130 million years ago, and then only as microscopic pollen grains. The first fossils of actual vegetation turn up in some of the same Pompeii-style volcanic deathbeds as feathered dinosaurs and furry mammals, in 125-million-year-old rocks in Liaoning. These are weedy things, with flimsy stalks and delicate leaves, resembling the thyme or oregano you might grow in your garden. Angiosperms remained this way—inconspicuous and perhaps largely restricted to swamps and wetlands—until about 100 to 80 million years ago, when they began to increase in size, evolve their signature flowers and fruits, diversify into everything from shrubs to lofty trees, and construct diverse forests. By the end of the Cretaceous, angiosperms composed some 80 percent of floras and had blossomed into familiar forms like palms and magnolias.

The angiosperm takeover was, like so many human revolutions, a combination of skill and timing. The skill component was their many adaptations that set them apart from other plants, like horsetails, ferns, and evergreen trees (pines and kin). Angiosperm flowers and fruits promoted pollination by insects and wide dispersal. Their greater leaf vein densities (which could transport more water) and increased number of stomata (the small openings on the leaves that bring in carbon dioxide) allowed them to take in more of the raw ingredients for making their own food during photosynthesis, which helped them grow faster and more efficiently. This might not have mattered, however, if their environment was not changing alongside them. As Pangea fractured, the arid belts on both sides of the equator—which were prominent in the Triassic and persisted into the Jurassic—decreased to small pockets in the mid-Cretaceous. More uniform, humid conditions came to characterize most of the new continents, and higher-latitude temperate regions were no longer separated from the tropics by deserts. These were ideal circumstances for angiosperms to spread, and to thrive.

The bounty of new angiosperms was manna for the multituberculates. Zofia's bataars and Mátyás's pea-brained *Litovoi* would have used their multicusped molars to chew backward into the leaves, stems, buds, fruits, flowers, roots, and other parts of these fast-growing angiosperm revolutionaries, which were generally more nutritious than things like fern fronds and pine needles. This is why multituberculates were proliferating in the Late Cretaceous: they were developing bigger and more complex teeth, with more cusps, to take advantage of the new greenery. As they did so, they diversified into scores of new species, specialized for feeding on different plants.

BUT THIS IS only part of the story. Evolving alongside the angiosperms, in a coevolutionary waltz, were their pollinators: many new groups of insects, particularly moths, wasps, flies, beetles, butterflies, and ants, plus spiders and other creepy crawlies. As we've seen, insects had long been the preferred grub of many mammals, going back to the very first ones like *Morganucodon*. With so many new insects about in the Cretaceous, it was no surprise that mammals took advantage. In doing so, one group of mammals brandished a special type of molar tooth, ideal for pulverizing the hard exoskeletons of insects to extract the juicy nutrients within.

These were the therian mammals: the major group that includes placentals and marsupials. Their new molar design was key to their—or, I should say, *our*—success, starting during the Cretaceous Terrestrial Revolution, and continuing today. As meager as a molar may seem, it helped spawn the splendor of us, and (nearly) all modern mammals.

Therian molars are "tribosphenic," a term that combines ancient Greek words relating to grinding (friction or "tribo") and shearing (wedge or "sphen"). This new molar was a marvelous evolutionary invention, because as its name implies, it serves two functions at once. It can grind *and* shear. Multituberculates, you might recall, advanced beyond other Jurassic and Cretaceous mammals via their derived dentition, which integrated premolars for cutting and molars for grinding. Therians did one better: they developed molars that could both cut and grind—and do both very well—at the same time, in one chewing stroke. While the entire jaw of a multituberculate was their Swiss army knife, therians had multiple tools packed into the same tooth.

The tribosphenic molar was constructed over many evolutionary steps. Its foundation was the molar shape of the first Triassic and Jurassic mammals like *Morganucodon*, which looked like a mountain with three peaks when viewed from the side. To these three cusps on the lower molars, therians added another three. The six cusps divided into two distinct regions: a "trigonid" set of three pointy spires at the front of the tooth and a "talonid" basin rimmed by three more subtle cusps at the back of the tooth. Meanwhile, the upper molars transformed too: they sprouted a big new cusp on their tongueward side, called the "protocone," which snugly fit into the talonid basin of the corresponding lower molar when the jaws closed.

There were many benefits to this new arrangement. Most importantly, when the upper and lower molars bit together, they

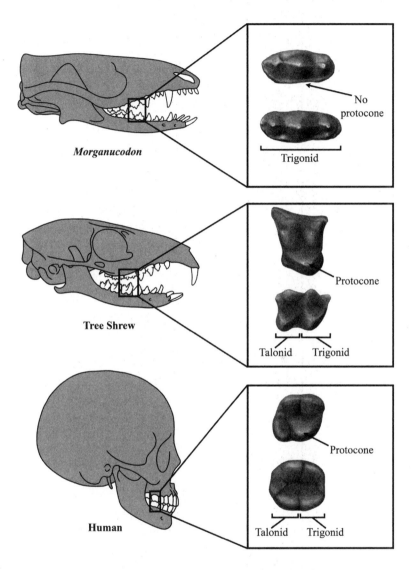

The evolution of tribosphenic molars. Inset boxes show the chewing
(occlusal) surfaces of the upper and lower molars for each species. The
simple three-peaked molars of early mammals (top) changed into the
more complex molars of tribosphenic therians, with a large protocone
on the upper molar that fits into a basin on the six-cusped lower molar
(middle). We humans have these teeth (bottom)! (illustrated by Sarah
Shelley)

could shear and grind at the same time. Shearing mostly occurred on the crests linking the trigonid cusps at the front of the lower molar, while grinding happened as the protocone of the upper molar bashed around in the talonid basin of the lower molar, like a pestle and mortar. These tribosphenic teeth were, therefore, much better bug-munchers than their predecessors. An insect shell could be sliced by the trigonid and macerated by the talonid—all in one jaw stroke—allowing therian mammals to eat more bugs faster, and to extract more nutrients from harder-shelled insects.

Tribosphenic teeth, with all their cusps and crests, were also more adaptable. Slight changes in the size, shape, and position of the cusps could fashion a bevy of new chewing and mashing tools, permitting therian mammals to enjoy a greater range of food choices and more quickly adapt as their environments changed. To read the technical literature on tribosphenic mammals is to become lost in the tongue-twister names of crests, cusps, and the bumps, ridges, and grooves associated with them. But we need so many terms, because tribosphenic teeth are so incredibly intricate and variable—much more so than the teeth of other mammals, or dinosaurs, or most any animals.

Therian versatility is remarkable. From their tribosphenic ground plan, they have diversified into a staggering array of modern species with innumerable diets, from insectivorous shrews that retain a basic tribosphenic molar shape, to carnivorous dogs and cats that slice and dice, fish-eating dolphins, and omnivorous primates, including us. If you gape into a mirror, you can see that your lower molars are divided into front and back sections, both outlined by low mounds. These are the trigonid and talonid, and the mounds are the cusps. You, and me, are tribosphenic! In humans, the cusps are gentle, and the trigonid and talonid subtle—an ideal setup for our generalist, omnivorous diet. But make no mistake, these are modifications from the ancestral tribosphenic design, pioneered by our distant insect-eating ancestors. Once again, innovation sprang from the insect-eating niche, the great incubator of mammalian experimentation and diversification.

Tribosphenic molars did not actually originate during the Cretaceous Terrestrial Revolution, but much earlier. The oldest fossil with unequivocal tribosphenic teeth is *Juramaia*, a shrew-size branch-climber from the approximately 160-million-year-old Middle-to-Late Jurassic Yanliao beds of Liaoning, described by Zhe-Xi Luo and his team. Therians, therefore, apparently developed their tribosphenic molars during that manic, running-to-stand-still phase of mammalian evolution in the Early-Middle Jurassic, when the docodonts, haramiyidans, and multituberculates were also evolving their own trademark dentitions and feeding styles. As these early mammals were competing with one another for resources in the Jurassic, the tribosphenic molar was a useful gadget for tiny insect-eaters, but not yet a game-changer. It was only during the Cretaceous Terrestrial Revolution, many tens of millions of years later, that the explosion of angiosperms sparked a diversification of insects, generating an all-you-can-eat bug buffet. After their long gestation in the humble, small-bodied insectivorous niches, the tribosphenic therians suddenly had the perfect utensil for catching, slicing, and crushing the many new insects. As therians prospered, mammals with the more primitive three-cusped teeth declined, and eventually went extinct.

Early on, long before they found success in the Cretaceous, the tribosphenic therians splintered into two tribes: the metatherians and eutherians. Metatherians include modern-day marsupials and their closest fossil relatives; eutherians encompass placentals— like us—and immediate kin.

There is no confusing marsupials and placentals today: marsupials, like kangaroos and koalas, give birth to feeble young that usually develop further in a pouch, whereas placentals give birth to well-developed young. But these reproductive differences likely arose later, long after metatherians and eutherians went their separate ways in the Jurassic. The founding metatherians and eutherians looked remarkably similar, as exemplified by the back-and-forth debate over the Liaoning *Sinodelphys*, originally described by Luo and his team as the oldest metatherian, but reidentified as a primitive eutherian by other researchers. Early therians are *really* hard to tell apart from each other, because they were all featherweight tree-climbers that chomped insects with their tribosphenic teeth. Only minor differences in the number of teeth, and the pattern of how baby teeth changed to adult teeth, differentiate the first metatherians and eutherians, and these are often difficult or impossible to discern in fossils.

Tens of millions of years later, during the Cretaceous Terrestrial Revolution, the eutherians and metatherians finally distinguished themselves. Members of both tribes were there in the Late Cretaceous dunes, oases, and riverbanks of Mongolia, alongside the still much more common multituberculates. Their first fossils were reported by Andrews's expeditions in the 1920s, but later Zofia's teams and the American Museum crews found much better skulls and skeletons. *Zalambdalestes* is a prototypical Gobi eutherian, a gerbil-size furball that strutted around on its long legs, stalking insects that it would snap up with its long snout and devour with its tribosphenic teeth. *Deltatheridium* is representative of the metatherians. It was about the same size as *Zalambdalestes* but looked and behaved much differently. Although its head was not much bigger than an apricot, *Deltatheridium* had bulging jaw muscles, big piercing canines, and large upper and lower molars, with reduced talonid basins but enlarged shearing cusps and crests, which resembled blades. They had taken their adaptable tribosphenic molars and turned them into guillotines, for slicing the flesh of small vertebrate

Zalambdalestes and *Deltatheridium*

prey—amphibians, lizards, maybe other mammals. Multituberculates, with their herbaceous diets, were probably particularly tasty.

By the end of the Cretaceous, multituberculates and therians had both succeeded where other mammals with more primitive dentitions had dwindled, and they advanced in a victory lap across the northern continents. But what of the southern lands: the emerging continents of Africa, South America, Australia, Antarctica, and India? Maybe multituberculates and therians never got there in the Cretaceous—a plausible scenario, because during the Jurassic and Cretaceous the northern and southern sectors of old Pangea increasingly diverged, as if they were two repelling magnets. At the time of the Cretaceous Terrestrial Revolution, north and south were separated by a wide equatorial waterway, called the Tethys Sea.

For decades, while the Gobi Desert, European islands, and North American floodplains spilled their Cretaceous mammal secrets, little was known about the austral faunas. Then, in the 1980s and 1990s, some tantalizing fossils appeared: only a few teeth and jaws, but they were weird. Or maybe they just seemed that way to paleontologists with a northern bias. Some of these southern teeth belonged to plant-eaters, but their tall molars with convoluted enamel folds looked more like those of a tiny horse than a multituberculate. Even more puzzling, other molars appeared tribosphenic, with a trigonid and a talonid, but the shapes and positions of the cusps were off-kilter. Were these multituberculates and therians or something else entirely?

JOHN HUNTER IS perhaps the second-most-famous dropout of the University of Edinburgh. It's hard to top Charles Darwin, who couldn't cut it as a medical student in the 1820s, his revulsion to blood a fatal flaw in his father's plan to make him into a doctor. Every autumn, I tell the students in my first-year evolution course in Edinburgh that, if they stick it through to graduation, they will have eclipsed Darwin.

Like Darwin, Hunter turned out all right. After quitting school in 1754, he did what so many aimless young men have done since time immemorial: he joined the navy. He quickly rose from servant to seaman to junior officer, and over the next few decades kept finding himself in the middle of the action, as Britannia ruled the waves. He battled the French during the Seven Years' War, made his way to both the West and East Indies, at least once circumnavigated the Southern Ocean, and participated in several battles during the American Revolution. After their surrender to Washington's forces at Yorktown, the British could no longer dump their convicts on the American colonies, so they needed a new plan. The First Fleet was thus launched, in 1787, to the distant shores of one of their newest colonies: Australia. Six ships carried prisoners, three transported supplies, and two provided naval muscle. One of these escort ships, the HMS *Sirius*, was captained by Hunter.

Hunter took a liking to Australia, but it was a complicated relationship. When the fleet arrived in early 1788, and found their new home considerably less inviting than promised, Hunter set out to find more hospitable land on Australia's treacherous southeastern coast. He explored the Parramatta River, the main tributary of a bay that was well-protected, easy to access, and surrounded by ample fresh water and rich soil. This would become the center of their penal colony—a place they called Sydney. After he was summoned back to England, and partook in more of that favorite Royal Navy pastime of fighting the French, Hunter applied to return to Australia. Not as a ship captain, this time, but as governor of the colony of New South Wales. His commission was approved and he began his tenure in 1795. It would not last long. At the end of 1799 he was relieved of his duties, unable to stop corrupt military officers from conspiring with convicts to traffic alcohol—a trade that made grotesque profits for the saboteurs, and later culminated in a rebellion.

Hunter proved so feckless a politician, at least in part, because his real passions were elsewhere. Again like Darwin, Hunter was a keen naturalist, who took advantage of his maritime sojourns to observe plants and animals all over the world. As governor, he regularly sent animal skins and botanical specimens back to England, and with each new shipment, it became more apparent that Australia sported a most peculiar flora and fauna, radically different from Europe, or the New World. The island continent—far removed from any other large landmass—was an ecosystem unto itself. Many of its mammals gave birth to helpless, minuscule babies that they raised in their marsupial pouches—a style of child-rearing unknown among the familiar foxes, badgers, bears, and mice of Europe. And some of the furry Australian fauna, it turned out, were even stranger.

One day in 1797, when he probably should have been quashing corruption, Hunter sat transfixed as an Aboriginal hunter eyed something lurking in the Yarramundi Lagoon, north of Sydney. The man watched this creature, about the size of a groundhog and covered in dense brown fur, come up for air, time and time again. Then, after about an hour of waiting, he decided the time was right. With a quick flick of the wrist, the hunter launched a short spear into the murky waters, as the animal paddled away with its big hands and feet. Governor Hunter was flummoxed. He later sketched what he saw and titled his drawing with the best description he could fathom: "an amphibious animal of the mole kind." Except it was much larger than a mole, and it had webbed hands and feet, sharp claws, and a fat and short tail like a beaver. Its face, though, looked like a duck. Instead of teeth, it had a beak. Really, it wasn't much like a mole at all. But what was it?

Hunter managed to capture one of the animals, which he sent back to England, pickled in a barrel of spirits. When the specimen reached Newcastle it caused a sensation. It seemed part mammal, part bird, part reptile, part fish. Its fuzzy pelt cried out mammal, but it seemed to lack mammary glands, and there were rumors that it laid eggs, a most unmammalian way of reproducing. The

A platypus paddles in a creek in Tasmania. (photo by Klaus via Flickr)

Aboriginal peoples swore it was true—that the animal they long knew about, and considered a duck-rat hybrid, nested like a bird—but the English scientific establishment refused to believe it. When he described Hunter's specimen in 1799, and gave it the name "platypus," the parson-turned-museum curator George Shaw didn't even have the confidence to consider it a real animal. He, and many others, felt that it was most likely a fake: a Frankenstein monster constructed from parts of many animals, stitched together by tricksters.

As eastern Australia became better explored by the British, settlers caught more glimpses of the platypus. They could no longer deny it: this was a genuine animal. It inhabited rivers and lakes, gliding through the water propelled by its webbed hands, and steering with its tail. Constantly hungry, it could dive for about a half a minute, emerging with a mouthful of shrimps, worms, and crayfish. Well over a century later, researchers would prove that its bill is an alarm system, packed with tens of thousands of receptors for sensing both movement and electrical impulses. In this way, the platypus can shut off its vision, smell, and hearing, dive into the water, sweep its bill along the muddy bottoms, and sense prey in stealth mode. Occasionally, the settlers noted, the platypus ventured onto the shore to burrow with its long, sharp claws, which stuck out of the webbing like the tines on a garden rake. The burrows were its resting places, and also, apparently, where female platypuses nurtured their young. The males—not really the fathering type—had spurs on their ankles, conduits for venom to assert dominance over other males during the mating season, before leaving the females to do whatever it is they did with the babies.

It was that mystery—how females reproduced and nurtured—that held the key to figuring out exactly what the platypus was: a strange mammal, or something else? For nearly a century, the arguments were ceaseless, and often nasty. Many of the leading

European naturalists and anatomists of the day became involved, including the petulant Richard Owen, whose name has already popped up so many times in our story. The two points of contention were whether platypuses fed their young with milk, which would be a sign of their mammalness, and whether they laid eggs, which would be a strike against a mammal identification.

Owen, as you might suspect, had no time for the Aboriginal peoples' insistence that the platypus laid eggs. He was convinced it was a mammal, and to him it was cut and dry: mammals gave live birth. Thus, when the British Army lieutenant Lauderdale Maule snatched a mother and two daughters from a nest in 1831 and reported milk oozing from small openings in the mother's abdomen, Owen responded with a cheer. But when Maule proceeded to mention eggshell fragments in the nest, and when other observers actually saw a female lay two eggs, Owen scoffed. This was no natural birth, he sneered; the female must have aborted her pregnancy due to fear. It was a sad example of a scientist letting preconceived ideas and personal grudges overrule common sense, not to mention thousands of years of local experience with an animal that he, a Londoner who cavorted with royalty, had never seen alive.

Eventually everyone had to concede that the platypus laid eggs. It was akin to a deathbed confession, when, in 1884, an eighty-year-old Owen gave the thumbs-up to a gruesome report from a young zoologist named William Caldwell. Over the course of two months, Caldwell conscripted 150 Aboriginal people to slaughter as many platypuses as they could find, along with another furry Australian oddity that was purported to lay eggs: the spiny echidnas. It was imperialistic science at its worst: some fourteen hundred animals were killed, and the Aboriginal peoples were horribly mistreated. During the massacre, Caldwell himself shot a pregnant female, as she was laying her eggs. One was found next to her corpse, the other still inside her uterus.

It was now obvious: the platypus and echidna were both eccentric species, which laid eggs like a bird or reptile, but lactated
like a mammal. They were given the name "monotremes"—the
"one-holed" animals in Greek—in reference to their single all-
purpose orifice for urination, defecation, and reproduction. By
this time, Darwinian evolution had become widely accepted, and
anatomists understood what the monotremes represented: unusual
mammals, less "advanced" than marsupials and placentals, which
retained many primitive features from their distant "reptilian" ancestors (like egg laying) but had also developed many of their own
peculiarities (like their sense-receptor bills). And they lived only
in Australia and New Guinea, isolated in one small patch of the
globe. No living monotreme, or other egg-laying mammal, has
ever been seen anywhere else in the wild.

One mystery was solved, but another arose: Where did the
monotremes come from, in an evolutionary sense? It was a riddle
made all the more difficult because platypuses and echidnas lack
teeth as adults, and it's by comparing all the elaborate cusps and
crests on teeth that anatomists have long built their family trees of
mammals. The secret, as it happened, was locked inside the jaws of
baby platypuses, which sprout teeth for a short time before they are
resorbed into the bill. The answer helped unravel another mystery:
What mammals were leaving those peculiar tribosphenic-like
teeth in the Cretaceous fossil record of the southern continents?

The dominoes fell in a chain of inferences. First was the discovery, in the 1970s, of a couple of upper molar teeth in South
Australia, with cusps linked by unusually thick crests called lophs.
Nearly identical lophed teeth appear, fleetingly, in baby platypuses
today. The fossil teeth—named *Obdurodon*—were 15 to 20 million
years old, at that time making them the oldest platypus in the fossil record. Subsequently, isolated lower teeth, and then a complete
skull, of *Obdurodon* revealed that the lower molars were lophed,

too. Paleontologists now had a search image: if they could find similarly lophed teeth in older rocks, they could trace monotreme origins further back in time.

The big break came in the early 1980s. In the dusty outback town of Lightning Ridge—population barely 2,000—a miner was sifting through circa 100-million-year-old Cretaceous sandstones and mudstones, on the lookout for anything with a blue-green sparkle. This was prime opal territory, and on this particular day, the prospector found what he was looking for. It was an opal, but not in the form of a sphere or a blob or another shape that could easily be cut into a gemstone. This most remarkable opal was about an inch long, consisting of a rod, flattened from side to side, with three craggy things sticking out from one end. The rod was a dentary bone; the jagged structures were three molar teeth, with cusps and crests. It was a mammal jaw, which had been buried in the sands of a shallow sea, dissolved, replaced with silica, and turned to opal—one of the most stunning, and unlikely, fossils imaginable.

It was also a tremendously important fossil. When described by Michael Archer and colleagues in 1985, and given the name *Steropodon*, it held several records. It was the first Australian mammal fossil older than the *Obdurodon* teeth and similarly aged fossils—and thus the first Down Under mammal from the Age of Dinosaurs. It was the first mammal specimen from the Early Cretaceous of the *entire* breadth of the southern continents, as nothing yet had been reported from that age in South America, Africa, Antarctica, or India. And most importantly for our story, it was the oldest monotreme. It, too, had the unmistakable lophed teeth of today's baby platypuses and the extinct *Obdurodon*. Not only that, but a void on one end of the jaw was later interpreted as the edge of a big mandibular canal—a tube that extends through the dentary of platypuses, transmitting the dense network of arteries and nerves that animate the electroreceptor bill. This tiny, shiny fossil, therefore, proved that monotremes went far, far back in time.

There was something else notable about the *Steropodon* molars, which is not immediately obvious from the transient and highly derived teeth of today's baby platypuses. They were tribosphenic, or at least they were described that way by Archer and his team. The lower molars had a trigonid region of sharp cusps and crests at the front, and a talonid-style basin at the back. The upper molars, however, were unknown (and still are), so it was not clear if they donned a protocone, the pestle that is integral to the tribosphenic design.

Over the next couple of decades, as paleontologists fanned across the southern continents on the mammal hunt, similar teeth began to turn up elsewhere. First in the state of Victoria, at the southeastern tip of Australia, where the husband-and-wife duo of superstar paleontologists Tom Rich and Pat Vickers-Rich reported the delicate, barely half-inch-long jaw of a species they named *Ausktribosphenos*. Like *Steropodon*, it was Early Cretaceous in age. Then from Madagascar, where John Flynn (one of my grad school mentors in New York) and his team somehow spotted a flake of lower jaw, about one-tenth of an inch long, with three teeth. They called it *Ambondro*, and it was much older than the Australian species; it was from the Middle Jurassic, nearly 170 million years ago. It also became a record holder: it more than doubled the age of the oldest mammal then known from Madagascar: an island realm, like Australia, that today flaunts a unique fauna but a terrible mammal fossil record. A few years later, another Jurassic jaw—named *Asfaltomylos*—came to light, this time from Argentina.

What were these mammals? Their fossils were almost comically meager—a handful of lower jaws, often broken, and some teeth, spanning thousands of miles of southern lands and over 70 million years of time. The all-important upper molars were, by dint of terrible luck, always missing. It was not a lot to draw on. Were they all related to monotremes, like *Steropodon* clearly was? Or maybe some of them were therians? After all, their lower molars did look tribosphenic. The Riches went as far as proposing a link between some of these ancient species and modern-day hedgehogs, as one of many possible genealogical explanations.

Enter Zhe-Xi Luo and Zofia Kielan-Jaworowska—our afore-mentioned mammal experts, veterans of studying so many Liaoning and Gobi fossils. They were now working together, with their American colleague Richard Cifelli. Combined, the trio had spent many decades scrutinizing the peaks and valleys of mammal teeth under the microscope. When they looked at the southern teeth, something did not sit right. The lower molars were divided into front and back portions, true, and it was easy enough to call them trigonids and talonids. When they forensically examined them, however, they seemed subtly different from the trigonids and talonids of therian mammals. To clarify this conundrum, Luo, Zofia, and Rich built a massive data set of mammal teeth, including the best-known representatives of the southern molars, along with everything from primitive species with three-cusped molars like *Morganucodon*, fossil and modern therians of the eutherian and metatherian lines, and monotremes. They assessed each tooth for dozens of nuanced features of the dental anatomy, relating to the presence, size, shape, and position of cusps, crests, and lophs. When they ran the data set through a computer algorithm, which builds family trees by grouping together species that share unique, derived similarities, the results were shocking.

The southern teeth did not group together with therians. Instead, the southern species like *Ausktribosphenos* and *Ambondro* all clustered with *Steropodon*, on the monotreme line, forming a broader "monotreme stem group" that Luo's team called Australo-sphenida, the "southern wedges."

The two types of tribosphenic teeth, in other words, are not equivalent. There is a therian version, and a monotreme version, which evolved independently, probably both during the Middle Jurassic, and for a similar reason: to increase shearing capability, and maybe to add a little bit of grinding. The therian version evolved in the north, enabled marsupial and placental antecedents to flourish during the Cretaceous Terrestrial Revolution, and persists today as a highly adaptable tooth design, including in our own mouths. The monotreme version evolved in the south, seems to have spread widely below the equator during the Jurassic and Cretaceous, but then essentially disappeared, its only remnant the ghostly teeth of the platypus that disintegrate as babies leave the nest.

There was something else notable in the family tree: the therians and australosphenidans were widely separated, with many other mammal lineages interspersed in between them. Australosphenidans are close to the root of the tree, not too far removed from docodonts, and a few steps up from Late Triassic–Early Jurassic species like *Morganucodon*. Therians, on the other hand, make the crown of the tree. Monotremes and therians both survive today, but nestled between them are countless extinct lines: many of the so-called dead end groups that no longer endure, like multituberculates.

What this implies, to me, feels a little chilling. Monotremes are the product of a long pedigree that goes back at least to the Middle Jurassic, the last survivors of a once gallivanting austral tribe. That they persist at all seems miraculous. You could easily envision an alternative history, in which the monotreme line was snipped back in the Cretaceous, and the platypus and echidna never held on, robbing us of our one good glimpse of the early vestiges of mammalian evolution, when furry burrowers laid eggs and leaked milk from their abdomens. You could also imagine another history, where somehow a single thread of the docodont or multituberculate bloodline survived to the modern day, in some remote corner of the world. As it stands, we should be thankful that the platypus and echidna remain—the zoological equivalents of the old couple who refuse to give up their apartment as their neighborhood gentrifies around them.

What I write about here is very much a developing story as fossils of Jurassic and Cretaceous mammals from the southern continents remain exceedingly rare. But the fossils are out there, to be found, as a few recent discoveries prove.

One of these is *Vintana*, from Madagascar, christened after the Malagasy word for luck by David Krause and his team in 2014. It was indeed an auspicious find, as it finally clarified the enigma of those strange horselike teeth I alluded to previously. For years, paleontologists had nothing but fragments, found across the south: pieces of tall, stout molars with complex folds of thick enamel on the occlusal surface, often worn down by repetitive grinding. They were shunted into their own group, called Gondwanatheria. These mammals were clearly plant-eaters, which employed a backward chewing motion like in multituberculates, as indicated by tooth wear. *Vintana* revealed the gondwanatherians as big-eyed, agile, strong-biting vegetarians, with wide cheekbones to support the muscles that scraped the molars against each other—an image confirmed by Krause's 2020 description of a close relative called *Adalatherium*, represented by a skull joined to a skeleton. These gondwanatherians were their own distinctive group, but positioned near multituberculates on the family tree. Essentially, they were the multituberculates' long-lost southern cousins and filled similar angiosperm-eating niches.

There was another group of successful southern mammals during the Cretaceous, called the dryolestoids. These are close kin of marsupials and placentals, but not quite therians, as they did not have tribosphenic molars. The oldest dryolestoids appear in

Jurassic rocks of North America and Europe, but it was in the Cretaceous of South America where they really prospered. In 2011, Guillermo Rougier and his colleagues announced the discovery of a new species—called *Cronopio*—represented by two skulls, with long muzzles, huge sweeping canines, and deep depressions for head muscles, which would have rotated the jaws as they bit down with their sharp-cusped teeth. They were probably insect-eaters, but of a different style than the northern therians, or the southern monotreme tribosphenic-mimics.

This is where things stood at the end of the Cretaceous, about 66 million years ago. Mammals were ubiquitous, although still small. *Vintana*, weighing in at about twenty pounds (nine kilograms), was by far the largest—regardless, an easy snack for a tyrannosaur or other meat-eating dinosaur. Therians and multituberculates were ensconced in the north, from the Asian heartlands to the North American mountains to the European islands. Among them were insectivores that demolished bugs with their tribosphenic teeth (eutherians), herbivores that feasted on flowers and fruits and other angiosperm parts (multituberculates), and the odd carnivore, which knifed through muscle and sinew with its sharp, modified tribosphenic molars (metatherians). To the south, across the crystal blue waters of the Tethys Sea, were other mammals that filled similar roles: the monotreme-line insect-eaters with copycat tribosphenic molars (australosphenidans); other insectivores with long snouts (dryolestoids); and herbivores (gondwanatherians).

They would have all been there when an incandescent spark lit up the sky, from north to south. All these major mammal groups would survive, at least for a while. But things would never be the same.

5

DINOSAURS DIE, MAMMALS SURVIVE

Ectoconus

THERE ARE CERTAIN INEXPLICABLE RULES of fieldwork. You always find the huge skeletons on the last day, when there's no time left to collect them. If you've been searching in vain for several hours, stop for a bathroom break and you'll inevitably spot a nice skull or jaw right where you're squatting. And, it's not the professors, but the students who often find the best fossils.

This last rule held true during our 2014 field season in the badlands of New Mexico. For about ten days in May, our team prospected the candy-striped hillocks and gullies of the Four Corners region, just north of Chaco Canyon, where the ancestral Pueblo people built a great city out of the rocks a millennium ago. Today this is sacred land to the Navajo, and one morning we found ourselves in a dry creek bed they call Kimbeto—the "sparrowhawk spring." Here, 65.6-million-year-old mudstones are bursting with fossils, which poke out of the parched ground like mushrooms, waiting to be plucked by anyone with a keen eye.

Carissa Raymond, just days out of her freshman year of college, was one of many members of our crew. She was part of the University of Nebraska contingent, recruited as a field assistant by her professor, Ross Secord. At that point she had yet to take any paleontology courses, but she performed superbly in Ross's geology class, so he gave her a shot at fossil collecting. As we all fanned across Kimbeto that sun-scorched morning, Carissa's scarlet Cornhuskers T-shirt stood out like a beacon against the crisp blue sky. You could see her, wandering with eyes to the ground, from at least a half mile away. She had yet to learn the veteran bone-hunter's trick of wearing muted tones when working in the desert. But how could she know? This was her first fossil-hunting trip.

The first few days hadn't gone so well for Carissa. Her eyes had yet to become accustomed to the way light sparkles off a fossil tooth, or the shape a jawbone makes as it erodes from the mud. These things take time and can't easily be taught. You have to earn those instincts through experience—which often means long spells of empty-handed frustration, until one moment, it clicks, and the fossils seem to levitate out of the rocks. Once a student— with their youthful eyes and exuberance—reach this state of nirvana, they can become ace fossil-finders.

Carissa's moment would be today. As she walked over a hill onto a flatter stretch of ground, she gazed at the bands of black and tan and red rocks alternating on the steeper bluffs on the horizon, scarred by erosion. She then peered down at the desert pavement, cracked into the polygonal shapes of dried mud, dusted with winddrifted stones that, if you're not careful, can send your feet flying out from underneath you. Her eyes scanned the surface. Stone, stone, another stone.

Then, something different. Shiny. Black—a rich, dark black. And a weird shape. Then another, and another. In sequence. These were not stones—they were fossils! Fossil teeth: a row of them, making up a jaw. The big ones looked like Lego bricks or corncobs, with three rows of kernel-like cusps, lined up in parallel, separated by sharp slits.

Carissa gave a siren call and our team assembled from each corner of Kimbeto. Tom Williamson, curator at the New Mexico Museum of Natural History and Science and the leader of our expedition, was farthest away, so he arrived last. Carissa handed him the teeth.

"Holy crap! I can't believe it!" he shouted, not knowing that I was recording with the video function on my Nikon camera.

Tom had been collecting in this area for some twenty-five years. With his encyclopedic knowledge and photographic memory, he can casually glance at nearly any fossil and tell you what it is. What type of tooth or bone, which species it belonged to. We awaited his word, as if listening to an oracle.

It was a multituberculate, he told us. A member of that group of plant-eaters that looked like rodents, but were not actually closely related, which Zofia Kielan-Jaworowska had discovered by the dozens in the Gobi Desert. The teeth gave it away. Only multituberculates had Lego-brick teeth with rows of big cusps, which they used to pulverize plants as the teeth of the upper and lower jaws slid against each other. As you may recall, these teeth were their secret weapon as they prospered during the Cretaceous, becoming more diverse as they fed on angiosperm fruits and flowers.

Something was strange about Carissa's multituberculate, however.

"It's big, really big!" Tom continued, with a mix of excitement and puzzlement.

Zofia's Gobi multituberculates were the size of shrews and rats, with molars that could easily fit on a penny. Most other Cretaceous multituberculates were in this size range, too. The molars of Carissa's fossil, however, were about twice the size of my fingernails, and approaching the length of my entire thumb. This implies a body weight of twenty to forty-five pounds (ten to twenty kilograms), about the heft of a beaver, which is the second-largest modern rodent.

For the next hour we scoured the area and gathered jaws from both sides of the head, containing molars and premolars, plus front incisors and part of the upper skull that surrounded the brain. When we returned to the lab in Albuquerque we got to work cleaning, gluing, photographing, and measuring the fossils, and about a year later we described them as a new species: *Kimbetopsalis*, named after where it was found. We thought that was a little cumbersome to pronounce, so we gave it a nickname: the Primeval Beaver.

"I knew it was cool, but not this cool," a shell-shocked Carissa told a reporter, during a media blitz that saw her interviewed on National Public Radio and profiled in the *Washington Post*.

Kimbetopsalis, the "Primeval Beaver": fossil skull and teeth (top) and Carissa Raymond and Ross Secord collecting the fossils a few moments after discovery in 2014 (bottom). (photos by Tom Williamson and Steve Brusatte, respectively)

Tom was also frequently quoted in these articles. "I wish I had found it," he conceded to one journalist, which didn't surprise me, given that Tom and his twin sons Ryan and Taylor—who he trained from a young age to sniff out fossils on weekend camping trips—still gather around the cooler at the end of each field day to argue over who found the best fossil, between chomps of tortilla chips and salsa.

Tom shouldn't have been so bummed, though, because the previous day he had found an impressive fossil himself—which would have been that year's prizewinner if Carissa hadn't spotted the *Kimbetopsalis* teeth. Within the first hour or so of our arrival at Kimbeto, during a quick afternoon reconnaissance survey after our morning drive from Albuquerque, Tom noticed a few cracked fragments that didn't look like rock, eroding out of the desert floor. Upon closer inspection, they notched together like puzzle pieces, forming part of something Tom immediately recognized as the humerus (upper arm bone) of an animal called *Ectoconus*, which had been described in 1884 by one of the first paleontologists to work in this part of New Mexico.

Tom called me over, and I shouted for Sarah Shelley, my PhD student in Edinburgh (remotely cosupervised by Tom), whom we met in the opening pages of this book, and whose art has been enlivening these pages. We got on our hands and knees, careful to both avoid crushing any bones and scuffing our skin on the ball-bearing desert stones, and used trowels and pin vises to scrape around the humerus fragments, deeper into the mudstone. The more we dug, the more bones we found.

The arm led to a skeleton!

We dug a trench around it and covered the bones in bandages soaked in wet plaster of paris, which hardened into a protective cast. Out came the hammers, chisels, and picks, and we pried the encased bones clean from the rock, with the sort of blue-collar

Sarah Shelley and Tom Williamson encasing the skeleton of the "archaic" placental *Ectoconus* in a protective plaster jacket in 2014. (photo by Steve Brusatte)

physical labor that academic scientists aren't usually called to do. It was fun. And it was also fun when I found some *Ectoconus* teeth nearby—probably from another individual, although since our skeleton was headless, who could be sure?

Once again—and, as we are seeing, is so often the case with mammals—the teeth tell us where *Ectoconus* sits on the family tree. It does not have the multiple cusp rows of multituberculates. Instead, its teeth are tribosphenic: the upper molars fit into the lower molars like a pestle and mortar. This means that it had those trademark teeth of therian mammals—the marsupial and placental group—that could shear and grind at the same time. The number of teeth, furthermore, tell us that it is a eutherian: a member of the placental line. Maybe it was even a true placental mammal like we are. Perhaps *Ectoconus* mothers were able to nourish and protect their developing babies with a placenta in utero, allowing the youngsters to be born in a well-developed state.

There's another, more obvious, way in which *Ectoconus* differs from Carissa's *Kimbetopsalis*. It is bigger—*considerably* so. The skeleton we collected is about the size of a pig, with bulky shoulder and pelvic girdles, and robust arm bones, which must have scaffolded powerful muscles, capped with claws that look like they are trying to morph into hooves. From the size of the bones, we can tell that it weighed about 220 pounds (100 kilograms) in life—substantially larger than any fossil mammal we've talked about so far.

Although we didn't find any of their fossils that trip, a third type of mammal is found in the same rocks. They are thus far known only from teeth, which are so tiny each one can sit on the tip of a ballpoint pen. These are metatherians—the marsupial line of the therian group—which also had tribosphenic teeth. Strangely enough, while the multituberculates and eutherians in New Mexico were so much larger than any of their predecessors, the metatherians seemed smaller, and meeker.

That's it. Multituberculates, eutherians, and metatherians. There are tens of thousands of mammal fossils, belonging to over a hundred species, from Kimbeto and adjacent badlands in New Mexico, which have been collected as far back as the 1880s. Every single one of them belongs to one of these three groups. There are none of the other mammals we've celebrated in the previous chapters: no *Morganucodon*-like scurriers, no docodonts or haramiyidans, none of those precursors of tribosphenic-toothed therians that had a line of three mountain-peaked cusps. No egg-laying monotremes, either, although that's not too surprising since they are a Southern Hemisphere group. But while all the New Mexico mammals are restricted to only three branches of the family tree, they are nonetheless remarkable. They are actually more diverse than ever before: there are more species, with a greater range of body sizes and diets and behaviors, compared to the mammals in any ecosystem that had yet existed.

And there's one more thing. The rocks immediately below the mudstones entombing the *Kimbetopsalis* teeth and *Ectoconus* skeleton are other mud rocks, deposited in a similar river floodplain and forest environment, which are ever-so-slightly older at about 66.9 million years old. These layers are stuffed with the bones of *T. rex*, horned relatives of *Triceratops*, monstrous sauropods like the epically named *Alamosaurus*, and duck-billed dinosaurs. Shards of busted dinosaur bones fall out of these rocks and litter the desert floor, and it's impossible not to trample on them. But nobody, ever, has found any sign of a non-bird dinosaur in the Kimbeto rocks, or in any rocks above it. Not a single bone, or scrap of bone. Not one tooth, or footprint.

It's as if the dinosaurs had evaporated, but not the mammals. And now, the mammals were bigger than they ever were during the Triassic, Jurassic, or Cretaceous.

KIMBETOPSALIS, ECTOCONUS, AND the other mammals we've collected at Kimbeto are Paleocene in age. The Paleocene is the time interval directly after the Cretaceous, but the two seem to be totally different worlds, two consecutive chapters in a novel with different characters—in this case, fossils of dinosaurs and mammals—that don't synch into a clear narrative. That's because the story line lurched dramatically as the Cretaceous ended and the Paleocene began. Separating them was the single greatest catastrophe in Earth history, literally the worst day our planet has ever endured.

The asteroid—or maybe it was a comet, we're not certain—

came from the far reaches of the solar system, beyond the orbit of Mars, perhaps even farther. It was about six miles (ten kilometers) wide, around the size of Mount Everest, and about three times as wide as Manhattan. A speck of cosmic dust in the grand scale of the universe, to be sure, but it was the largest celestial body to approach our corner of the solar system in at least half a billion years. As it hurtled through the heavens, on the random trajectory of a drive-by gunshot, it was traveling more than ten times faster than a speeding bullet.

It could have gone anywhere, but as fate had it, the space rock made a beeline for Earth. It could have whisked by, a close call that rustled the upper layers of the atmosphere before disappearing into the blackness of space. It could have disintegrated as it neared the earth, pulled apart by gravity. Or it could have delivered a glancing blow. But it did none of those things. It collided with what is now the Yucatán Peninsula of Mexico, impacting with the force of over one billion nuclear bombs, punching a hole in the crust 25 miles (40 kilometers) deep and over 100 miles (160 kilometers) wide. That scar is still visible today as the Chicxulub Crater, which straddles the Gulf of Mexico coast not far from the tourist town of Cancún.

Once asteroid met Earth, about 66 million years ago, nothing would ever be the same again.

It first was a matter of physics. The energy released by the collision had to go somewhere, and it was converted to heat, light, and noise, of unimaginable fury. Almost instantaneously, everything within about six hundred miles (one thousand kilometers) of ground zero was vaporized. Many dinosaurs, mammals, and other animals met their end this way, turned to ghosts.

The New Mexican species were slightly luckier, as they lived about fifteen hundred miles (twenty-four hundred kilometers) from the Yucatán. They merely had to contend with hurricane-

force winds, earthquakes far larger than any humans have experienced, and bullets of hot glass that rained down from the
sky—made of the dust and rock liquefied during the impact
that solidified while falling back to Earth. As the molten pellets
whizzed down, the sky turned red and the atmosphere warmed like
an oven. This was enough to make forests spontaneously combust,
and wildfires raged across the land. Each of these calamities was an
agent of murder, and they were more potent closer to the impact
site. It's hard to gauge how many of the New Mexican animals died
during these chaotic few hours, but probably a lot of them, maybe
most of them.

Anything that survived the immediate effects of the asteroid
then had to cope with the longer-term repercussions. Soot and
smoke from the wildfires streamed into the atmosphere, mixing
with the residual grime that hadn't yet condensed into glass bullets. This toxic cocktail choked the high atmosphere currents that
circulate air around the earth, plunging the entire planet into cold
darkness. It was a nuclear winter that lasted several years. Whatever plants spared by the wildfires were now starved of the sunlight
needed for photosynthesis, so they withered and died. As forests
collapsed, ecosystems fell like houses of cards. But that wasn't
all. Volcanoes in India, which already had been spewing out lava
and gases for thousands of years, went into hyperdrive. Nitrogen
and sulfur vapors combined with water to make acid rain, which
leached out of the etched land and poisoned the oceans. These
were all global agents of doom, and for the years and decades after
that fateful day, nothing was safe, no matter how distant from the
impact crater.

Then, in one final stroke of cruelty, the asteroid found a way to
keep killing for generations. As if its raw destructive power wasn't
bad enough, the asteroid just so happened to smash into a carbonate platform: an expanse of rocks, formed in the shallow ocean

by corals and shelled creatures, made up of calcium, carbon, and oxygen. As the carbonate rock was annihilated, the carbon and oxygen were liberated and escaped into the atmosphere as carbon dioxide. We've seen it before at the end of the Permian and Triassic, and we're experiencing it now: carbon dioxide is a greenhouse gas, which warms the atmosphere, the surface of the earth, and the oceans. After a few decades at most, nuclear winter turned to global warming. For several thousand years, the roasting temperatures made it difficult for ecosystems to recover.

There is not a hint of doubt in my mind: this was the most dangerous time to be alive in the entire four-billion-plus-year history of the earth. The asteroid was the ultimate serial killer, so effective because it had so many murder weapons at its disposal: the energy pulse that acted within seconds, the wildfires and hot glass rain during the next few hours and days, the nuclear winter over decades, and a few millennia of global warming. It would take quite the combination of talent *and* luck to survive so many obstacles, and not many animals did. Approximately 75 percent of species died out, making this one of the worst mass extinctions ever.

Dinosaurs didn't make it, except for a few birds, which is why their fossils vanish so suddenly from the rocks in New Mexico. Also succumbing were many groups of large reptiles that ruled the oceans, like the long-necked plesiosaurs. Gone, too, were the pterosaurs: the family of flying reptiles, commonly called pterodactyls, which until the end of the Cretaceous had kept birds out of some of the aerial niches. Other animals made it through, but wounded: crocodiles, lizards, turtles, and frogs, to name a few. Many plants perished, as did a huge percentage of the microscopic plankton in the ocean, forever changing the base of the food chains on land and in the water, and ensuring that entirely new ecosystems would have to be constructed in the Paleocene.

And what about mammals? We know they lived on, of course,

or else we wouldn't be here. But it's a much more complex—and fascinating—story than the textbook fable of "dinosaurs died, mammals survived." For mammals, the asteroid impact was both their moment of greatest peril, and their big break.

MAMMALS *ALMOST* DIED out. They nearly went the way of the dinosaurs. All they had accomplished—their entire evolutionary legacy, of hair and milk, jaw bones turned to ear bones, and all those varieties of teeth—was almost lost forever. All they would accomplish—woolly mammoths, whales the size of submarines, the Renaissance, you reading this page—was almost rendered a nonstarter. It was a close call, and it all hinged on what happened in the days, decades, and millennia after the asteroid impact— a pittance compared to the vast depths of geological time.

We have a good handle on what transpired during this most precarious time in mammalian history. In the cattle country of northeastern Montana, where the Missouri River and its tributaries sculpt the plains into sagebrush-scented badlands, there is a fossil archive. It is held within the mudstones and sandstones that make up the hilly topography, crisscrossed by barbwire fences and covered in cow pies. These rocks were formed by rivers that drained the ancestral Rocky Mountains and flowed eastward into a seaway that cut North America in half, during an approximately 3-million-year stretch spanning the very end of the Cretaceous and beginning of the Paleocene. Layer by layer, the rocks and their fossils provide an unparalleled record of how a single ecosystem changed after the asteroid impact.

For nearly half a century, Bill Clemens (who sadly passed away in late 2020) worked this land, befriending the ranchers who control access. Year by year, he and his teams of students gathered fossils from rocks of the Cretaceous-aged Hell Creek Formation and Paleocene-aged Fort Union Formation, building a collection of tens of thousands of teeth, jaws, and bones, which is still growing. But Bill might not have gotten the opportunity, if fate acted differently. While he was becoming an expert on extinction, he faced his own bizarre brush with violence.

A native of the Bay Area, Bill joined the faculty of the University of California at Berkeley in 1967—a dream job for a local boy. That same year another young hotshot was made professor, in the mathematics department. His name was Ted Kaczynski, although he's better known by the nickname the FBI gave him when he was still a mystery-man posting mail bombs: the Unabomber. When Kaczynski was captured in 1996, a hit list of targets was found in his dilapidated shack, deep in the woods of far western Montana, some 350 miles from Bill's field sites. Bill's name was on the list, along with other new hires from way back in 1967. The FBI interviewed Bill, who confirmed that he never knew Kaczynski. His spot on the list, it seems, was a matter of wrong place, wrong time, nothing more. Sometimes that type of coincidence can be deadly, but thankfully, Kaczynski was incarcerated before he reached Bill's name.

"Bill didn't seem too perturbed by it, but I opened boxes with a stick for about ten years after that," Anne Weil told me. I've spent many late-spring weeks with Anne in New Mexico, where she has been the resident multituberculate specialist on our field teams. A former ice hockey player at Harvard and a stellar writer who often pens summaries of new mammal discoveries for *Nature*, Anne today is a professor at Oklahoma State University, but at the time of the FBI interview, she was Bill's grad student. She is one of dozens of students Bill has mentored over the years—a who's who of leading paleontologists, including several of the most prominent women in the field today. His students, even more than his fossils, will end up being Bill's greatest legacy.

Greg Wilson Mantilla was another of Bill's students. Greg

grew up in Michigan and went to college planning to be a medical doctor—that is, if a career in soccer didn't pan out, which wasn't an absurd dream, as he was captain of Stanford's team. Then his brother, Jeff—a paleontologist who's made his name studying some of the biggest sauropod dinosaurs—took Greg on a fossil dig. It was an intoxicating experience, and Greg decided to become a paleontologist, too—but one who studied the tiny mammals that seized the crown from Jeff's colossal dinosaurs. Greg is now a professor at the University of Washington, and over the last few years he assumed the reins of Bill's Montana fieldwork, as Bill transitioned into a much-deserved retirement.

The fossils collected by Bill, Greg, Anne, and their colleagues paint an evocative picture of latest Cretaceous Montana. It was a

Greg Wilson Mantilla (back) and Bill Clemens (front) collecting mammal fossils in Montana. (photo by Diane Clemens-Knott and courtesy of Greg Wilson Mantilla)

dinosaur-dominated world, no question about that. Some of the most famous dinosaurs of all are found in the Hell Creek rocks: *Tyrannosaurus rex* and the three-horned *Triceratops*, the duck-billed herbivore *Edmontosaurus*, the armor-covered tank *Ankylosaurs*, and close cousins of *Velociraptor*, too. But while *T. rex* was the undisputed heavyweight champion of the Hell Creek forests and river floodplains, the lightweight class was ruled by a mammal: a marsupial-line metatherian called *Didelphodon*. Pound for pound, it was probably a *more* ferocious fighter than the tyrant dinosaur king, which as a longtime devotee of *T. rex*, pains me to admit.

Didelphodon was massive as far as Cretaceous mammals were concerned, tipping the scales at about eleven pounds (five kilograms)—about the size of its distant cousin, the modern opossum. Its molars were cutting blades, its premolars were bulbous crushers, and its canines were thick pegs. When Greg described gorgeous new fossil skulls of *Didelphodon* in 2016, he used measurements of the skull and teeth to estimate its bite forces, based on math equations for today's mammals. The results were astounding: *Didelphodon*'s canines were stronger than those of dogs and wolves, and its bite force, when standardized for body size to make a "pound-for-pound" measure, was stronger than *any* modern mammal Greg studied. Stronger than a wolf, a lion, or a Tasmanian devil. *Didelphodon* probably filled a hyena-like role in the Hell Creek ecosystem: a ferocious predator and scavenger, which killed live prey and devoured dead carcasses, crushing bone to immobilize its victims and extract every ounce of nutrition. On the menu was any small animal it wanted: other mammals, turtles with hard shells, even baby dinosaurs.

Didelphodon is one of thirty-one mammal species thus far known from Hell Creek rocks. They would have filled many ecological roles near the base of the food chain, ranging from specialized meat-eaters like *Didelphodon*, to various angiosperm-munching herbivores and omnivores, down to minuscule shrew-size insect-eaters. The vast majority of these mammals were metatherians like *Didelphodon* (twelve species) or multituberculates (eleven species). Eutherians—the placental-line mammals—were less common, as only eight species have been identified, and these exhibit nowhere near the range of body sizes or rich variety of diets as the other mammals. These eutherians were a fringe group, subsisting in the forest undergrowth where metatherians were the top predators and multituberculates the main herbivores.

This situation is stable if you follow the stack of Hell Creek rocks upward, moving through the layers that record the final two million years of the Cretaceous. There are some dips and rises in mammal diversity, as new species came and went, probably in response to small changes in climate wrought by the Indian volcanoes and shifts in the shoreline of the adjacent sea. Overall, though, the latest Cretaceous mammals were doing well, particularly the metatherians and multituberculates. They were still generally small, but they were diverse, and they were ensconced in many niches. There was no sign of any trouble.

Then everything changes. There is a thin line in the rocks, saturated with iridium, an element that is rare on the surface of the earth but common in outer space: the chemical fingerprint of the asteroid. All the dinosaurs abruptly disappear. The Hell Creek Formation gives way to the Fort Union Formation. The Cretaceous has turned to the Paleocene.

The first Paleocene rocks present a dire scene. There is a fossil locality dated to approximately 25,000 years after the asteroid hit, called the Z-Line Quarry. It reeks of death. Not only are all the dinosaurs gone, but most of the mammals are, too. There are only seven species, all of them represented by tiny teeth that you need a microscope to see properly. Three of them—one multituberculate called *Mesodma*, one metatherian called *Thylacodon*, and one eutherian called *Procerberus*—are exceptionally common. They are "disaster" species. We've seen their kind before, earlier in our story, after the great extinction at the end of the Permian. They are the types of animals that revel in chaos, the mammalian equivalents of cockroaches that prosper in the darkness and filth. These three mammals or their immediate ancestors were all survivors, which managed to endure the heat pulse, the wildfires and scalding rains, the nuclear winter and global warming. They carried the torch of mammalness through the long night of the end-Cretaceous extinction, but make no mistake, their proliferation in the earliest Paleocene was not a sign of recovery. It was a sign that ecosystems were unhealthy and unbalanced.

Several other fossil sites in Montana reveal what was happening during the next 100,000 to 200,000 years. It is only by observing this broader span that we can appreciate the true destruction of the asteroid. If you pool together all the mammal fossils from this time, there are twenty-three species. Nine of those are multituberculates, meaning they suffered a modest extinction. Only one, however, is a metatherian: these marsupial-line mammals, so abundant and diverse in the Cretaceous, were very nearly wiped out, saved locally by *one single species* that managed to hold on. Taking their place were eutherians: these previously marginal placental-line mammals increased from their Cretaceous roster of eight species to thirteen in the earliest Paleocene.

One of these Paleocene eutherians would have been our ancestor. Maybe it was one of the Montana species, or maybe it was living elsewhere. Simply put, we wouldn't be here had this one plucky forebear not gutted things out.

Where did the Montana eutherians come from? Most of them, it seems, must have immigrated from afar, as they don't have obvious ancestors in the underlying Cretaceous rocks. Maybe they came from Asia, which was connected to North America via a land bridge at the time. Perhaps Asia—much farther from ground zero than Montana, and thus less severely affected by the destruction of the first days and weeks after the impact—helped replenish the decimated North American mammal communities. In fact, it seems like many species were on the move during the millennia after the impact. Some were like Tom Joad and his family in *The Grapes of Wrath*, leaving their destroyed homes in search of a better life. Others were more like gold prospectors or land speculators in the American West after the horrors of the Native American clearances: they rushed in to fill the void, sensing opportunity. For one reason or another, it is likely that our eutherian ancestor was one of these migrants.

All told, if you compare the Cretaceous and Paleocene mammals of Montana, the numbers are grim. Three out of every four species living in the latest Cretaceous disappeared, either failing to make it through the environmental destruction themselves, or failing to leave any descendants. If you consider all Cretaceous and Paleocene fossil sites from western North America, the statistics get worse. A mere 7 percent of mammal species survived. This number is even more devastating than it seems, because it takes into account migrants: if a species died in Montana but lived in Colorado, it would be included in the survivors category. Imagine a game of asteroid roulette: a gun, with ten chambers, nine of which hold a bullet. Take your shot. Even those odds of survival—10 percent—are slightly better than what our ancestors actually faced in that brave new world after the asteroid hit.

This begs the question: What allowed some mammals to endure? The answer is apparent when looking at the victims and survivors. The Paleocene survivors were smaller than most of the Cretaceous mammals, and their teeth indicate that they had generalist, omnivorous diets. The victims, on the other hand, were larger species, with more specialist carnivorous or herbivorous diets, like *Didelphodon*. They were supremely adapted to the latest Cretaceous world, but when the asteroid threw everything into anarchy, these adaptations became handicaps. The smaller generalists, however, were able to take advantage of their flexible palate to eat whatever was on offer—probably little more than seeds, decaying vegetation, and rotting flesh. It also seems like species that lived across broader areas in the Cretaceous, and which were more abundant in their ecosystems, had a better chance of survival.

It adds up to a hand-of-cards scenario. I've used that analogy before, to explain why mammal ancestors were able to survive previous mass extinctions. It is particularly apt here. When the asteroid so suddenly and unexpectedly toppled the Cretaceous world, Earth became a casino. Survival came down to a game of probability. Dinosaurs had a terrible hand of cards—the dead man's hand. Most of them were big, they couldn't easily escape into burrows or hide underwater, and most had highly specialized diets. Many mammals had a bad hand, too: species like *Didelphodon*, in particular, with their larger sizes and fussier eating habits. But some mammals—only a slim fraction of them, but including, thankfully, our eutherian ancestors—were holding much better cards: they were small, they could hide easily, they could eat many things, they lived over wide areas, there were lots and lots of them. No one feature secured victory, but together they won the pot.

It's not just about winning the evolutionary poker game, though. It's also about what you do with your prize. After all, crocodiles and turtles and frogs survived, too, but never reached the heights mammals would achieve. That's because those few mammals that had a royal flush didn't squander their luck. There was something about them—their versatility, their evolvability, their wanderlust—that allowed them to quickly surpass other groups that survived. Within a few tens of thousands of years at most, some of these mammals were prospering as disaster species.

Others moved around, immigrants filling jobs vacated by the extinction. The local survivors and immigrants interacted with each other and their environments, they evolved, they split into new species—and most importantly, they got bigger. Within about 375,000 to 850,000 years of the asteroid, as temperatures stabilized and ecosystems recovered, mammals were thriving in Montana. There were more species than ever existed during the Cretaceous, and there were entirely new groups, including a variety of stocky creatures with hooves, and a bevy of tree-climbers with long limbs.

Fossils of these new mammals from Montana are nice, but the ones from New Mexico are even better.

ON JULY 25, 1874, an exploratory party set out from the railway terminal in Pueblo, Colorado, and headed south. They were on horseback, trailed by a team of mules loaded with enough provisions to last at least several weeks in the sparsely populated mountains, high desert plains, and badlands they were set to traverse—still the domain of Navajos and other Native Americans. Among the six men were two scientists and one assistant, a mapmaker, a foulmouthed teamster, and a cook. Their mission: to map the topography of the region around the San Juan River, where Colorado meets New Mexico—neither of which were yet states. It would be one small contribution to the Wheeler Survey, commissioned by the US Congress to chart the expanse of land west of the one-hundredth meridian. While making their map, the crew was also expected to take a census of the Native American tribes, appraise sites for railroads and military installations, and seek out mineral resources.

Nominally leading the team was a zoologist named H. C. Yarrow, but the man really in charge—through the brute force of his personality—was a paleontologist from Philadelphia,

Edward Drinker Cope in 1876, two years after discovering the "Puerco marls" in New Mexico. (photo by AMNH Library)

Edward Drinker Cope. In his midthirties, with shrewd eyes and a goatee that draped below his chin, Cope was one of the nation's eminent fossil experts, and a veteran of other land surveys in Colorado, Wyoming, and Kansas. A year prior, he caught wind of some intriguing new mammal fossils from New Mexico and was determined to explore the area himself. When he heard that the cartographer George Wheeler was organizing a mapping trip, he begged to be part of it. Wheeler was hesitant, knowing Cope's reputation as a lone wolf immune to authority, who would follow the fossil trail wherever he wanted, orders be damned. Cope continued to grovel and borrowed money from his father to contribute to the team's finances. Finally, Wheeler relented, under the condition that Cope act as the team's geologist, and nothing more. When Cope agreed, surely both men knew that it was a promise never intended to be kept.

Less than three weeks after departing Pueblo, Cope had already rebelled. He found mammal teeth and refused to move northward with the team before he finished collecting. Yarrow conceded, which only served to embolden Cope. A month later he overheard tales of sensational fossil fields to the west, in a stretch of badlands far off the team's route, at a place called Arroyo Blanco. This time, Cope simply left. He took three of the crew, a mule, and a week's worth of rations, and headed into tribal territory. The rumored fossils materialized—a bounty of crocodiles, turtles, and sharks, and at least eight types of mammals. Cope recognized these as early members of some of today's mammal groups, such as horses, and deduced that they came from the Eocene, a time interval that we now know began approximately 10 million years after the dinosaur extinction, and stretched from 56 to 34 million years ago. "It is the most important find in geology I ever made," Cope wrote to his father a few days later, an ironic turn of phrase given that, by this time, he had given up all pretenses of being the team's geologist.

After collecting his fill, a triumphant Cope retreated to meet Wheeler personally, ostensibly to apologize. The boss, however, had more serious concerns. He relayed the shocking news: after Cope had deserted the team, the mapmaker was accidentally killed, and Yarrow had been recalled to Washington. An exasperated Wheeler told Cope that he was now on his own. Inconceivably, Cope escaped a formal reprimand, but he was never invited back on another Wheeler expedition again.

A page from Cope's 1874 field notebook, depicting the fossil-rich rocks of New Mexico. (photo by AMNH Library)

It was mid-September. Cope, suddenly free of any constraints, would have at least another month to explore before the weather turned sour. He went south, and in late October passed a tent city called Nacimiento, where thousands of roughnecks were mining Triassic-aged petrified logs—not because they were fossils, but because they were infused with copper. Cope crossed the dry channel of Rio Puerco and noticed some petrified wood sticking out of gray and black clay. The wood was different from the logs being mined, and Cope could tell that the clay was located in between the Triassic rocks and the Eocene rocks bearing his "most important" fossils. He made a note of it, calling the clay the "Puerco marls" after the river. He suspected it might hold mammal fossils, too, but he was unable to find any before turning back to the railway station in Pueblo, in advance of the coming snows.

The Wheeler survey continued the next year, and with Cope no longer welcome, they hired a local frontiersman named David Baldwin. Baldwin was an elusive character; there is no record of his birth or death, and he seems to have spent most of his life as a loner. It's remarkable that he agreed to join the team, as he usually ventured into the hinterlands accompanied only by his donkey, preferably in the depths of winter, when he could melt snow for drinking water. Legend has it that he dressed like a Mexican cowboy, with a pickax balanced on his shoulder, and he subsisted on sacks of cornmeal.

After getting his bearings on the Wheeler survey, Baldwin returned to the San Juan area of New Mexico in 1876, under a new commission. He was to personally collect fossils for a young, demanding East Coast paleontologist who despised authority.

Not Cope, though, but Cope's rival: Othniel Charles Marsh of Yale University.

The "Bone Wars" feud between Cope and Marsh is the stuff of scientific infamy, and it was once set to get the Hollywood treatment as a film starring Steve Carell and James Gandolfini as the dueling scientists (which was put on hiatus when Gandolfini suddenly died). Read any book on dinosaurs and you're bound to hear the sad tale of these once friendly bone-hunters who let greed, ego, and fame turn them into bitter rivals, who sabotaged each other's work, destroyed each other's fossils, and slagged each other off in the press. Their feud is largely remembered today as a fight over dinosaurs, probably because some of the most famous names in the dinosaur lexicon—*Brontosaurus*, *Diplodocus*, *Stegosaurus*—were found during that manic time in the 1870s and 1880s when Cope and Marsh were desperately one-upping each other.

Really, though, much of their bickering was about fossil mammals. Each man was gunning to find the oldest and most primitive fossils of horses, primates, and other modern groups. With that in mind, it's puzzling how Marsh reacted to the fossils Baldwin sent him, between 1876 and 1880. He ignored them. And refused to pay up, too. Baldwin responded by doing the obvious: switching his allegiance to Cope. It was the biggest mistake Marsh ever made, because soon after he changed his colors, Baldwin found mammals in Cope's "Puerco marls." These bones and teeth—sandwiched between the last-known dinosaurs and the more modern-flavored Eocene mammals Cope found in 1874—were the first records of the transitional fauna that swept out the Age of Dinosaurs and ushered in the Age of Mammals.

Over much of the next decade, Baldwin and his donkey meandered across the New Mexican desert, tracing the "Puerco marls" and collecting thousands of fossils. Among the places Baldwin found mammals was the dry creek bed the Navajo called Kimbeto—the opening scene of this chapter, where our 2014 field team worked. Baldwin sent all his finds to Cope in Philadelphia; Cope paid Baldwin not only in money, but also with respect. Whereas Marsh was apathetic, Cope enthusiastically described and named the mammals as rapidly as Baldwin could provide them. From 1881 to 1888, Cope published some forty-one papers on Baldwin's mammals, describing them as nearly one hundred new species. The work was often rushed and sloppy, but it was only the start.

For the next 125 years and counting, discoveries have continued in this part of New Mexico, which is now called the San Juan Basin. The mining town of Nacimiento was abandoned long ago, replaced by the city of Cuba, not far from the Navajo Reservation, where so many of the Native Americans were forcibly settled. The "Puerco marls" are now considered part of the Nacimiento Formation, widely recognized as the world's premier record of mammals from the Paleocene: the first 10 million years after the end-Cretaceous extinction. Today, fieldwork in the area is led by Tom Williamson, who moved to New Mexico as a grad student in the early 1990s, enticed by the legend of Cope and Baldwin. Tom also studies the Cretaceous dinosaurs found underneath the mammal fossils, and it was through his research on *T. rex* that I met him when I was an undergraduate. We struck up a friendship, and after many years of wearing me down, Tom convinced me to study fossil mammals, too. It is solely because of him that I now wear the hat of a mammalian paleontologist—and I am forever grateful.

Through Tom's research, we now understand how the Paleocene mammals of New Mexico fit into the broader story of mammalian extinction, survivorship, and diversification. The fossils from Kimbeto date to around 65.6 million years ago, which means they lived, at most, 380,000 years after the asteroid impact, and a couple hundred thousand years after the "disaster" faunas of Montana. Geologically speaking, that is not much time. But it was more than enough for the Kimbeto mammals to eclipse all mammals that had come before, in terms of their overall number of species, their array of behaviors and food sources and habitats and ways of moving, and most notably of all, their body sizes.

"Finding mammals in the Cretaceous is a very rare thing, and we have to crawl around to find them, and collect dirt to wash through screens to gather their teeth," Tom explained during a team Skype chat, as we digitally reminisced after the Covid-19 pandemic canceled our planned May 2020 fieldwork. "And in the earliest Paleocene in Montana they have screenwashed everything too. But in the Puerco rocks of New Mexico things change dramatically! We all of a sudden find big jaws of mammals everywhere!"

We've already met two of these Kimbeto mammals. There is Carissa Raymond's "Primeval Beaver" *Kimbetopsalis*, a plant-grinding multituberculate substantially meatier than any multituberculates from the Cretaceous. It was Cope who created the name "multituberculate," in 1884, based in part on Baldwin's discovery, in the Puerco marls, of another approximately beaver-size species called *Taeniolabis*. Baldwin picked the rocks of Kimbeto clean, as did many other field crews over the past century, which makes Carissa's discovery of *Kimbetopsalis* all the more impressive.

And then there's *Ectoconus*—represented by the skeleton Tom found, but which was also originally named by Cope in 1884, from another of Baldwin's hauls. *Ectoconus* was the biggest animal in the Kimbeto ecosystem, about the girth of a pig. It is a condylarth: a term—once again—invented by Cope, which refers to a nebulous assortment of hard-to-classify Paleocene and Eocene mammals, with generally primitive skeletons and stocky builds. Sarah Shelley, who excavated the skeleton with Tom and me, did her PhD on *Ectoconus* and other condylarths, and with her typical sense of humor, she describes *Ectoconus* as a "really fat, sheep-pig thing, with a long tail, and a tiny head for its plump size." Its big premolars and molars sported low, rounded cusps, and they would have worked like one of those bumpy rolling balls for massaging your back, but in this case, the cusps tenderized tough plant material. *Ectoconus* was at home on the ground, where it ambled from shrub to shrub, not too slow or too fast. You can sense, though, that it was evolving in the direction of more speed, as its fingernails and toenails were starting to look like miniature hooves.

Ectoconus is one of dozens of placental-line eutherians from Kimbeto. They—not the multituberculates or the few remaining, tiny metatherians—were firmly in control. I could spend several chapters trying to describe all these eutherians, as they are outrageously diverse, and they formed complex food webs. During the Paleocene, they lived in a swampy jungle, with a dense canopy of palms and other trees with huge leaves and long, tapering tips, a

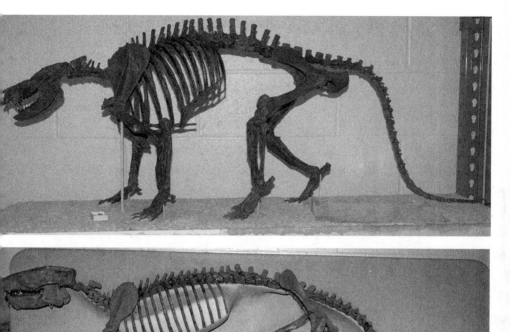

Two "archaic" placental mammals: *Ectoconus* (top), and *Pantolambda* (bottom). (photos by Tom Williamson)

sign they were constantly shedding rainwater. The forest was thick with vegetation—from the tallest palms down to the ferns and flowery shrubs that engulfed the damp soils. All these plants grew so well because it was hot, humid, and wet year-round. On top of this, there was intense seasonality in rainfall levels, and monsoons would have drenched the forests even further during parts of the year. Rivers draining the Rocky Mountains—which were still rising—cut through the jungles, often flooding their banks and

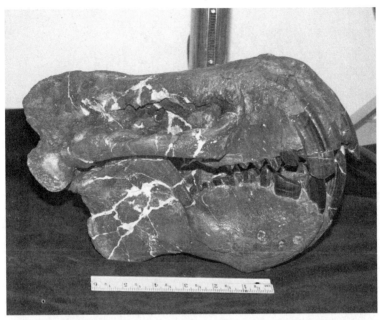

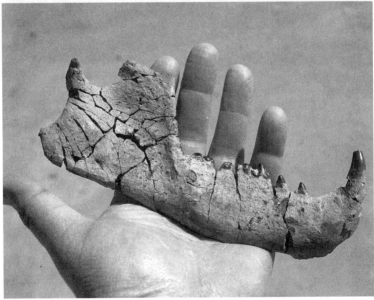

Skull of *Stylinodon* (a taeniodont cousin of *Wortmania*) and jaw of *Eoconodon* (bottom). (photos by Steve Brusatte and Tom Williamson, respectively)

pooling into ponds, where mammals were occasionally trapped, their bones and teeth turning to fossils.

Another of the eutherians living in this jungle world, *Eoconodon*, has a name similar to *Ectoconus*. It, too, is classified as one of Cope's stocky condylarths—although the similarities end there. *Eoconodon* was the terror of Paleocene New Mexico, a brute the size of a wolf—but more buff—that topped the food chain. Its jaws hinged open extra wide, so its cylindrical, fang-shaped canines could grab onto prey. As its paralyzed victim bled out, *Eoconodon* would slice through skin and muscle with its premolars, whose sharp cusps pointed backward, and then smash through bones with its big, crushing molars, which resembled those of bears. *Ectoconus* would have been a delicious meal, but it was probably fast enough to escape much of the time.

An easier target was another eutherian, called *Wortmania*, about the size of a badger. It would not have won any Paleocene beauty pageants, or as Sarah puts it less diplomatically, "it was probably *really* ugly." *Wortmania* was a muscular digger, which used its huge clawed forearms to tear through the dirt and its massive jaws and enlarged canines to root out tubers. Sturdy and plodding, *Wortmania* would have been safe in its burrows, and probably a nasty opponent in one-on-one combat, but if *Eoconodon* could chase one down in an open forest clearing, then it was game over. *Wortmania* was one of several so-called taeniodonts that lived in the Paleocene of New Mexico. These were among the first mammals to develop high-crowned teeth, a godsend for eating tough vegetation, like dirt-encrusted roots and tubers. Because mammals can't replace their teeth throughout their lives, eating hard foods is risky, as a broken tooth can mean a death sentence. Evolving supertall teeth, which can gradually wear down over many years of abrasive chewing, is a clever end-around strategy.

About a million years after the Kimbeto fossils were formed, a new type of eutherian enters the New Mexico fossil record. Its name is *Pantolambda*, and it is the archetypal pantodont: a mysterious group that flourished in the Paleocene and Eocene. They were named by, you guessed it, Cope in the 1870s, but their first recognized fossils were a couple of teeth, dredged from the clays around London, described in the 1840s by another recurring character in our story: the Victorian villain Richard Owen. *Pantolambda* was a beast for its time, around twice the size of *Ectoconus*, and about the bulk of a Vechur cow (today's smallest cattle breed). When it was lumbering across New Mexico, some 64 million years ago, it was the largest mammal that had ever lived there. It was, however, a gentle giant, which languidly snipped and swallowed leaves, probably much like a giraffe, albeit with a chunky neck, not a long one. With a barrel-shaped chest, wide hips, and enormous hands and feet—which recall those giant foam hands so popular with sports fans—it would have cut a comical profile from a distance. And it would look goofier if you saw it up close, with a tiny head, deep jaws, eyes forward on its snout, and oversized conical canines that were probably used to attract mates and intimidate rivals. *Pantolambda* skeletons have been found jumbled together, suggesting they lived in herds, and they were probably highly social animals.

All these Paleocene eutherians were likely endowed with one of the marvels of our own reproductive biology: a placenta, a temporary organ that exists only during pregnancy, connecting fetus to mother. Placentas are not unique to mammals; they have evolved around twenty times, in a variety of species that traded egg-laying for live birth, including even some fishes. It's easy to see why. An egg is essentially its own care package, with a yolk that includes all the nutrients a growing embryo needs to develop. Once a mother lays her eggs, she can protect them, but she can't really break through the egg shell to provide additional nourishment. Live birth, however, requires that an embryo (and then fetus) grows inside its mother until it emerges into the world. It needs access to food and oxygen during this time, plus a way to expel waste. The placenta does the trick. It is the ultimate multitasker, acting as the baby's pantry, lungs, and excretory system at the same time. Then, after childbirth, it is simply discarded—as what we unceremoniously call the "afterbirth."

Mammal placentas, compared to those of baby-birthing fishes and reptiles, are special. They're so sublime that the most diverse major division of today's mammals bears their name. We are a "placental mammal," as are other derived eutherians like rodents, bats, whales, horses, bears, dogs, cats, and elephants. Every mammal alive right now that isn't a monotreme or marsupial is classified as a placental. This is a slight misnomer, though, as marsupials actually do have a placenta, for a very short time, as part of their curious way of making babies. In these marsupials, the fertilized egg is briefly encased in a shell, then *hatches* in utero, is implanted in the mother's uterus and nourished by a placenta, and born as a naked pipsqueak, which will keep developing from the safety of its mother's pouch. Marsupial placentas are small, made of only a single membrane that provides nutrition. Placental mammals like us, on the other hand, have enormous and complex placentas, with separate membranes for feeding and waste disposal. These elaborate placentas can sustain pregnancy for a long time, allowing placental mammals to give birth to large and well-developed young.

As anyone who has witnessed childbirth can attest, placentas are cake-shaped masses of soft tissue, covered in blood vessels and connected to the umbilical cord. Stuff like this doesn't normally fossilize. But the placenta makes its presence known on the skeleton. Monotremes and marsupials have a triangular bone called the epipubis that sticks forward from the pelvis, into the abdominal cavity. These were once called "marsupial bones," because it was thought they supported the pouch, but we now know they serve other functions, like anchoring limb-moving muscles and providing support for the suspension of lots of small, lactating offspring (of either the hatched or birthed variety). Epipubis bones are found in many fossil species, including some of the cynodont precursors to mammals, multituberculates, and even Cretaceous eutherians discovered in the Gobi Desert. This suggests that the earliest eutherians—the immediate ancestors to placental mammals—reproduced either like monotremes, by laying eggs, or marsupials, by birthing many minuscule offspring.

We, however, do not have epipubis bones, nor does any other modern-day placental mammal. Our placentas and larger, longer-developing offspring take up too much space in the abdomen, so there is no room for these bones. Plus, we don't need epipubis bones to support hordes of tiny babies constantly latched to the nipples. Epipubis bones are absent in *Ectoconus* and *Eoconodon*, and *Wortmania* and *Pantolambda*, and every other Paleocene eutherian from New Mexico and elsewhere, which is a strong indicator they had evolved large and complex placentas, and thus, had become true placental mammals like us. It probably was one of their secrets to success after the extinction.

These Paleocene placental mammals were not, however, particularly intelligent. CAT scans studied by my team, led by my colleague Ornella Bertrand—a French paleontologist who is a whiz at taking the x-ray images of the brain cavity and building 3-D models of the brain—show that pantodonts and most of these Paleocene mammals had unusually small brains. Now, their brains were indeed big compared to those of lizards and frogs and crocodiles—after all, these Paleocene species were mammals, and as we saw earlier, as soon as they started feeding milk to their young, the first mammals developed enlarged brains with a novel structure, the neocortex, for sensory processing. But, compared to modern-day mammals of similar body size, the Paleocene species had remarkably tiny brains, with much smaller neocortices. It doesn't seem to make sense, at first thought. Shouldn't the mammals that survived the end-Cretaceous extinction, and prospered afterward, have used keen intelligence and sharp senses to succeed? Alas, that seems not to be the case. Instead, these Paleocene mammals so rapidly ballooned in body size that their brains could not catch up, and it was only more than 10 million years later, in the Eocene, that the signature gargantuan brains of modern placental mammals appeared, with voluminous neocortices that swell over much of the surface of the cerebrum.

Brawn, not brains, explains how Paleocene mammals were able to prosper. After more than 100 million years of constraint, imprisoned in small-bodied niches and unable to get any larger than a wolverine, mammals were suddenly free. It's no mystery why: the dinosaurs were now gone. There was nothing holding mammals back, and literally within a few hundred thousand years—a hiccup of Earth history—placental mammals were moving into the roles once occupied by *Triceratops*, duck-billed dinosaurs, and raptors. By the time of the Kimbeto fossils, and definitely by the time of *Pantolambda*, the dinosaurs were a distant memory, and it was as if they had never existed at all. Mammals formed an entire food chain—of puncture-toothed flesh-eaters and giant leaf-munchers, piggish plant-grinders and muscle-bound burrowers, plus so many other species scurrying over the ground, ambling up trees, and clambering across the branches. It had become a mammal world.

Except that's not entirely accurate. One type of dinosaur did survive: birds. They held their own winning hand of cards: they were small, bred fast, could fly away from danger, and had beaks perfect for eating seeds, a nutritious food source that would have remained in the soil long after the forests collapsed. The delicate, paper-thin bones of Paleocene birds are found alongside the mammals in New Mexico, and these postextinction pioneers enjoyed

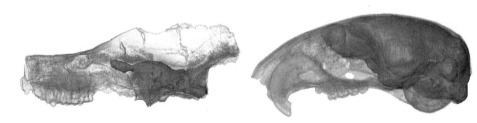

Ornella Bertrand studying CAT scans of fossil mammal skulls (top). Digital models showing the small brain of the "archaic" placental *Arctocyon* (bottom left) and the much larger brain of the modern ground squirrel (bottom right). Scales = 1 cm. (photo by University of Toronto Scarborough; *Arctocyon* specimen curated at Royal Belgian Institute of Natural Sciences).

their own run of success—culminating in the ten-thousand-plus species of birds that live today, around twice the number of mammal species! But numbers can be deceiving, and although birds are an undeniably diverse part of our world, they are not dominant in the same way as mammals. Even the very biggest birds of all time, the extinct elephant birds of Madagascar—which weighed between 1,000 and 1600 pounds (500 to 730 kilograms)—are dwarfed by real elephants of the mammalian kind, which cause small tremors as their 12,000-pound (6,000-kilogram) bodies thunder across the African savanna. Most birds, though, are tiny, easily able to fit in your hand or nest on your windowsill. They're the culmination of a long-term evolutionary trend of miniaturization, which began before the extinction but was accelerated afterward.

The evolutionary roles had thus reversed. Birds were getting smaller, mammals were getting bigger. The mammals had not only replaced the dinosaurs, but they had, in a sense, become the dinosaurs. The Age of Mammals had begun.

6

MAMMALS MODERNIZE

Eurohippus

IN CENTRAL GERMANY, A SHORT drive southeast of Frankfurt at a place called Messel, there is a big hole in the ground. It spans about one hundred acres and is some two hundred feet (sixty meters) deep, a gouge in the otherwise flat and wooded landscape. In the 1700s, locals here discovered black shale rock saturated with kerogen, which they could turn into oil. For nearly two centuries, through falling empires and two world wars, they mined the oil shale, until it became unprofitable in the early 1970s. The mine closed, but the pit remained.

The hole was an eyesore. The government wanted it gone and proposed to make it into a landfill, where the garbage of Frankfurt could be dumped. Locals objected, and after twenty years of legal wrangling, the authorities backed down.

Instead, the pit became a UNESCO World Heritage Site.

This designation—bestowed by the United Nations on only about eleven hundred places worldwide, deemed to have outstanding cultural, historical, or scientific impact—has nothing to do with the mining, or the human history of the area. It has everything to do with the other stuff found inside the black shale: fossils, which tell the story of a much deeper history, some forty-eight million years ago during the middle part of the Eocene, the time interval after those first placental mammal communities were flourishing in the Paleocene times of New Mexico.

We can imagine this world, and its inhabitants, and how they lived their lives. Here's one story, fictional, but based on real Messel fossils.

One spring evening in the Eocene, when this part of Germany was an island in an archipelago much like Indonesia today, a mare felt a craving. During the waning days of the previous summer she had become pregnant, and by now she had been carrying her baby for over two hundred days. Her abdomen was swollen, as were her wrists and ankles, and she was finding it hard to walk without pain—a strange sensation for an animal accustomed to running through the forest undergrowth. This was to be her first child, but she instinctually knew: the birth would be soon.

At that moment, the mare was consumed by hunger. A particular type of hunger: she yearned for the sweet, white and purple flowers of the water lilies that peeked out from the shallows of the nearby lakeshore. She had been fasting all afternoon, through the hottest and dampest part of the subtropical day, but not by choice. She just couldn't muster enough energy to do anything other than lie around. As the sun receded, and cooler evening temperatures descended on the forest, she felt a surge of ambition. Now would be a good time to eat. But there was one flaw in her plan: the lake was a good kilometer away, and could be reached only by walking through the thickest groves of the jungle, right as the predators would be starting their sunset hunts.

The mare glanced around, adjusting her eyes to the encroaching dusk. She could see about twenty other members of her herd, fanned across the small meadow within the forest. They all looked the same: furry creatures, each the size of a terrier dog, mostly covered in coarse brown fur, but with long stripes of black hair following the contours of their arched backs. They stood proudly on all fours, balancing on their hooves, many of them swatting away mosquitoes with their stubby, bristle-haired tails. Some were already nose to the ground, nipping leaves and berries with their incisors, their tongues sweeping across their mouths as they chewed. All of them had their ears perked up, listening both for the rustle of predators in the shrubs and for the sounds of their compatriots. They were a social species, and discipline was strongly enforced through high-pitched whinnies from the herd leaders—which were both male and female.

Although she was not one of the herd's elite, the mare felt the need to call out. Perhaps some of the others would join her on the quest for water lilies? She let out a loud neigh, which screeched through the stillness of the meadow. Some of the herd looked up annoyed, then returned to their dinners. The others just ignored her. If she wanted to satiate her pregnancy craving, she would have to do it alone. So she got up off her haunches, steadied herself on her tiny hooves, and ambled into the jungle, leaving the herd behind. Within moments she was engulfed by the foliage.

The forest was a riot of weeds and shrubs and trees. There were ferns and pines, but most of the greenery was of the angiosperm variety, and because it was spring, the flowers were in bloom. The pink blossoms of magnolias sparkled in the twilight, and the aroma of bay laurels and roses wafted through the air, which was heavy from the afternoon's downpour. There were nutmegs and palms, mistletoes, dogwoods, and heathers, tea-trees, sweetgums, beech, and birch, and citrus trees, whose tart fruits were starting to peep out from their flowers. Grapevines snaked up the tree trunks, choking the midlevels of the forest in a dense blanket of big, flat leaves, dripping with rainwater still making its way down from the canopy. Bean pods, walnuts, and cashews hung above. A few months later, these would cover the forest floor, providing a bounty of food for the herd, and so many other animals that called the jungle home.

As the mare pushed her way through a laurel patch, she heard a crackling in the branches overhead. She locked eyes with another furry animal, about a meter long but more tail than body, climbing up the trunk with its hooked claws. It balanced on a branch, sticking its bushy tail backward for stability, propped its head forward, and began gnawing on the leaves with its enormous chisel incisors. Meanwhile, higher up in one of the sweetgums, another animal moved through the canopy, with considerably more skill. It was a little smaller than the leaf-gnawer and clung to the branches with confidence. Its hands and feet were huge and ended in flat nails instead of claws. With one hand, it wrapped its opposable thumb and long, spindly fingers around a branch, and with the other it grabbed a fruit, which wasn't quite ripe, but was apparently edible.

As the mare was transfixed with the activity in the treetops, it took her a moment to notice an even more peculiar animal, parked on the ground in front of a nest of ants. Stocky and muscular, it grunted as it swung its arms forward, its shovel claws shredding the nest. As the ant colony dispersed in a panic, spilling across the decaying leaves of the forest floor, the nest-destroyer licked its lips with its long, snakelike tongue, and then began to suck up the ants with its narrow, toothless snout. The mare had never seen anything like this before. The ant-muncher had some hairs sticking off its back at odd angles, but most of its body was covered with scales, each one looking like a guitar pick, which imbricated together to sheath the creature in a strong, but flexible, coat of armor.

The mare was so distracted by the insect pandemonium that, for a few minutes, she forgot about her hankering for water lilies. She let her guard down just long enough for one of the night-stalking carnivores to notice her. From the camouflage of some ferns, this mongoose-size meat-eater watched with intense, red eyes and read the scent of the breeze with its whiskers. It had already devoured several lizards and frogs, but it was still hungry, and its blade-shaped cheek teeth were ready for the evening's next course. The hunter weighed its options. The scale-covered animal might be delicious, but biting into its armor would take too much effort. Much better to go for the plump mare that seemed completely oblivious to her surroundings.

As the predator shot out of its hiding place, teeth bared, it realized that it had another option. To the left of the mare was another animal with nails that resembled hooves, but with four toes instead of the mare's three. It, too, seemed oblivious as it stood on its long, toothpick arms and legs, gulping rotten fruit encrusted with fungi. This thing seemed more clueless—and defenseless—than the mare, so the hunter made its choice and pounced on the fruit-eater. The mare heard the commotion and snapped back to sanity. It was a close call, and she had better be on her way.

With her wits about her, and her mother's instinct in full force, the mare kept moving. No distractions this time; the water lilies beckoned. A few minutes later she untangled her way through a heather patch, and suddenly darkness turned to shimmering moonlight. The forest had cleared, and she now gazed across the vista of the lake.

Concentric waves rippled across its dark blue surface, washing through blooms of algae that gave the water a dirty appearance. Bubbles streamed up from fish swimming in the depths, and turtles poked out their heads for air. In the sky above, the silhouettes of flying animals fluttered against the last hues of a fiery orange sunset. Some were clearly birds, of the nightjar family, hunting moths and dragonflies on the wing, their chirps and squawks ushering in the impending darkness. But others were different—they had broad wings of skin stretching between their fingers, were covered in hair, and emitted ultrasonic clicks as they swooped between the nightjars, piercing bug cuticles with their sharp-cusped molar teeth.

And then the mare saw it. The light of the moon, nearly full, danced across the line where lake met land, illuminating the colorful petals of the water lily flowers. Her stomach rumbled. She felt her baby kick. Having learned her lesson in the forest, she inched toward the shore, careful to check for predators. A few crocodiles paddled in the water, but she saw them, and they were too far offshore to pose a threat. Another furry creature was stirring, but it was nothing to fear: a wee thing, about a foot long, with dense, spiky hair covering its back and sides. It, too, was after a sundown lakeside meal, but of fish, not flowers.

The mare sunk her hooves into the mud of the shore, waded into the warm water, and immersed herself in a lily patch. A playful whinny of delight, and then it was down to business. She ate voraciously, first nipping at the choicest flowers, but then giving up and biting whatever she could, whether flowers, leaves, or stalks. It was bliss, exactly what this expectant mother needed for the strength to birth new life. No doubt her time was soon, so with a full belly, she turned to the land, to make her way back to the safety of the herd.

But something didn't feel right. As she took her first step out of the water, she wobbled. Disoriented, she fell onto her haunches, and tried to get up again, but couldn't. In an instant, all went black, and the mare—and her baby, in utero—slid into the algal waters of Lake Messel.

EOCENE LAKE MESSEL was no ordinary lake. It was forged from a volcanic explosion, when magma oozing up from the deep earth came into contact with groundwater, detonating a steam blast that formed a crater. Rivers draining the surrounding rain forests filled the crater with water, which over time became deep and stratified. Every so often, without warning, the lake would gurgle and a cloud of invisible gas would ascend from its anoxic abyss. Sometimes the gases were volcanic, other times a by-product of bacteria or algae. Regardless, the fumes were toxic and quickly asphyxiated anything in their path: animals swimming near the lake surface, loitering by the shore, clinging to branches over the water, flying overhead, or eating water lilies in the shallows.

Felled by a killer that left no physical calling card, the animals would slip into the stagnant depths of the lake, looking like they suffered no trauma. Without oxygen to catalyze their decay, the carcasses would sit on the lake bed and were slowly buried by mud, transforming them into the most exquisite fossils: not merely skeletons, but things that look like actual animals, with hair, and last meals in their stomachs, and in some cases, unborn babies in their bellies. In this way, entire ecosystems were encased in shale, permeated with kerogen distilled from algae that sunk to the bottom alongside the animals. The thousands of Messel fossils include everything from flowers to bugs, fish to turtles, lizards to crocodiles, birds to mammals.

Although I didn't name them in my story, you might be able to identify the types of mammals I'm writing about. The mare is a horse: a wee species called *Eurohippus* that would barely come up to the ankle of a thoroughbred. There are several Messel *Eurohippus* skeletons with a single late-term fetus coiled inside the uterus, as if captured in an ultrasound photograph, surrounded by traces of the placenta. The bushy-tailed leaf-gnawer is a rodent called *Ailuravus*, which would have looked like a squirrel. Its fellow tree-dweller, with opposable thumbs, is a primate called *Darwinius*—an early cousin of ours. The anteater is *Eomanis*, one of the first known pan-

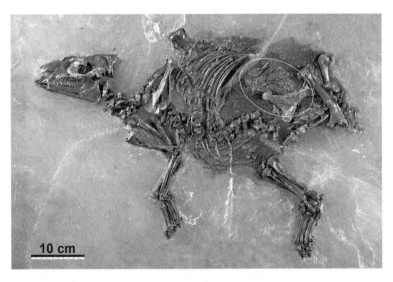

10 cm

The Messel mare *Eurohippus* with a preserved fetus (circled). (image from Franzen et al., 2015, *PLoS ONE*)

golins, which today are endangered because their scales are popular ingredients in traditional Asian medicine. The carnivore is called *Lesmesodon*, and although its relationships are disputed, it is probably a primitive carnivoran: part of the dog and cat group. The stilt-legged frugivore that it ate, *Messelobunodon*, is an artiodactyl related to cows, sheep, and deer, with an even number of toes. The spike-haired fish-scavenger is *Macrocranion*, of the hedgehog lineage. And of course, the winged insectivores echolocating over the lake are bats—several species are known from Messel, and they are some of the most common fossils in the oil shale.

This mammal community had more species, with a greater prosperity of ecologies, diets, body types, and behaviors, than the Paleocene faunas of New Mexico, or any other mammal ecosystem of Paleocene age. Just as the Paleocene was more diverse than the Cretaceous, the Eocene was richer than the Paleocene. Beyond that, two things about the Messel mammals immediately stand out.

First, all the mammals in my story are placentals. There were metatherians—members of the marsupial clan—living in the Messel jungles, but they were such background characters that they are barely worthy of mention. Of the thousands of animal skeletons pulled from the ancient lake, a meager five are metatherians. These were omnivores that used their prehensile tails and strong feet to hang from branches, like an opossum, but they were clearly being crowded out of the arboreal niches by rodents and primates. This is reflective of a bigger evolutionary story: after metatherians were hammered by the end-Cretaceous extinction, they managed to hold on in Europe, Asia, and North America for a few tens of millions of years, but then vanished across the north. Their legacy was saved by dispersal to South America and Australia, where they would blossom again—a story we will return to later. And what of multituberculates, that major group that was so common in the Cretaceous, endured the extinction, and got bigger

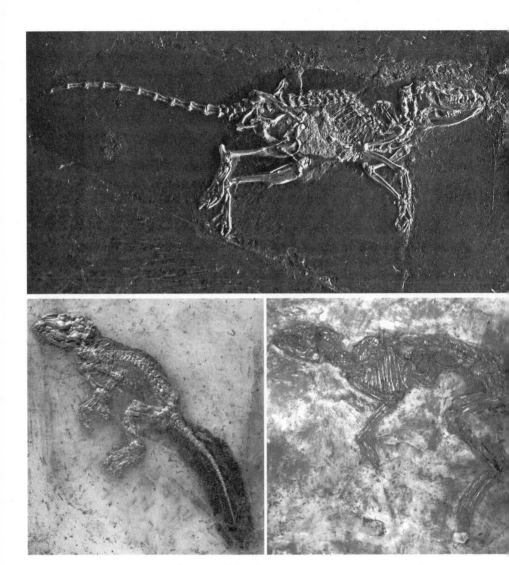

Montage of Messel mammal fossils: *Macrocranion* (top), *Lesmesodon* (bottom left), and *Messelobunodon* (bottom right). (photos by H. Zell, Norbert Micklich, and Ghedoghedo, respectively)

in the Paleocene? There is not a whiff of them in Messel: not one skeleton, or one jaw, or even one of their trademark Lego-brick molars. As Lake Messel entombed its treasures, multituberculates were in the process of going extinct, and by the end of the Eocene, some 34 million years ago, they were gone.

The second important aspect of the Messel fauna is that these placentals are recognizable to us. We can classify them into the main subgroups that exist today. Our hero, the mare *Eurohippus*, is a horse. The bush-tailed *Ailuravus* is a rodent, the branch-grabber *Darwinius* is a primate, and so on. This is not the case with Paleocene mammals from the first 10 million years after the dinosaur extinction. If you recall from last chapter, there were plenty of placentals in Paleocene New Mexico, but they seem strange and are hard to categorize. What the hell is a condylarth, a taeniodont, or a pantodont—those three keystone groups that raised the curtain on the Age of Mammals a few hundred thousand years after the asteroid? They don't obviously share features with today's placental groups, like the gnawing incisors of rodents or opposable thumbs of primates. Instead, they appear generalized, primitive—so much so that they are often dismissed as "archaic" placentals. Ever since the first discoveries of their fossils in the late 1800s, Edward Cope and legions of paleontologists thereafter have struggled to place them on the mammal family tree.

Scientists have been making family trees of mammals for well over a century. The consensus view was solidified by George Gaylord Simpson's landmark 1945 classification scheme, underpinned by a 350-page manifesto. Simpson was one of the giants of twentieth-century paleontology, who brought fossils to the "high table" of evolutionary biology by showing that ancient species were governed by the same laws of natural selection that operate today. After dabbling in door-to-door sales, Simpson went to college and became enamored with fossils, and at age twenty-two blustered his way onto his first field crew as a driver—even though he didn't know how to operate a car. Within a few years he was curator at the American Museum of Natural History and doing regular fieldwork in New Mexico, focusing on the Eocene rocks above Cope's Paleocene beds. Simpson's career was interrupted during World War II, when he served as an intelligence officer in North Africa and Italy. Not only was he awarded two Bronze Stars, but he also claimed to have defied General George Patton's order to shave his beard by appealing directly to Dwight Eisenhower, the US Supreme Commander in Europe who later became president.

Simpson was a man of authority, and his family tree was gospel for decades. It was updated in 1992 by one of his successors at the American Museum: Mike Novacek, who we met earlier collecting Cretaceous mammals in the Gobi Desert. By and large, Novacek's genealogy was similar to Simpson's. Egg-laying monotremes like the platypus were the most primitive mammals, followed by pouched marsupials and placentals. Among placentals, elephants were placed in a cluster with hoofed mammals like the odd-toed horses and even-toed cattle; bats nestled close to primates in a congregation of big-brained tree dwellers; ant-slurping pangolins perched next to anteaters and sloths; and there was a master group of bug-eaters called Insectivora, which included many small, fast-growing vermin from around the world. Most of the relationships made intuitive sense, because Novacek's tree, like Simpson's, was based on anatomy. Mammals sharing particular features—such as hooves, or molars with pointy cusps for puncturing insects—were grouped together. Logically it seemed sound: evolution fashioned mammal bodies through natural selection, so if a bunch of species all had hooves, it was a sign of their common ancestry.

There's one serious problem with this approach to family trees: convergent evolution. Two organisms may independently evolve the same feature if they are faced with similar environmental pressures. Take hooves, for instance. There's no reason that hooves had to evolve only once, in the common ancestor of all of today's hooved species. Instead, they could have developed a few separate times, when different species—distantly related to each other—each had to survive by running faster on open land. Ditto for pointy-cusped molars: maybe many different mammal groups got a taste for insects, so evolution did the same thing over and over: produced teeth that were mechanically suited to breaking insect cuticles. Because form follows function, these teeth would naturally look similar to each other even though they were the product of many separate evolutionary acquisitions. It would be easy to mistake these similar molars as a sign that their possessors were closely related, when in actuality, they just had similar diets and ecologies. Simpson and Novacek knew of this pitfall, but they didn't have the tools to untangle common ancestry from convergence.

DNA came to the rescue, but only later in the 1990s. This was the era of the Human Genome Project, one of the greatest achievements in scientific history, which mapped our genetic code and illuminated the shared foundation of all humanity. This was also when DNA fingerprinting became a common technique in law enforcement, putting many murderers behind bars. Behind it all was improved technology for sequencing genes—essentially putting human tissue into a machine, which uses chemical reactions to read out the string of gene components, which go by the letters A, C, G, and T. These same techniques can be applied to animal tissue, and before long biologists were awash with new evidence perfect for building family trees.

You can think of each A, C, G, and T as a characteristic—the molecular version of a hoof or pointy molar. If you sequence the genomes of a bunch of animals, line them up, and compare them, you can make a family tree by grouping species with the most similar DNA. In practice, this means constructing family trees by uniting species based on shared mutations in the DNA, which they possess but other species do not. Each mutation is a discrete evolutionary event, just like the development of a hoof or tooth. Like anatomical features, DNA can be affected by convergent evolution. But the problem is not so acute, because there are potentially billions of base pairs of DNA to compare, so a few convergent mutations can more easily be filtered out. Plus, DNA can unmask cases of anatomical convergence: if two animals evolve a hoof independently, they will probably do it through different genetic pathways, the same way that two historians would use their own words to describe the physical appearance of the same person—say, Charles Darwin.

Finally, paleontologists had a way to tell whether an anatomical feature shared by two mammals was likely due to common ancestry—and thus could be used to build family trees—or was a misleading effect of convergence. Even better, paleontologists could work with molecular biologists to use DNA to actually construct the family trees, bypassing the problems of anatomy altogether.

When the first DNA-based genealogies of mammals were published in the late 1990s and early 2000s, by molecular biologist Mark Springer and his network of collaborators, paleontologists were shocked. Many of the relationships among placentals championed by Simpson disintegrated, revealed as illusions of anatomical convergence. Genes showed that pangolins are not closely related to anteaters and sloths, but group with dogs and cats. Bats are not next-of-kin to primates, but part of a larger assemblage with dogs, cats, and pangolins, plus the perissodactyls with an odd number of toes (like horses) and even-toed artiodactyls (like cattle). These latter two groups both have hooves, but there are other hoofed mammals dispersed around the family tree—like the cute little hyraxes, which group with elephants. Hooves, therefore, really did evolve multiple times. But that is nothing compared to the madness of the insect-eaters. Once thought by Simpson and Novacek to compose a single group, they were scattered all across the DNA tree. Some, like the golden moles and tenrecs, are closely related to the hyraxes and elephants—a most unusual union that nobody had ever predicted from anatomy. Insectivory, and the distinctive molars that enabled it, was thus reinvented numerous times by numerous different mammal lineages.

Springer's family tree has now supplanted Simpson's as the standard. On it, the placentals are divided into four fundamental groups. Near the base of the trunk, two lineages branch off from the others. One is that unexpected cluster of golden moles, tenrecs, hyraxes, and elephants, plus aardvarks and manatees. This group was given the name Afrotheria, because most of its members live in Africa today and have a fossil record showing they have been there for quite some time. The second early-splitting group is Xenarthra, which incorporates mostly South American species like anteaters, sloths, and armadillos. Making up the crown of the tree are two diverse groups of northern species, distributed widely across Europe, North America, and Asia, but with members south of the equator, too. One is called Laurasiatheria—this is the clique of dogs, cats, pangolins, odd and even-toed hoofed mammals, whales, and bats. The second is Euarchontoglires, of which we are a member, along with our primate cousins, and rabbits and rodents.

The overall structure of the tree, therefore, reflects geography more than anatomy or ecology. The histories of the major placental subgroups have largely played out on certain continents or landmasses, and while living apart, these subgroups often converged on one another in diet and lifestyle. This suggests that the subgroups split from one another back when the continents were closer together, and then Afrotheria and Xenarthra became largely isolated on Africa and South America, respectively, as the continents moved farther apart. The northern groups, on the other hand, were able to range more freely across the high-altitude land bridges that have intermittently linked North America, Europe, and Asia since the Cretaceous. Overprinted on this general pattern were dispersal events, which brought some afrotherians (like manatees and mammoths) and xenarthrans (like armadillos) up north, and boreal species like primates and rodents down south. We'll discuss these geographic patterns and dispersals throughout the remainder of this book.

If the geographic structure of Springer's genealogy was not surprising enough, paleontologists were stunned by another implication of the DNA evidence. As discussed in the opening pages of this book, in recounting the story of when the mammal lineage split from the reptile line back in the Pennsylvanian-aged coal swamps, DNA can be used as a clock. You can line up the DNA of two species, count the number of differences, and if you know the rate at which DNA mutations normally accumulate (which can be estimated using lab experiments and other techniques), you can back-calculate to figure out when those two species last shared a common ancestor. Think of it like one of those elementary school math problems: if Jack and Jill are five hundred miles apart, and we know they move apart from each other at the rate of one hundred miles per week, then they must have gone their separate ways five weeks ago. When Springer's team applied this rationale to their DNA trees, they were in for another shock: many of the modern placental lineages—not only the fundamental groups like Afrotheria and Laurasiatheria, but also individual lines like primates and rodents—must have originated back in the Cretaceous or the earliest Paleocene. In many cases, this is long before their fossils first appear, hinting at a vast unrecorded history.

This raises an intriguing possibility: maybe some of those Paleocene "archaic" placentals like condylarths, taeniodonts, and pantodonts are the missing fossils from the ghostly early histories of the modern groups. We just haven't been able to easily link them to the modern groups because they had yet to develop the signature anatomical features that define those groups today. This is not a new idea; going back to Cope, paleontologists have speculated along these lines, and there is reasonable evidence that some condylarths were early members of the odd-toed and even-toed hoofed lineages, and that some of the tree-swinging mammals that lived soon after the end-Cretaceous asteroid impact were primeval primates. Our biggest handicap is that, with fossils, usually all we have is anatomy, and as we've seen, anatomy can be misleading. If only we had DNA samples from these bizarre Paleocene species, that would settle things as quickly and conclusively as a paternity test identifies the real father on one of those afternoon talk shows—something else made possible by the 1990s DNA revolution.

More work needs to be done, especially in combining evidence from the anatomy of fossils with the anatomy and DNA of today's species, to build a master family tree. This is the big project that my lab is working on right now. After several attempts, I was awarded a grant by the European Research Council to build that master family tree and attempt to slot in the "archaic" Paleocene species. We have a top team working on it—including my New Mexico comrades Sarah Shelley and Tom Williamson, the mammal brain expert Ornella Bertrand, mammal anatomy specialist (and one of my favorite mammal mentors) John Wible, and several outstanding PhD students, all shown in the accompanying photo—so if anybody can figure it out, it should be us. As I write, I don't know what we will find.

For now, though, what we do know is that, by the time Lake

Messel was burying its dead in the Eocene, all of today's key placental mammal groups had emerged, and many were thriving. As the Paleocene transitioned to the Eocene, the archaic faunas modernized. Once again, a spurt of environmental change was the trigger.

THE PALEOCENE WAS a greenhouse world. The "archaic" placentals of New Mexico lived in a rain forest jungle, a lush biome so unlike the dry scablands that cover the same area today. Back then, much of the midlatitudes were carpeted with subtropical forests, composed of new flowering trees that had evolved after the asteroid impact. Crocodiles basked in the sun in the high latitudes, which were ice-free and shrouded in temperate woodlands. Any snow would have been limited to the very tallest mountain peaks—like those of the rising Rockies. All because the atmosphere was awash in carbon dioxide, which kept the earth hot.

Then, as the Paleocene turned to the Eocene 56 million years ago, the greenhouse got even hotter. Yet more carbon dioxide was injected into the skies, and global temperatures increased between five and eight degrees Celsius (nine to fourteen degrees Fahrenheit). The average land temperature in the Arctic skyrocketed to 25 degrees Celsius (77 degrees Fahrenheit), and crocodiles and turtles now lounged *above* the Arctic Circle, in the shade of palm trees. The equatorial regions breached 40 degrees Celsius (104 degrees Fahrenheit), turning large swaths of low-latitude waters into no-go zones, too roasted to support much—if any—life. This was the hottest Earth had been since the dinosaur-killing asteroid, and it has yet to get any hotter since. It all happened very fast: the carbon release took about 20,000 years at most, and the surge of global warming peaked and abated within 200,000 years. Still, it

My team of students and colleagues studying mammal genealogy. Back row: Hans Püschel, Sarah Shelley, Sofia Holpin, Paige dePolo, Zoi Kynigopoulou, Tom Williamson. Front row: Jan Janecka, me, John Wible (and his favorite pangolin). (photo courtesy Steve Brusatte)

was enough to disrupt environments around the world, and change the course of mammalian evolution.

This brief interval of climate change, called the Paleocene-Eocene Thermal Maximum, or PETM for short, is *the* exemplary global warming event in the geological record. It has been studied by legions of scientists who wish to better understand contemporary climate change and predict how our earth may respond. No doubt it is the most suitable ancient parallel to our modern predicament. Its cause, however, was different. Modern warming is on us, fueled by the carbon dioxide released from burning oil and gas. The PETM, as with so many prehistoric heat waves, was instigated by volcanoes.

As you read this, magma is snaking its way through the mantle and crust underneath the North Atlantic, scabbing into basalt rock as it hits the cool ocean waters. The still-growing basalt blob has a name: Iceland. It marks the spot where Europe and North America began to separate in the late Paleocene. Until then, Greenland was linked to Europe. Then the magma plume began to rise, pushing the two landmasses apart and opening the North Atlantic corridor—one of the final acts of the disassembly of Pangea, which had begun some 140 million years earlier, when the first mammals were scurrying.

As the magma percolated through the crust, on its way to the surface, it fanned out into thousands of horizontal sheets called sills, which literally baked organic matter they came into contact with. Like an engine burning gasoline, this released greenhouse gases: carbon dioxide, and the more potent methane. Trillions of tons of carbon leaked into the atmosphere, raising the carbon dioxide level between two and eight times beyond that of the already scorching Paleocene. Temperatures spiked, leaving a telltale chemical fingerprint in the rocks: a sharp decline in the ratio of the heavier isotope of oxygen (18O), which has more neutrons than the lighter isotope (16O). From lab experiments, we know that the ratio of these two isotopes is a paleothermometer—and it pinpoints the temperate rise, of five to eight degrees Celsius, right at the Paleocene-Eocene boundary.

Such intense global warming had major repercussions, on ecosystems and their mammals. This is best seen by studying the world's premier record of mammal fossils spanning the Paleocene-Eocene boundary, from the Bighorn Basin of northern Wyoming, west of the imposing Bighorn Mountains that tourists traverse on the way to Yellowstone. Philip Gingerich and his litany of students and colleagues—among them Ken Rose, Jon Bloch, Amy Chew, and Ross Secord—have made their careers documenting these fossils, excavating thousands of skeletons, jaws, and teeth of the mammals that endured the PETM.

The Bighorn Basin is badlands today, but during the PETM it was a wet, verdant forest not unlike that of Paleocene New Mexico. Before the temperature rise, the forests were a diverse mixture of evergreen conifers and flowering trees like walnuts, elms, and laurels. As the Icelandic volcanoes belched their carbon, and the planet warmed, earliest Eocene Wyoming got drier. The conifers wilted and were replaced by trees that could better withstand heat, particularly those of the bean family, which migrated some four hundred to nine hundred miles (six hundred to fifteen hundred kilometers) northward from the tropics. Then, the magma plume slowed to the dribble that it remains today—the source of Iceland's geysers and airplane-canceling ash eruptions. The carbon gusher was turned to a trickle. Temperatures stabilized, it started raining heavily again, and the conifers returned.

These 200,000 years of upheaval—the temperature swings, the changing vegetation, the aridity and return of the rains—fashioned an entirely new mammalian community. The Paleocene of Wyoming was similar to that of New Mexico, dominated by "archaic" placentals. They were prospering when a change in the carbon composition of the rocks signals the onset of volcanism. Then, within the next 10,000 to 27,000 years, as the oxygen composition of the rocks recorded the warming spike, scores of new mammals suddenly appeared in the Bighorn Basin. Chief among these were the first members of three modern groups, what we call the PETM Trinity: primates, even-toed artiodactyls, and odd-toed perissodactyls.

This same Trinity appeared at the same time in Europe and Asia, too. The PETM, it seems, unleashed a mass migration. Fossils of the Trinity materialize so rapidly—like a swarm of locusts—that it's difficult to tell exactly how they migrated. Did they arise in

Asia, and then make their way to Europe and North America? Or take the opposite journey, or another route altogether? Did they evolve earlier, perhaps from "archaic" placentals like the New Mexico condylarths, but were restricted to an isolated valley or mountain range, before the warming climates helped them break free and spread northward, across the polar corridors? Or did they arise during the PETM interval itself, in a frenzy of evolution catalyzed by the temperature and environmental changes? We're not yet sure. All we really know is that things transpired quickly, and by the time the volcanism subsided, three of the most canonical modern mammal families were broadly distributed across the northern continents.

The arrival of the PETM Trinity was transformative. In the Bighorn Basin, the immigrants claimed the forests as their own. Almost instantaneously, newcomers made up about half of the individuals in the ecosystem, and they brought with them their own customs: they were, on average, larger than the natives, and their diets more neatly fell into leaf-eating, fruit-eating, and meat-eating categories, compared to the more omnivorous and insectivorous palates of the locals.

The migrants also boasted new adaptations. The first Wyoming primate, *Teilhardina*, had big eyes, nails on its fingers and toes to grip branches, and lithe ankles that allowed it to move gracefully through the canopy. The earliest-appearing artiodactyl, *Diacodexis*, looked like a deer, albeit one the size of a rabbit. Its body was customized for speed: its limbs were long, thin, and capped with hooves. Its main ankle bone, the astragalus, had a deep groove at each end, ensuring that the foot could extend and flex in a fore-aft direction without rotating laterally. This "double pulley" condition is a hallmark of today's artiodactyls—from cows to camels—and permits them to run fast without dislocating their ankles. The pioneering Wyoming perissodactyl, the tiny horse *Sifrhippus*, was fast in another way. It, too, had hooves perched on long limbs, but its more flexible shoulder and pelvic joints provided greater maneuverability as it galloped through the dense undergrowth—like its close cousin, the Messel mare *Eurohippus*, which lived later in the Eocene, after the warming pulse. And in general, all these migrants had bigger brains than the comparatively dim-witted Paleocene "archaic" placentals.

Something peculiar happened to many of these immigrants—and some of the locals—as they weathered the global warming. They dwarfed. Then, as the climate cooled, they got bigger again. Phil Gingerich first noticed this pattern, and then one of his graduate students, Ross Secord, figured out why it was occurring. Ross, now a professor at the University of Nebraska, is a member of our New Mexico field crew, and he brings discipline instilled from the Bighorn Basin camps to our more spontaneous bunch. His tents are always neatly assembled, and dinner—cooked in a spotless kitchen tent—is always on time, and usually includes some variant of hot dogs, whether on the traditional bread rolls, or chopped up inside burritos, or served over pasta (which, I must admit, always upsets my Italian American sensibilities). I respect Ross deeply: he is a rare paleontologist who combines expertise in the nuances of mammal anatomy with the know-how to read carbon and oxygen isotopes in the rocks, which allows him to put ancient mammals into an environmental context and understand how they changed as temperatures and climates shifted.

In a landmark study published in 2012, Ross surveyed the Bighorn Basin mammal fossils. He found that about 40 percent of the Paleocene locals got smaller during the PETM, most of which then bounced back and grew larger. Even more striking was the fate of the immigrants, particularly the little horse *Sifrhippus*. The first equine settlers entered the Bighorn Basin right as the volcanoes began releasing carbon, and they were small: about 12 pounds

(5.6 kilograms) on average. Then, as temperatures increased, the horses got even smaller: they shrunk by about 30 percent, to a measly average of 8.5 pounds (3.9 kilograms), making them some of the teeniest horses that have ever existed. For about 130,000 years they stayed this way, before rapidly beefing up some 75 percent to an average of 15.5 pounds (7 kilograms) as climates ameliorated. The trend of size change almost perfectly matches the trend in temperature, as shown by the oxygen isotope paleothermometer in the rocks. These horses got progressively smaller as the world went hot, and then got bigger as the world cooled.

We see something similar today, albeit on a spatial scale instead of a temporal one: animals living in warmer areas are often smaller than their contemporaries in colder climes—an ecological principle called Bergmann's rule. The reasons are not entirely understood, but it is probably, in part, because smaller animals have a higher surface area relative to their volume than plumper animals and can thus better shed excess heat. What's neat about Ross's finding is that we can predict that, as temperature rises today, many mammals might get smaller. Including, perhaps, humans. After all, we are mammals, too—subject to many of the same ecological and evolutionary pressures as minihorses and our primate kin in the PETM Trinity. And as we'll see later, humans have dwarfed before.

As the Trinity colonized Wyoming, and oscillated in size to the dancing temperatures, the forests became more diverse than ever before. And they stayed that way, because the migrants managed to move in without exterminating the natives. "Archaic" placentals like condylarths, pantodonts, and taeniodonts persisted alongside the new species, for well over 10 million years. Paradoxically, the PETM global warming—unlike the volcanic climate changes at the end of the Permian and Triassic periods, which we learned about earlier—did not cause a mass extinction. But over time, the PETM migration had a slower drip of consequences. The "archaic" placentals survived for a while, but their death warrant was sealed. The future belonged to monkeys, cattle, and horses.

For the remainder of the Eocene, it was the horses and their perissodactyl relatives that really thrived. Today, perissodactyls are a paltry lot compared to their more diverse artiodactyl cousins—there are fewer than twenty species of odd-toed horses, rhinos, and tapirs, in contrast to the nearly three hundred species of even-toed cows, camels, deer, pigs, and whales (which evolved from land-living artiodactyls, as we will learn in the next chapter). While these groups are defined based on their feet, they also differ in their digestive systems. Perissodactyls are hindgut fermenters, which break down the cellulose in plant matter in their intestines, after it has already passed through the stomach. So are we, and most mammals. Many artiodactyls, on the other hand, do most of the work in their stomach, which is a multiplex of four chambers. This is why cows "ruminate," or chew their cud: they swallow their food, process it in the first two stomach chambers, then regurgitate it, chew it some more, and pass it back through the entire stomach. By doing so, they extract maximum nutrition out of every morsel they eat—a nifty trick when eating tough or lower-quality vegetation, like grasses. The Eocene, however, was still a time of forests, and grasslands wouldn't spread until much later. In this world of abundant fruits and leaves, perissodactyls were primed to prosper.

Horses, rhinos, and tapirs all got their start in the Eocene, but the most remarkable perissodactyls of the time were two now-extinct groups, with stupendous body types that make them among the most fantastic of the fantastic beasts that fire our imaginations. One of them, the brontotheres, never made it out of the Eocene. The other, the chalicotheres, held on until less than a million years ago in Africa, where they would have made contact with—and possibly been hunted by—our hominin ancestors.

Brontotheres—the "thunder beasts"—were the largest mammals of the Eocene, and the first mammals that truly tried to imitate the colossal dinosaurs of ages past. The "thunder beast" moniker works on two levels. These furry, horned behemoths really would have shaken the ground as they walked. Plus, their name alludes to the story of the "Thunder Beasts," exalted characters in Sioux legend that would leap from the clouds during thunderstorms and shepherd herds of buffalo toward the Native American hunting parties. The story may sound a little far-fetched to us, but it was not merely a myth. The Sioux lived on the plains of the American West, before they were forced onto reservations so settlers could steal their land and gold. They were surrounded by fossils. They noticed these fossils, and collected them, and tried to make sense of them, as we do today. Othniel Charles Marsh—Edward Cope's great rival in the Bone Wars—was a friend of the Sioux chief Red Cloud, and in what may seem an uncharacteristic act of decency for such a prickly bone-hunter, he spent considerable time lobbying the US government about the plight of the tribes. Marsh's crew was shown the fossil jaw of a brontothere by a band of Sioux, who recounted the tale of the "Thunder Beasts." It was Marsh, in 1873, who formally proposed the brontothere name for these extinct perissodactyls.

The very first brontotheres were humble sprinters that looked a lot like the miniature Eocene horses. Then, evolution went insane. Over the course of the Eocene, brontotheres ballooned in size, the largest ones measuring some eight feet (two and a half meters) tall at the shoulders and sixteen feet (five meters) from snout to tail, and tipping the scales at two to three tons. That's approximately the size of a modern African forest elephant. As

Right: Bizarre extinct perissodactyls: a brontothere (opposite, top) and chalicothere (opposite, bottom) in classic exhibits at the American Museum of Natural History. (photos by AMNH Library)

the brontotheres got bigger, their bodies grew fat and stocky, their limbs turned to Greek columns, and their snouts began to sprout horns. Many brontotheres had horns that bifurcated into a fork at the top, whereas others had a single, terrifying, three-foot-long (meter-long) battering ram that arched upward. These were instruments of intimidation, used in wrestling matches and head-to-head battles, like many horned mammals do today. Brontotheres were social animals, which traveled in herds, as shown by the discovery of mass death assembles where scores of skeletons are preserved together. You can imagine a pack of hundreds of these ogres, grunting as they plowed their way through the forests of the Eocene, trampling on ferns and shrubs, clearing their own meadows as they went, on the search for tasty leaves and fruits that only they—and not their diminutive horse cousins—could reach.

As extraordinary as they were, brontotheres had nothing on the chalicotheres. These were surely some of the most implausible mammals that ever lived, looking like what might result if a horse was able to mate with a gorilla. When their bones were first noticed in the 1830s, they were thought to belong to two different animals: one with the head of a horse, whose hooves were unknown, and the other a strange anteater-like species with long, curving claws, which was missing a skull. A half century later the puzzle pieces were put together, and it was realized that the head and claws belonged to the same animal: a hallucinogenic oddball with long arms and short legs, which walked in a stupor, on its knuckles, so it could keep its sharp claws from rubbing against the ground. The claws were not for defense or catching prey. Instead, chalicotheres would sit on their butts, lean up against a tree, and use their claws to pull down branches. They lost their front teeth as adults, probably to make way for a long prehensile tongue—like that of a giraffe—to strip leaves off branches. Imagine what our ancestors must have thought when confronted with such a creature—and lament that we are not able to experience them today, because if they managed to survive extinction, surely they would have taken their place alongside elephants and pandas as the most popular exhibits at our zoos.

As chalicotheres, brontotheres, and other perissodactyls were diversifying during the Eocene, they were joined by other groups. Not only other members of the PETM Trinity, but two additional families that are hugely important today: rodents and carnivorans. Both originated during the Paleocene, before the temperature spike, but it was afterward when they migrated far and wide, alongside the Trinity.

The early rodents—like *Paramys*—looked like a mix between a squirrel and a prairie dog and mostly lived in the trees. They possessed two signatures of modern rats, mice, beavers, and kin: the ability to chew food by sliding the jaws forward and backward and ever-growing incisors for gnawing. X-ray a rodent skull, and you will see that the incisors are massive: only the tip protrudes from the gums, but most of the tooth is hidden inside the jaw, where it curves far back, its root often reaching beyond the molars. These superior eating adaptations may have helped rodents outcompete the previously diverse multituberculates, which ruled the gnawing and plant-chewing niches during the Age of Dinosaurs, and for much of the Paleocene. By the end of the Eocene, the multituberculates were extinct, and rodents were on their way to their astounding modern diversity. Today, there are over two thousand species of rodents—around 40 percent of all mammals.

Eating these rodents, and the wee horses and maybe the brontotheres, if they wanted to risk it, were carnivorans: members of the dog and cat group. For much of the Paleocene, predator niches were filled by the sharp-canined "archaic" condylarths like *Eoconodon* from New Mexico. Carnivorans did them one better, by evolving a new dental utensil for cutting flesh and breaking bone:

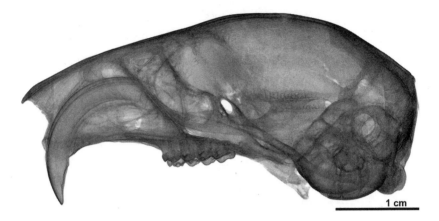

1 cm

X-ray image of a red squirrel skull showing the extremely long and looping incisor, whose root extends far into the jaw. (photo by Ornella Bertrand)

enlarged cheek teeth (either premolars or molars) that resemble knife blades. There are four of these so-called carnassial teeth in the mouth, one on each side of the upper and lower jaws, and the corresponding upper and lower pairs work against each other as carnivorans bite. Dare to peek into your cat's mouth, and you'll see the menacing carnassials, which have crowded out many of the other teeth. Watch a dog chew a bone, and you'll see it use the carnassials on the sides of its mouth, rather than its front canines and incisors, to get to the marrow. With their new razor teeth, carnivorans supplanted the "archaic" meat-eaters, and they have remained at the top of the food chain—as lions, tigers, hyenas, and wolves—ever since.

Not too long after the PETM, and certainly by the time the Messel mammals were getting suffocated by lake gases, ecosystems would have been familiar to us. If we were suddenly transported to the mid-Eocene, things wouldn't have seemed *too* strange. Sure, a brontothere or chalicothere might alert us that something was amiss, but there would have been horses, and primates, and rodents and flesh-eaters that resembled small dogs. We must remember, however, that this was the situation on the northern continents: Asia, Europe, and North America. The southern lands were separated from the north by oceans, as they had been since the Cretaceous, and they had very different ecosystems. South America, in fact, was an island continent at the time—and its mammals were undergoing an evolutionary drama all their own.

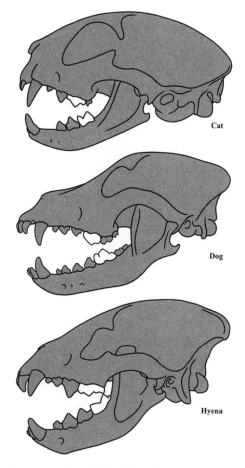

The blade-shaped carnassial teeth (in white) in carnivoran mammals. (illustrated by Sarah Shelley)

INDIGENOUS PEOPLE IN South America, like the Sioux of the North American plains, occasionally came across large petrified bones. Revered as "Thunder Beasts" by the Sioux, such bones were reviled by southern tribes, who considered them primal giants smitten by God for romantic improprieties. These stories were conveyed to the Spanish conquistadores who brutally colonized much of South America beginning in the 1500s, and then later to Catholic missionaries.

In 1832, another Christian traveler turned up on the Atlantic coast of what is now Argentina and Uruguay—a twenty-three-year-old Englishman of blue-blooded bearing, recently graduated from Cambridge. After quitting medical school in Edinburgh, he was pressured by his father to study theology. His new degree fated him to a quiet life as an Anglican county parson, to him a most dreary destiny. When presented with the chance to sail around the world on a ship called the *Beagle*, the young man jumped at it. It wasn't a glamorous post: he would be the dining companion for the ship's captain, who wanted another posh person to talk to, to save the indignity of eating with his working-class sailors. As the ship left Plymouth in late 1831, nobody could have predicted how the next five years would unfold, and how the aimless preacher-to-be would one day shake the core of Western civilization by writing about what he observed on the journey.

Charles Darwin's voyage on the *Beagle* has become mythical. In most tellings, the heroic climax occurs on the Galápagos Islands, off the west coast of Ecuador, where Darwin is portrayed as undergoing an awakening, where the many species of finches—each one living on a different island, with its own unique beak, specialized for eating a certain food—revealed to him that species evolve through natural selection. But curiously, Darwin didn't lead with that story when writing the *Origin of Species*. Instead, his opening paragraph alludes to something else he saw, on the mainland of

South America: fossils. As the *Beagle* glided along the coast and dropped anchor at port after port, Darwin would get out and venture inland. On several of these trips, he collected the fossilized bones of large mammals. Not an expert in anatomy, he sent them back to England, where they were studied by another young naturalist who was at the time a friend, but would later become Darwin's fiercest critic: our recurring villain, Richard Owen.

Many of these mammals were immediately recognizable even to a nonspecialist like Darwin. They were not giant humans, as the local legends held, but the bones of sloths and armadillos—two peculiar placental mammal groups that live across South America today, but which Darwin knew did not live in Europe. They were not, however, identical to modern sloths and armadillos: they were in many cases larger, the sloths strikingly so, and they clearly belonged to different species. These things excited Darwin. Here were strange extinct mammals that were obviously similar to the current inhabitants of South America, but unknown practically everywhere else, alive or dead. They pointed to, in the words of Darwin's opening paragraph, "relations of the present to the past inhabitants." More than the finches, Darwin saw this continuity of mammal descent—"this wonderful relationship in the same continent between the dead and the living"—as the keystone evidence for his theory that species changed over time.

Some of his other mammals, however, flummoxed Darwin. One of these, which Owen named *Macrauchenia*, was about ten feet (3 meters) long and weighed over a ton, with a long neck and stretched legs. It kind of looked like a camel, but its feet were much bigger and sturdier, more like a rhino. Even weirder was a creature Owen called *Toxodon*, which also weighed over a ton. Its stocky body shouted out rhino or hippo, but it had high-crowned and ever-growing teeth like a rodent, and backward-shifted nostrils that recalled an aquatic animal like a manatee. Darwin called it

Charles Darwin's bizarre South American Ungulates: *Toxodon* (above) and *Macrauchenia* (opposite). Drawings from William Scott's classic 1913 monograph. (photos by Hans Püschel)

"perhaps one of the strangest animals ever discovered," and he speculated that its bewildering combination of anatomical features meant that groups of mammals today classified into separate families were once "blended together." It was yet another observation from his voyage that got him thinking about how species changed over time.

All philosophical considerations aside, what exactly were these mammals, and how should they be classified? A few decades after the *Beagle* voyage, a band of brothers attempted to answer these questions. The Ameghinos were born in Argentina to poor Italian immigrants, and unlike wealthy gentlemen of Darwin's breed, they had to support their work through honest labor. Juan ran a bookshop, whose profits funded Carlos's trips across Patagonia, where he collected fossils that Florentino studied. Carlos discovered thousands of mammals, including many that Florentino recognized as similar to Darwin's *Macrauchenia* and *Toxodon*. These were not rarities; rather, there was an entire group of strange South American mammals that couldn't be pigeonholed into any of the

familiar northern groups. Florentino thought they were Cretaceous in age and regarded them as a primordial stock ancestral not only to today's South American species, but to all mammals. Others, though, started to see them as exotics unique to South America, a view that solidified once it became clear that they lived after the Cretaceous, from the Paleocene to the very recent, a time when South America was an island unto itself.

These mammals became known as Darwin's South American Ungulates, because many of them walked on hooves—a hoof being the modified ungual, or last bone, on each finger or toe. Darwin's Ungulates were astoundingly diverse. Hundreds of species have been found, ranging in size from things that would make a nice lap pet to ones approaching three tons in weight. Various species boast features of antelopes, camels, horses, rhinos, hippos, elephants, rodents, and rabbits, often in unexpected combinations, as if evolution had cut up a bunch of northern species and glued them back together in new arrangements. Darwin's *Macrauchenia* and *Toxodon* are prime examples, as is the massive *Pyrotherium*, which has upper and lower tusks and a trunk, like an elephant, on the body of a hippo. Others resemble northern species, but in a superficial way, by possessing a singular feature of a northern group, but not sharing the rest of the skeleton. *Homalodotherium*, for instance, has scythe claws on its hands like those hallucinogenic knuckle-walking chalicotheres. Yet others took the specializations of northern species to whole new levels. The dainty *Thoatherium* walked on only a single hoofed toe, like a modern-day horse, except that horses retain traces of the two digits on each side of their main toe, whereas *Thoatherium* does not.

Darwin's Ungulates persisted for well over 60 million years. They nearly made it to the present, but the last survivors were among the victims of the Ice Age extinctions around 10,000 years ago, of which more later. Many were runners that sped across the land on their hooves, others were jumpers or diggers, or flat-footed plodders, or semiaquatic waders. All of them, it seems, ate plants, but some were specialized for eating soft leaves, others grittier and tougher vegetation. A few of them managed to hop over to Antarctica during the latest Paleocene or Eocene, when spindly fingers of land from each continent ever so slightly touched before inching away. But otherwise, the Ungulates are known only from South and Central America. Despite a few false alarms, not a single bone or tooth of theirs has surfaced anywhere else (with one exception, which proves the rule, to which we will return).

None of this answers the most fundamental question: What are Darwin's Ungulates? This has nagged at paleontologists ever since the *Beagle* returned to England. There has been huge debate about where they fit on the family tree, and who their ancestors may be. Might they be related to the hoofed mammals of the northern continents, like the odd-toed perissodactyls or even-toed artiodactyls? They do have hooves, but as we've seen, hooves evolved many times in mammal history, so they alone are not reliable indicators of genealogy. Might they instead be related to other heavy mammals, like elephants, or odd offshoots of another group, like rodents? Or maybe they don't have anything to do with northern placentals, but occupy their own branch on the family tree that diverged long ago.

Only in 2015 was this mystery put to rest. Two groups of molecular biologists were able to extract proteins from Darwin's own *Macrauchenia* and *Toxodon*. Usually such soft tissues are difficult, if not impossible, to find in fossils, but both of Darwin's mammals made it to the Ice Age, so their bones are preserved in much more pristine conditions than older skeletons from the Paleocene or Eocene. When these proteins were entered into a database, and used to construct a family tree, both species grouped together with the northern perissodactyls. Two years later, this was corroborated by an even stronger type of evidence: DNA from *Macrauchenia*.

The paternity test had spoken, and Darwin's Ungulates—at least most of them—were close cousins of horses, rhinos, and tapirs. They likely evolved from "archaic" placental ancestors, like the New Mexican condylarths, which jumped across islands spanning North America and South America in the Paleocene. Then, as the Paleocene turned to the Eocene, South America became properly isolated from North America, and the northern and southern tribes went their separate ways. Those in the north migrated during the PETM, dwarfed and then got bigger in North America, and spread across Asia and Europe, where some caught a taste for water lilies along the Messel lakeshore. Those in the south were free to evolve on their own, and in doing so, they acquired some unusual features, but also converged on the hooves, tusks, trunks, single-toed-feet, and ever-growing teeth of their brethren to the north. Because they were disconnected from the northern bloodline—a separate experiment in evolution—Darwin's Ungulates developed slightly different versions of their convergent features and put them together in different permutations. As they continued to evolve from the Eocene until the Ice Age, Darwin's Ungulates became more and more different from their long-lost northern cousins.

Darwin's Ungulates are one part of an insular South American mammal community, which developed, in the words of George Gaylord Simpson, in "splendid isolation." Some of these mammals, like the Ungulates, are now extinct. Others remain characteristic components of the Amazon rain forests, Andean meadows, and Patagonian pampas grasslands today.

Among them are Darwin's other fossils: sloths and armadillos, which along with anteaters are part of the broader group Xenarthra. Recall that Xenarthra is one of the four main subdivisions of the placental mammal family tree. They split from the tree near the base, making them some of the most primitive placentals. Their tongue-twister name refers to the extra joints—"xenarthrous articulations"—between their vertebrae, which strengthen and stabilize their backbones, probably an adaptation retained from their digging ancestors. This ancestor may have skipped over from North America during the latest Cretaceous or early Paleocene, perhaps with the ancestor of Darwin's Ungulates, and then diversified on the South American island, leading to today's around thirty xenarthran species.

Xenarthrans are some of the most photogenic mammals of all. Sloths are chastised for their low-energy lifestyle, but there's no denying how adorable they are, hanging upside down from trees with their gangly clawed limbs, munching on leaves, their fur teeming with symbiotic green algae that helps them blend into the canopy and escape the glare of jaguars. Armadillos are memorable in another way: they have distinctly unmammalian bodies, covered in bony tiles called osteoderms, which grow in the skin and fit together like the panels of a soccer ball. If a jaguar tries to attack, some armadillos can roll up into a rock-hard ball and tough it out.

The jaguars that pester sloths and armadillos today are more recent arrivals to South America—immigrants two or three million years old. There were no cats, dogs, or bears prowling South America when it was an island continent. Rather, for tens of millions of years, the xenarthrans and Darwin's Ungulates were tormented by a totally different type of mammal, which filled predatory niches from the Paleocene to a few million years ago. Their identity is surprising: they weren't even placentals. They were metatherians: members of the marsupial line, which raise their microscopic babies in pouches. We nearly dropped metatherians from our story when they went extinct on the northern continents, but in South America (and later Australia) they were given new life, reborn on outposts that placentals had yet to fully seize. On their new island kingdoms, they were able to earn back the prominence they started to enjoy in the Late Cretaceous, before the asteroid changed everything.

The South American metatherian carnivores are called sparassodonts and were first described by Florentino Ameghino in the late 1800s. If you saw them alive, you probably wouldn't recognize them as metatherians, at least without checking below the belt for a pouch. They didn't look like kangaroos or koalas—marsupials we are familiar with today—but were dead ringers for weasels, dogs, cats, hyenas, and bears, all placentals. One of them, *Thylacosmilus*, had enormous saber canines on its upper jaws, which it used to rip open the bellies of its prey, to feast on the internal organs. You would swear it was a saber-toothed tiger, that famous giant Ice Age cat. This is another example of convergent evolution—one of the most striking in the entire fossil record. True carnivorans like dogs and cats weren't able to reach South America in the Paleocene and Eocene, but metatherians were, so they imitated the placentals. Or maybe it was the placentals up north who were mimicking the metatherians? Regardless, sparassodonts like *Thylacosmilus* were eventually replaced by jaguars and other actual placental predators that moved southward, but many of their cousins remain in South America, as about one hundred species of opossums and other marsupials.

The marsupial sabertooths and other sparassodonts ate Darwin's Ungulates; we know this from bite marks matching their teeth, found on their prey's bones. Sloths also would have been tasty targets, as would armadillos—if they could crack one open. And there were other things the pouched predators snacked on: big-brained tree-swingers with gangly limbs, and bucktoothed chubbers that burrowed in the leaf litter and paddled in the jungle streams.

Primates and rodents. Real ones—placental ones—and not strange marsupial versions. Where did they come from?

The paternity test gives us the answer, and it's astonishing. They came from Africa. Family trees made from both DNA and fossil evidence nest the South American primates and rodents within diverse African groups. Thus, they were immigrants, from a continent that had split from South America tens of millions of years earlier during the Cretaceous, and which during the Eocene—

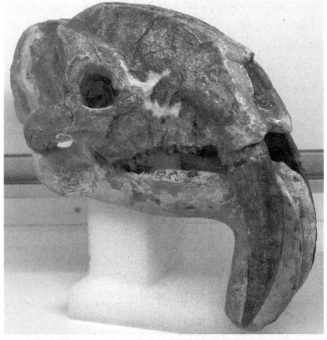

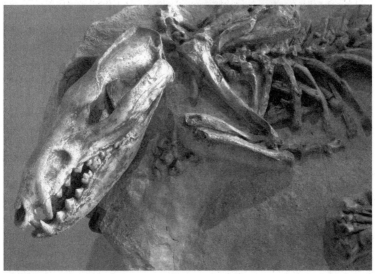

Predatory sparassodont marsupials: the saber-toothed *Thylacosmilus*
(top) and *Lycopsis* (bottom). (photos by Jonathan Chen and
Ghedoghedo, respectively)

when the migration happened—was separated by at least a thousand miles (fifteen hundred kilometers) of open Atlantic Ocean.

Africa was also an island continent in the Eocene, home to its own bespoke placental fauna of elephants and other afrotherians. Sometime after the PETM global warming, primates and rodents journeyed to Africa from Asia. This makes sense: only a narrow Tethys seaway, the forerunner of the Mediterranean, separated the two, and the European islands would have made stepping-stones in between. What makes little sense is how the primates and rodents could then move westward from Africa to South America. There simply were no land routes, and the refugees must have dispersed across the water. They probably floated on rafts of rotting vegetation, flung off the African coast by storms, which then washed up in South America. Maybe they hopscotched across a few islands on the way, or maybe they stayed on their lifeboats the entire time. Either way, they would have endured weeks bobbing in the waves, exposed to the sun, with little or no food. Like so many immigrants, they must have been hearty and resilient—traits that primed them for success in their new, distant home.

It all seems so improbable. And it was. But we've observed today that small mammals can cross water on foliage rafts and colonize new lands. Biologists have a term for these long-distance exoduses: waif dispersals, a waif being a slightly derogatory name for a homeless or orphaned child that leaves their wretched life for a faraway place. I prefer to use an analogy from one of my favorite sports, football (or American football for my international readers): a Hail Mary dispersal. At the end of the game, with only seconds to go and needing to score to win, but with most of the field in front of them, the losing team is desperate. The quarterback really has just one option: heave the ball all the way down the field and hope for the best. Usually the pass falls short, but every once in a while, it reaches the arms of a receiver in the end zone.

Touchdown! Maybe it works one time out of a hundred. But given enough time and enough chances, the improbable can become reality. The football team scores; the plant raft, and its mammals, reaches the other side of the ocean.

The migration of primates and rodents from Africa to South America was one of those fluke events that changed the course of mammalian history. It is because of this unlikely journey that we have New World monkeys and caviomorph rodents today. These monkeys—some sixty-plus species—are part of the fabric of Central and South American jungles. Some, like the howler monkeys, drown the rain forests in their cacophony of screams. Others, like the spider monkeys, are the only primates that hang from trees with prehensile tails, while the pygmy marmosets, half a foot long and weighing less than half a pound, set the record for primate smallness. The caviomorphs are even more diverse, with hundreds of species burrowing, climbing, running, and swimming across basically every environment South America can throw at them. Among them are the velvet-furred chinchillas and the largest rodents alive today, the dog-size capybaras, plus extinct ones that were much bigger, like the cow-size *Josephoartigasia*, named after the father of Uruguayan nationhood. And guinea pigs. My childhood pets—and maybe yours too?—were descendants of the Eocene rafters.

Traversing an ocean was an act of fortitude and grit, of which I can scarcely comprehend. But it's not even the wildest thing that mammals did during the Eocene. While monkeys and rodents were riding the waves, other placental mammals were inflating to gargantuan sizes, still others were evolving wings and taking to the skies, and some were turning their limbs into flippers, so they could make a much shorter but more remarkable migration: from running on land to living fully in the water.

7

EXTREME
MAMMALS

Deinotherium

HIKERS FIND BONES OF BIGGEST ANIMAL EVER?
July 25, 2831

NEW MIAMI (wire services)—A family backpacking in the Florida Desert stumbled upon giant fossilized bones, which might belong to the largest animal ever to live on Earth.

Paleontologists called to the scene estimate that the beast was over 100 feet long and may have weighed over 100 tons.

"It's astounding," said Professor Lola Bricker of the New Miami Institute of Climate and Environment. "I've never seen anything so huge. It breaks the rules for how big we thought animals could get."

The scattered bones cover an area larger than a football field. They belong to a still unnamed creature with a long, tubular body that resembled a submarine.

"If my calculations are correct, it is at least twice as big as any animal living today," Bricker told reporters, as she stood next to one of the animal's ribs, which dwarfed her by at least a foot.

Scientists say they have found pieces of the animal's front limbs, which look like the fins of a fish, but so far have not recovered any trace of the hind limbs. Its massive head is nearly intact, with bow-shaped jaw bones that lack teeth.

"The family, all four of them, really did fit inside of its mouth," Bricker said, referring to a widely circulated photograph taken by the hikers when they found the bones, which many pundits dismissed as a forgery.

What the beast ate and how it moved remain a mystery. Scientists say that the bones were found inside of mud rock, formed on the floor of a seaway about 5,000 years ago when

Florida was a lush semitropical peninsula separating the Gulf of Mexico and Atlantic Ocean.

"I'm confident that it lived in the water, and I can tell from its ear bones that it was a mammal, but that's all we really know right now," Bricker said.

She plans on excavating the bones, and reassembling them in her lab, but worries she will not have the resources.

"I'll need a team of at least twenty people, and we'll need probably six months to dig up all the bones, and then I'll need to find a space big enough to study them at my university," she cautioned, pleading for a wealthy benefactor to fund her research.

Such effort will be worth it, she argues, because finds like these inspire children to become interested in nature.

"Could you imagine seeing one of these things alive? Sharing the same world with them? It boggles the mind," she said.

Before removing the bones, Bricker hopes to narrow down the identification of the behemoth.

"My colleagues may scoff at this, but sagas from a millennium ago mention leviathans that lived in the oceans, called whales. Some stories talk about a blue whale more than 100 feet long. We've always thought they were a myth, but maybe not."

THIS MOCK NEWS ARTICLE, OBVIOUSLY, is fictional, and a bit facetious. Hopefully Florida does not become a desert, and we don't lose touch with our recorded history, and whales do not go extinct! But imagine if they do, and future generations find their fossilized bones. They would surely marvel at them like we do at gargantuan dinosaurs: as superlative animals from ages past that we only wish we could have seen in the flesh.

What many of us—me included, to be honest—don't often appreciate is that there are many superlative animals alive right now, which share the earth with us. Many of these are mammals. The blue whale is the most extreme of these "extreme mammals." It is not merely the largest mammal alive today, but the largest living animal, period. Nobody has ever found a fossil of anything bigger, which means that the blue whale is the all-time record holder, the heavyweight champion of the history of the world.

It's a simple but profound statement that bears repeating: the biggest animal that has ever lived is alive *right now*. Of all the

The blue whale, the largest animal to ever live on Earth. Skeleton
on display at Natural History Museum London (opposite) and whale
paleontologist Travis Park posing next to a skull (above). (photos by
Jan Beránek and Travis Park, respectively)

billions of species that have lived during the billions of years of Earth history, we are among the privileged few that can say such a thing. How glorious is it that we breathe the same air as a blue whale, swim in the same waters, and gaze at the same stars?

As you read this, blue whales are cruising the oceans—every ocean, as their range is nearly worldwide, except for the very north tip of the Arctic. The bulkiest elders are a shade over one hundred feet (thirty meters) long and usually weigh between a hundred and one hundred ten tons, a good twenty tons *more* than the maximum takeoff weight of a Boeing 737 airplane, and probably thirty to forty tons more than the most colossal dinosaur. Blue whale mothers birth three-ton calves the length of a speedboat, which bulk up by about fifteen tons during their half year of nursing. Adults can dive to depths of more than 1,000 feet (315 meters) and hold their breath for well more than an hour and expel a two-story column of water from their blowhole when they come up for air. With one gulp of their expandable mouths, they can take in a backyard swimming pool's worth of water, which they do several times a day, in order to gather the two tons of krill—little shrimpy crustaceans—they need to power their metabolism. They're smart and social, and their low-pitch vocalizations are the most powerful sound in the animal kingdom, able to reverberate for over nine hundred miles through the abyss.

But all is not well. It is estimated that 99 percent of the blue whale population was exterminated by whaling over the past couple of centuries. Of a community that once numbered in the hundreds of thousands at any given time, only a few tens of thousands of individuals—at most—remain. At the risk of sounding trite: let's celebrate them before it's too late, and let's do all we can to conserve and protect them while we still have the chance, before the blue whale goes the way of the *Brontosaurus*.

It's impossible to talk about blue whales without indulging in hyperbole. Same with other extreme mammals, whether they be other whales, or giant land species like elephants, or smaller mammals that have refashioned their bodies to do remarkable things, like bats—the only mammals that can fly using powered flaps of their wings, and one of only three types of backboned animals to ever figure it out (along with pterodactyls and birds). All these extreme mammals—elephants, bats, and whales—began to make their mark in the Eocene, and only through long evolutionary journeys did they reach their present splendor.

ELEPHANTS ARE THE largest mammals—in fact, the largest animals of any kind—that live on land today. The biggest species, the African bush elephant, is about ten feet (three meters) tall at the shoulders, the same height as a basketball rim. The largest males weigh between five and seven tons, roughly twice as much as a Ford F-150 pickup truck. If a single bush elephant was perched at the end of a playground seesaw, it would take about a hundred humans to balance it out. Elephants are not as big as blue whales, true, but they have to deal with a handicap that whales do not: gravity. While a blue whale can passively float in the buoyancy of water, elephants need to hoist themselves up, and move and mate and give birth, on their arms and legs.

Only three species of elephant survive today, scattered across sub-Saharan Africa, India, and Southeast Asia. They are but a sad remnant of what was once a flourishing family, with species like woolly mammoths and mastodons spread across much of the world, some of which were more than double the tonnage of the African bush elephant—making them the largest mammals to *ever* inhabit the land (well, probably, as we shall see). But before they went cosmopolitan and then declined, elephants were confined to Africa for tens of millions of years, as part of the great afrotherian radiation.

Afrotheria, as we learned in the last chapter, is one of the four main subdivisions of placental mammal genealogy. Like Xenarthra—the family of sloths and armadillos—Afrotheria branched off close to the root of the family tree, making it one of the most primitive placental groups. Also like Xenarthra, most of the early history of Afrotheria in the Paleocene and Eocene played out on a single landmass, sequestered from the rest of the world. Sloths and armadillos, along with Darwin's strange Ungulates and the saber-toothed marsupials, lived on the island continent of South America. Elephants and other afrotherians also lived on an island continent; as their name implies, this was Africa.

For tens of millions of years, Africa was an isolated fortress. After it split from the rest of Gondwana—the southern half of old Pangea—some 100 million years ago during the Cretaceous, Africa was on its own. To its west was the ever-widening Atlantic Ocean, occasionally crossed by rafting monkeys and rodents. To its south and east was the Indian Ocean, which lapped up against Antarctica and Australia and was rapidly being traversed by a smaller island—India—on its way to an Eocene collision with Asia. And north of the African island was the Tethys, the warm equatorial seaway that separated north from south. The Tethys was not an impenetrable barrier; every so often, northern animals could bounce southward across the constellation of European isles that approached the North African coast, but this would have been a challenging journey. The long night of African seclusion finally ended only about 20 million years ago, when Arabia docked with Eurasia, reducing the Tethys to its western arm—what we now call the Mediterranean.

Elephants are one of many afrotherians. Others in the family album include the aquatic manatees, the little hoofed hyraxes, the ant-eating aardvarks, the subterranean golden moles, the tenrecs of Madagascar (and nearby islands), and the elephant shrews, which look like a tiny rodent with an elephant's trunk. If this seems an odd mix to you, you're not alone. There aren't many—if any—obvious anatomical features that unite afrotherians. Maybe, some scientists have argued, they share a single peculiar tooth cusp, or nuances of how the backbone develops. Maybe. The prime evidence for Afrotheria, however, is genetic, and it is strong. The identification of an elephant-manatee-hyrax-tenrec group was one of the most startling revelations of the first DNA family trees in the late 1990s and early 2000s, and it's stood the test of time and more sophisticated whole-genome analyses. Much to the chagrin of old-school anatomists, the paternity test is conclusive, and Afrotheria is real.

The reason modern afrotherians look so different from one another is because they diversified long ago, from a primitive common ancestor, to fill a variety of ecological niches on the African island continent. This ancestor was probably a dog-size "archaic" placental that would have looked similar to the New Mexican condylarths, and it likely migrated across the Tethyan islands from the north, either during the latest Cretaceous or soon after the asteroid impact, in the early Paleocene. Once the dinosaurs were gone, many jobs in the food chain were suddenly open, and the afrotherians took advantage. They adapted to niches on the forest floor and in the meadows, up in the canopy and along the Tethyan coastlines.

This was parallel to what was happening independently up north, where another of the four main placental subgroups—Laurasiatheria—was diversifying. In the north there were hoofed perissodactyls and artiodactyls; in Africa there were hyraxes. In the north there were ant-eating pangolins and burrowing moles; in Africa there were aardvarks and golden moles. Up north there were shrews and hedgehogs, while in Africa there were elephant shrews and tenrecs—sometimes called "Madagascar hedgehogs."

And as we'll see shortly, whales evolved in the north, while in Africa sirenians (the manatee group) underwent a similar transformation from runner to swimmer. This story is another beautiful example of convergent evolution. Primitive placentals found themselves sequestered in different places in a suddenly dinosaur-free world. They were estranged from each other—unable to mix or share genes—yet as they diversified, evolution took similar paths in adapting both afrotherians and laurasiatherians to similar niches.

It is difficult to appreciate the afrotherian diversification from today's vantage point. Much of it has been overwritten by the more recent immigration of northern mammals—the ancestors of zebras and wildebeest, lions and hyenas—after Africa and Eurasia connected. But in the Paleocene, and especially in the Eocene and subsequent Oligocene interval, Africa was a kingdom of afrotherians. It would have been something to behold.

Hyraxes are mentioned in the Bible—as "feeble folk" called conies that "make their houses in the rocks"—but those of us from the northern continents would be forgiven for never having heard of them. Five species persist, only in Africa and the Middle East, either as stumpy woodchuck-like herbivores that clamber across rocky outcrops on their hooves, or tree-climbers with suction pads on their feet. In the Eocene and Oligocene, however, there were dozens of African hyraxes. They ranged from a few pounds to the rhino-size *Titanohyrax*, which tipped the scales at 1.3 tons. Some were omnivores, others had convoluted teeth for shearing plants, and some specialized on seeds and nuts. *Antilohyrax* would have mimicked an antelope as it sped through the forests on its stilt limbs, while others rooted in the dirt with their long snouts, like pigs. These hyraxes were filling niches later occupied by wildebeest, antelopes, warthogs, hippos, and rhinos. Essentially, they were hyrax versions of the familiar hoofed mammals of modern Africa.

Meanwhile, along the coast, the first sirenians were dipping their toes into the water. They began as land animals with sturdy arms and legs, which they also found useful for dog-paddling in the shallows. During the Eocene, evolution molded them fully to the aquatic realm, turning their arms to flippers and erasing their legs, substituting a big paddle tail whose up-and-down motions provided the propulsion necessary to swim, albeit awkwardly, through the currents. Sirenians were the first afrotherians to go global, as they migrated along the Tethyan shorelines and beyond, turning up as fossils everywhere from North Carolina to Hungary to Pakistan. But like the hyraxes, little of their greatness remains. There are only three modern species of manatee, restricted to the Caribbean, western Africa, and the Amazon, and a single dugong that lives in the Indian Ocean and southwestern Pacific. They're cute and cuddly emblems of endangerment: their languid lifestyle, of slow swims and seagrass feasts, makes them easy victims of fishing nets and boat collisions.

The outrageous extinct afrotherian *Arsinotherium*. (photo by Aram Dulyan)

Manatees and dugongs might succumb to extinction, if we are not careful. If so, they will sadly join a legion of other afrotherians that were once diverse and dominant, but then disappeared. The most memorable of these is *Arsinoitherium*, one of those extinct mammals in the pantheon of fantastic beasts. Looking something like a rhino on steroids, it had two enormous horns sticking out of its forehead, each one considerably taller than the rest of its head, like Marge Simpson's hair. Unlike a rhino—which is an odd-toed perissodactyl—the *Arsinoitherium* horns were made of bone, not keratin, and they were hollow, and they were angled so far forward that they surely would have restricted forward vision while walking. Maybe that wasn't too much of a problem, as *Arsinoitherium* was a hulking plant-eater that probably didn't fear many predators. It's hard to fathom any purpose for these ridiculous horns other than an overcompensating way to attract mates and intimidate rivals.

Then, of course, there are elephants. The iconic afrotherians, and even more than hyraxes and manatees and *Arsinoitherium*, a group that was once magnificent, but is now a trifle of its former glory.

Like all giants, elephants started small and humble. A transitional sequence of fossils from the phosphate mines of Morocco, studied by the French paleontologist Emmanuel Gheerbrant and his Moroccan colleagues, shows how they supersized. The oldest of these extinct elephants, called *Eritherium*, lived in the middle part of the Paleocene, about 60 million years ago. It was not much to look at: with a shoulder height of about eight inches (twenty centimeters) and a weight of around ten pounds (five kilograms), it would have been cowed into submission by most lapdogs and could have easily been pancaked by a single stomp of a modern elephant. Yet its molar teeth were beginning to show signs of an elephant staple: transverse crests, extending across the face of the

Eritherium

Daouitherium

Numidotherium

Sequence of elephant evolution. (illustration by Todd Marshall)

tooth from the tongue side to the cheek side, linking the cusps. These so-called lophs give the grinding surface of the tooth a corrugated appearance, perfect for pulverizing plants.

As the Paleocene transitioned to the Eocene, and the PETM global warming came and went, the teeny Moroccan elephants got bigger, and their tooth lophs became more prominent. First came *Phosphatherium*, about three times larger than *Eritherium*, with full-on corrugated teeth. Then *Daouitherium*, the first elephant that was properly bulking up, weighing about 450 pounds (200 kilograms). When it sauntered through the forests about 55 million years ago—long before they turned to savanna and grasslands—it was the largest African mammal that had ever existed. Both *Phosphatherium* and *Daouitherium* had procumbent incisors, a hint of the tusks that would develop at the next stage of evolution, in species like the Algerian *Numidotherium*, more than three feet (one meter) tall and a stout 650 pounds (300 kilograms). It's at this point that elephants also acquired their high foreheads, and their nostrils shifted backward, to fasten a small proboscis like that of a tapir. With a little more tinkering, evolution turned the proboscis into a trunk. Although they may look silly, trunks were game changers: they allowed elephants to procure food and water without having to move their entire bodies, which unlocked the potential of even vaster sizes.

Bigger and bigger they grew. By the early Oligocene, around 34 million years ago, species like *Palaeomastodon*, all two and a half tons of it, exceeded current-day African forest elephants (the smallest extant species) in mass. *Palaeomastodon*'s upper tusks were downturned, while its lowers stuck out from the jaw horizontally, to extend beyond the uppers in a goofy underbite. These were but one of a countless variety of tusk shapes that elephants developed throughout the Oligocene (34–23 million years ago) and subsequent Miocene (23–5 million years ago). There were species with

The extinct elephants *Palaeomastodon* (top) and *Deinotherium* (bottom). (photos by Egyptian Geological Museum and Alexxx 1979, respectively)

lower tusks resembling spatulas or shovels, others with long sets of matching upper and lower tusks that projected from the mouth like a giant set of tweezers, and a species called *Deinotherium* that lost its upper tusks but deformed its lowers into a backward curving pair that resembled a bottle opener. This assortment of shapes probably served a dual purpose: they allowed different species to specialize on different plants—for instance, the bottle-opener tusks of *Deinotherium* could hook branches off trees—and they were wielded as display features, to advertise strength or attractiveness to the herd.

Some of the Miocene elephants became giants, surpassing all of today's species in tonnage, and in doing so, they spread out of Africa when Arabia made contact with Eurasia. *Deinotherium* stood about thirteen feet (four meters) tall at the shoulders, and weighed up to fourteen tons. That's double the weight of an African bush elephant. But *Deinotherium* wasn't even the grandest elephant of them all. That title goes to a later species called *Palaeoloxodon*, which is known from fragmentary bones of preposterous dimensions, indicating a body over sixteen feet (five meters) tall and around twenty-two tons in mass. If these numbers are valid—and it's always a bit iffy when extrapolating whole-body size from isolated bones—that would make *Palaeoloxodon* the most enormous mammal that has ever lived on land. It would wrestle the record from the mammal most textbooks consider the all-time biggest: a hornless rhinoceros called *Paraceratherium*, which lived around the Eocene-Oligocene boundary, thought to be about 15.5 feet (4.8 meters) tall and weighing about seventeen tons.

Ultimately, it doesn't really matter which one holds the crown: elephants like *Palaeoloxodon* and rhinos like *Paraceratherium* were both gigantic. They were probably fairly similar to each other in overall dimensions, which speaks to a broader pattern in mammal evolution: by the Eocene-Oligocene boundary, 34 million years ago, land mammals reached the largest sizes they ever would, and from that point until the present, rhinos and then various groups of elephants have taken turns with the title belt. This suggests there is some limit to overall land mammal body size, probably dictated by a combination of factors. First is diet: the bulkiest mammals are always herbivores and are generally about ten times heavier than the largest predators they live alongside. To get epically big, you need a steady source of calories, and the best way to do this is binge on plants, which are much more readily available than meat. Second has to do with temperature: big animals are at risk of overheating, and the problem becomes more acute the bigger you get. It may be that the sizes of *Palaeoloxodon* and *Paraceratherium* are near a functional limit. Any larger, perhaps mammals couldn't take in enough food or shed body heat quickly enough. And there might be other limiting factors, too. But then again, the discovery of a single bigger land mammal fossil would cause us to reconsider.

You might be asking another question: Why didn't land mammals get as big as dinosaurs? As huge as they were, *Palaeoloxodon* and *Paraceratherium* were still not even half as heavy as the most enormous long-necked dinosaurs. There's no easy answer to this conundrum, but I suspect it has to do with the lungs. Mammal lungs are tidal; breaths go in and out, as the lung expands and contracts. We feel this every time our chest goes up and down as we breathe. Birds are different: they have a flow-through lung, as air can go through it in only one direction. This feat of engineering is choreographed by balloonlike air sacs, which connect to the lung and funnel air through it in a precise sequence. When a bird breathes in, some of the oxygen-rich air goes directly across the lungs, while the rest is shunted into the air sacs. Then, when the air sacs contract, the still-oxygenated air inside is passed across the lung during exhalation, meaning birds—and the giant dinosaurs with the same lungs—take in oxygen while breathing in *and* out. This means dinosaurs got more oxygen with each breath than a simi-larly sized mammal. And there's more: the air sacs extend through the body and even into the bones, acting as an air-conditioning system, and lightening the skeleton. The end result: large dino-saurs were more efficient breathers, could cool their bodies easier, and had lighter and more limber skeletons. This, I think, is why no land mammal has been able to approach their titanic sizes.

While they may not be dinosaurian in their proportions, today's African and Indian elephants are—by any objective measure—still really, really big. Their entire body is tailored to their size. Their limbs are beams, like Greek columns, to support their girth. Their comically floppy ears are cooling panels, helping them discard excess heat. They have turned the traditional lineup of mammalian teeth into a simplified set of elongate tusks and shoe-size molars and premolars, which are so big that only one or two can fit in each jaw at a single time. This necessitates a whole new way of growing their teeth, called serial replacement. Their jaws are conveyor belts: new molars erupt at the back and grad-ually move forward, getting worn down as they chew and then falling out of the front of the jaw, replaced by the next tooth be-hind. All so they can eat lots and lots of plants—a few hundred pounds each day—to maintain their stature. And their largesse has consequences: they eat so much, they can uproot trees and turn savannas into grasslands, and when they dig for water, they can create waterholes, which become the lifeblood of new mini-ecosystems.

Don't, however, mistake elephants as dim-witted goliaths. What makes them truly impressive is that they combine brawn with brains. Their brains are massive in absolute terms, as would be expected of a giant animal. But they are also huge in relative terms: the volume of an elephant's brain, in comparison to its body size, is in the same zone as primates. With primates and whales, elephants are members of the Big Brain Club: the mammals with the proportionally largest brains compared to their weight, capable of many feats of intelligence. Elephants have exceptional long-term memory, manufacture tools with their trunks, exhibit complex social behavior and problem-solving abilities, and can recognize themselves in a mirror. They converse with each other over long distances by using either low-frequency infrasound vocalizations or seismic communication: they talk by making small earthquakes! Perhaps, biologists debate, elephants even display a form of empathy, showing concern for sick or dying members of the herd, and taking an interest in the bones of their ancestors and cousins.

For all their skills, though, there's one thing elephants can't do. Although the silver-screen hero Dumbo was able to soar through the air by flapping his ears, the real-life elephants of Africa and Asia, obviously, cannot fly. Nor can any other mammal that has ever lived, except for one group: bats.

WHEN I DID my PhD in New York, I had an office at the American Museum of Natural History, on the west side of Central Park. I was there for the dinosaurs: the AMNH houses one of the world's largest collections, including some of the most famous skeletons of *T. rex*. As I walked to the dinosaur storehouse, through the high-vaulted hallways lined with wooden cabinets, I would occasionally run into Nancy Simmons.

I didn't know Nancy well at the time and probably spoke to her on only a couple of occasions during my nearly five years at the museum. And I regret it, because as I've gradually shifted my focus from dinosaurs to mammals, and have become more familiar with her research, she's come to be one of my scientific heroes.

Nancy is a paleontologist, too, although only part of the time. As a graduate student, she began her career by becoming one of the world's experts on multituberculates, those bucktoothed, flower-munching mammals that lived underfoot of *T. rex* in the Cretaceous. Then, remarkably, she switched gears and became a world expert on bats. A few months before I began my PhD, she published a sensational discovery, which earned the cover of *Nature*: the world's oldest and most primitive bat, called *Onychonycteris*, from the early Eocene, about 52.5 million years ago. But most of the time she isn't describing old bones; she is bushwhacking through the jungles of Southeast Asia and the Neotropics, searching for new species of living bats, and collecting samples of bat blood and other tissues, to harvest DNA to make family trees.

Bats need no introduction. Whether you adore them or are creeped out by them, they're instantly familiar as the only mammals that properly fly. Some other mammals can soar or glide on membranes of skin, like flying squirrels and dermopterans (the inaccurately named "flying lemurs," which are not primates), or the extinct haramiyidans of the Jurassic and Cretaceous, like the secret fossil I saw in that Chinese workshop a few chapters ago. Bats, however, are the only mammals that employ *powered* flight: by actively flapping their wings, they generate the lift and thrust needed to propel themselves through the air.

Flapping flight is no easy task, which is why evolution has managed to do it just three times in the entire history of vertebrate life. Each instance was a different experiment in how to get airborne. Pterodactyls stretched their ring finger to support a giant sail of skin. The dinosaur ancestors of birds lengthened their entire arms to anchor an airfoil of feathers. Bats, on the other hand (pun intended), elongated most of their fingers to create a hand-wing. This bat wing is an ingenious design: the skin extending between the fingers is thin and flexible, and it flaps as big muscles attaching to the breastbone contract. This allows bats to fly quickly—some have been clocked at 100 miles per hour (160 kilometers/hour)—

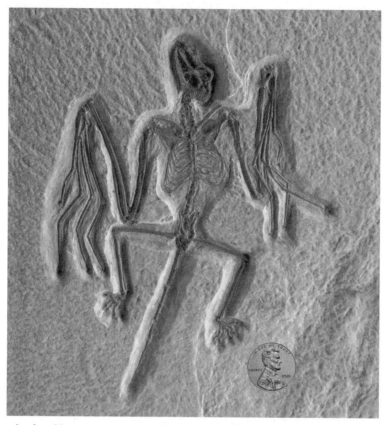

The fossil bat *Onychonycteris*, described by Nancy Simmons. (photo by Matthew Dillon)

and to maneuver seamlessly around obstacles, a useful talent for animals active mostly at night.

Flight is a bat superpower. It gives access to habitats and food inaccessible to earthbound mammals and is surely the overriding reason why bats are so diverse and abundant today. One out of every five mammal species alive right now is a bat—some fourteen hundred total—a diversity exceeded only by rodents. Not only are there many bat species, but they are aces at coexisting: in the tropics, more than a hundred species are known to inhabit the same ecosystems. And some of these bat species are enormously

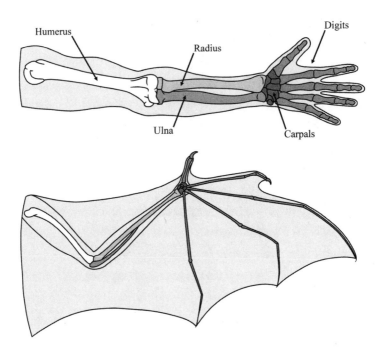

The wing of a bat next to the arm of a human. (illustrated by Sarah Shelley)

abundant. One reason bats seem so spooky is that they often live in dense colonies, where innumerable individuals roost by hanging upside down by their feet, in caves or under bridges, piling upon one another so that from a distance they look like a vast blanket, their animal nature betrayed by their stench and copious amounts of dripping guano. I will always remember the spectacle of visiting one of these colonies, where some *one and a half million* bats room together under the Congress Avenue Bridge in Austin, Texas, all emerging at dusk, in a flowing wave of wingbeats, for their evening insect hunt. It was a biological performance unlike anything I had ever seen. But it's because of this communal living that bats are so susceptible to infection and disease, and why colonies are notorious incubators for viruses that later jump to humans.

How did bats evolve their aerial superpower? Surprisingly, we know very little about how they developed wings and flight. What we do know, from the DNA paternity test, is that bats are members of the northern group Laurasiatheria, and they cluster near carnivorans (dogs and cats) and hoofed mammals (odd-toed perissodactyls and even-toed artiodactyls) on the family tree. It goes without saying, but a bat looks nothing like a horse or a dog, so there must have been a transitional sequence of extinct species that morphed from a ground-dwelling mammal with walking limbs to a hand-winged flier. The problem is, we don't have many fossils depicting this evolutionary transformation. The first bat skeletons that turn up in the Eocene, like Nancy Simmons's *Onychonycteris*, already look like bats. They have the unmistakable outline of the Batman logo: small head, compact body, wee tail, and broad wings reaching out from both hands. These were featherweight animals, whose delicate skeletons could be preserved only in exceptional circumstances. *Onychonycteris*, for instance, was buried in the peaceful depths of a Wyoming lake, similar to how bats were entombed in Germany's Lake Messel later in the Eocene. The forebears of these fortunately preserved Eocene bats were probably even smaller and more liable to break apart after death, and we haven't been lucky enough to capture them as fossils . . . yet.

While we are still missing fossils of the actual bat ancestors that first sprouted wings and took to the air, there are things we can infer from *Onychonycteris* and the other early bats. When Nancy and her colleagues described *Onychonycteris*, they realized it was much more primitive than current-day species. Sure, it had a hand-wing and a large attachment ridge for flapping muscles on the breastbone, so it could fly. Its wings, however, were uniquely shaped: they were short and stout, compared to the broader and more graceful wings of most modern bats, meaning that *Onychonycteris* was less maneuverable and would need to fly very fast to create enough lift to stay airborne. It probably had a bizarre flying style, alternating between flapping and gliding, fluttering about like a drunken butterfly. There were other anatomical oddities, too. Intriguingly, if you took away its hand-wing, *Onychonycteris* had the trademark body proportions of a glider, like a flying squirrel. Plus, it had sharp and curved claws on all digits—unlike modern bats, which lack claws on their wing fingers—a sign that it was an agile climber that scrambled up trees on all fours.

Put the clues together, and this all suggests that bats evolved from a tree-living glider ancestor, which became a flapping flier by lengthening its fingers into a hand-wing. The new wing changed everything: compared to the skin flaps of gliders, it had about twice the surface area for lift and thrust generation and was thus capable of more precise and longer flights. Even so, the first powered flights would have been awkward. *Onychonycteris* had lower aerodynamic efficiency than extant bats—it needed to work harder to produce lift, and it never could maneuver as easily around branches or other obstacles. But a threshold had been crossed, and now bats were flapping, and from that point forward natural selection could mold them into better and better fliers, with more expansive wings that endowed more lift, longer flights, and greater maneuverability.

It's a good story, and it makes sense based on what we currently know, but ultimately we need the fossils of intermediate species—spanning the ground-to-gliding-to-flapping transition—to prove it.

Regardless of exactly how bats acquired their hand-wings and started to flap, it's clear that once flightworthy, they quickly spread around the world. By the end of the early Eocene, just a few million years after the PETM global warming, bats were leaving their fossils in North America (like *Onychonycteris*), Europe, and Africa. Fossils even pop up in Australia, already an island at the edge of the world like today, and India, then an island yet to couple with Asia. Bats thus became the first cosmopolitan placental mammals—the first to break the shackles of geography that had dictated the evolution of Paleocene and Eocene species up to that point. There's no mystery as to why. Because bats could fly, they could easily disperse across the ocean barriers that hindered, or outright prevented, the migrations of land mammals. They advanced so rapidly across the globe, like an invasive horde, that they reached South America and Africa—the two great island continents—before the natives were able to produce their own endemic flying forms. Alas, that's likely why there are no weird marsupial bats or afrotherian bats—just bats, of the laurasiatherian kind.

As bats dispersed over the Earth—reaching their present distribution of living on every continent except Antarctica, and on pretty much any stretch of land that isn't freezing—they also diversified their body sizes, wing shapes, flight styles, diets, and ecologies. In the middle Eocene, about 48 million years ago, hundreds of bats fell into the toxic muds of Lake Messel, and their stunning fossils are a roster of variety. There are seven known species, and their wing shapes and petrified stomach contents are testaments to their diverse lifestyles. Among them were narrow-winged highfliers that soared in the open air above the lake, broader-winged daredevils that darted through the dense vegetation of the forest understory, and others with intermediate wing shapes that hovered in the open spaces between trees. Some species ate moths and other flying insects; some picked beetles and more stationary bugs from the branches.

All the Messel bats, it seems, employed a second superpower wielded by many of today's species: echolocation. It is a high-powered biological sonar system, totally incomparable to anything in our own sensory repertoire. Bats emit high-pitched squeaks from their throat or click their tongue, listen to the echoes, and create a "picture" of the sound landscape in their brain. In this way, they can "see" in the dark—a sixth sense that unmasks lurking predators, tasty bugs to eat, and tree branches to avoid flying into. To do it, throat-echolocating bats need two things: first, a big coiled cochlea in the ear to hear the echoes; second, a firm linkage between the throat and the ear—provided by an expanded throat bone called the stylohyal that wraps around the ring bone of the eardrum—so that the nervous system can compare outgoing squeaks to incoming echoes, and to support the prominent larynx muscles that do the squeaking. These anatomical hallmarks can be observed in fossils, which is how we know the Messel bats could echolocate, but that Nancy's *Onychonycteris* could not. Echolocation, therefore, likely evolved after flight.

As with flight, the first attempts at echolocation were modest. Some of the Messel bats had cochleas barely larger than those of nonecholocating species, and were probably capable only of low-duty echo detection, useful for general orientation and avoidance of glaring obstacles. Over time, the system was fine-tuned by natural selection, turning a 1990s Game Boy, of coarse graphics and limited capability, into a modern gaming console. Larger cochleas allowed more advanced bats to "see" clearer in the dark, allowing them to do more than simply avoid things, and to actively use their sonar to hawk insects on the wing. Bats today are some of the most sophisticated aerial hunters, able to hear the fluster of a bug in pitch-darkness *and* pinpoint its position precisely enough to snatch it, while both are midflight. Echolocation was destiny for bats, their ticket to mastering the night skies, thereby excluding birds—which had evolved earlier from dinosaurs, but aside from a handful of species never developed echolocation—from conquering the nocturnal niches.

Not all echolocating bats use their sonar to find insects; some use it to source other types of food. The most infamous of these are the most feared bats of all: the vampires, which really do drink blood. The three vampire species, which live in Central and South

America, are the only mammals that feed exclusively on blood, a most unusual diet called hematophagy. Nature doesn't get much spookier than these bats, which find their sleeping victims with the one-two punch of echolocation and a brain tuned to detect breathing rhythms. They'll quietly fly toward their unknowing target—in the dark of course—and land nearby, creep slowly on all fours, use a heat sensor in their nose to locate a place on the skin where blood is flowing underneath, and then attack with their pointy teeth, lapping up the oozing blood with their tongue. Their drinking binges last for about thirty minutes, and they're careful not to remove too much blood, so as to keep their host alive for the next feed. The victims are usually birds and cows and horses, but vampires have been known to attack humans. As if they're not frightening enough on their own, vampires live in colonies, where hundreds or thousands hang together from cave roofs during the day before stalking the night. In one year, a hundred-member colony can drink the blood of twenty-five cows.

Did any of the Eocene bats suck blood? We're not sure—and that's one of so many open questions about the evolution of these animals. If you want to make a mark as a young paleontologist, go and find fossils of Paleocene or earliest Eocene bats. And if you do, may I suggest that you name one of them after Nancy, the way that our New Mexico team formally named the "Primeval Beaver"—discovered by Carissa Raymond in the chapter-opening story several pages back—*Kimbetopsalis simmonsae*, in honor of Nancy's early-career work on multituberculates before she turned to bats. Until transitional fossils show how bats changed from runner to glider to flier, considerable mystery will remain. But there was another mammal group that underwent a very different but equally remarkable transition, and in this case, there are fossils to show us how it happened, step by step.

THE PYRAMIDS OF Giza have stood tall for over four millennia, battered by Saharan sun and wind. They, like many ancient Egyptian monuments, are built from hearty material: limestones, with a pedigree much older than the pharaohs. These hard, calcium-rich rocks were formed in the warm and quiet waters of the Tethys seaway more than 40 million years ago, during the Eocene. The creatures living in this lost world—long before northern Africa became land, and then a desert, and then a cradle of human civilization—left their marks as fossils. The very blocks of the Giza pyramids are heaving with bits of algae, shells of microscopic plankton, and petrified snails. About one hundred miles southwest, near the palm-lined Fayum Oasis irrigated by the Nile, much grander fossils protrude from the Eocene strata.

Wadi al-Hitan, in Arabic, translates as "Valley of the Whales." It is no metaphor. Whale skeletons litter the desert pavement, as if an Eocene seafloor has been thrust onto the land and turned to stone. The scene is jarring, a clash of reality: thousands of whales, sitting on the sand, roasting in the heat, in one of the driest parts of the world, more than a hundred miles from the nearest sea where whales now swim. They are whales out of water, indeed, almost as unfathomable as if they were plonked onto the craters of the moon. Many of the skeletons are as pristine as ancient bones can be: enormous bodies preserved in perfect arrangement, with toothy heads connected to gently arching backbones, ribs sticking out sideways. If you follow the serpentine contours of the trunk, flattened flippers emerge from the shoulders, then the back vertebrae transition to the tail, and finally, when the tail has started to taper, some small bones appear, unmoored from the rest of the skeleton.

A pelvis, and a leg.

Right: Fossil whale skeletons in the Egyptian desert at Wadi al-Hitan. (photos by Ahmed Mosaad, top, and Mohammed ali Moussa, bottom)

It's weird. You won't see legs on any modern-day whales, which have no need for them, because they use their fore-flippers to steer and fluked tail to swim.

The Wadi al-Hitan skeletons—belonging to the fifty-foot-long *Basilosaurus*, its smaller cousin *Dorudon*, and a few other species— are no normal whales. They harken to a time when whales once walked. Although they themselves lived in the ocean, the Wadi al-Hitan species retained the legs of their land-bound ancestors, which over some 10 million years of the Eocene ventured into the water, modified their bodies from the long-limbed frame of a sprinter into a submarine-shaped swimming machine, and never returned to terra firma. In doing so, they became ever bigger and ever more adapted to the waves, gradually losing body parts typical of land mammals and acquiring those necessary for a fully aquatic existence.

This is a prime example, heralded in biology textbooks everywhere, of a major evolutionary transition: the change of one type of organism into something that looks and behaves radically different, with a refashioned body suited to its new lifestyle. It is not hypothetical; we have a chain of actual fossil skeletons, including *Basilosaurus* and *Dorudon*, showing the metamorphosis of whales, stage by stage. If anybody tries to claim there are no "transitional fossils" or "missing links" in the fossil record, tell them about the Walking Whales.

Before we dive into the history of whales, let's get something out of the way, and state the obvious. Whales look like fish. There is no shame in mistaking them for giant fish; I was many years into school before I grasped that whales are mammals. They resemble fish because convergent evolution has molded their bodies to the same way of life: swimming, eating, and reproducing in the water. This means whales don't much resemble other mammals, because the Eocene transition from land to sea erased or repurposed many of the features and behaviors of the stereotypical "mammal." Whales are thus the most unmammalian of all mammals—but mammals they are. If you look closely, you can see the signs. They have the single lower jawbone and three middle ear bones that define mammals, they have mammary glands and feed their young with milk, and although their skin is smooth, they maintain vestiges of hair—as whiskers around the mouth, which in some species are present only in babies. Furthermore, whales are placental mammals: they birth large (often *very* large) and well-developed offspring, nourished by a placenta. If you don't believe it, the proof is in their belly buttons—where the umbilical cord, extending from the placenta, attached in utero.

So whales are mammals, but what type of mammals are they? Or put another way: What sort of land mammals did they evolve from? This has flummoxed thinkers for thousands of years. Aristotle recognized that whales were not fish, but in those days long before evolutionary theory, he had no way of conceptualizing that they arose from other species. In a rare moment of embarrassment, Darwin speculated that whales might have evolved from bears that swam, mouth open, skimming insects from the water surface—a notion so ludicrous he removed it from later editions of the *Origin of Species*. When he drew his famous mammal family tree in 1945, George Gaylord Simpson punted on the question, sequestering whales on their own branch, far separate from other groups. Finally, during the latter half of the twentieth century, a promising candidate for whale ancestor materialized in the fossil record: Paleocene "archaic" placentals, called mesonychids, whose thick, sharp teeth were similar to those of fossil whales. This was always tenuous, however, as relationships drawn on dental similarities often are, because of the problem of convergent evolution.

The real answer emerged at the end of the twentieth century. The riddle was solved with both fossils and the DNA paternity test, which in this case—unlike so many other cases we've talked about previously—came to the same conclusion. Whales are artiodactyls, members of the even-toed hoofed mammal family.

Initially, DNA showed that whales cluster on the family tree with cows, sheep, hippos, camels, deer, pigs, and other cloven-toed herbivores. Then, fossils proved it. In 2001, several primitive Walking Whale skeletons from the Eocene were found in possession of the most trademark artiodactyl feature of all: the double-pulley astragalus bone of the ankle, with a deep groove at both ends. As we learned in the last chapter, artiodactyls developed this unique ankle when they originated at the beginning of the Eocene, in the burst of PETM global warming, as a way to run fast without dislocating their ankles. No other mammals, not even speedsters like horses or dogs, have it. Of course, modern whales don't have it, either, because they've lost all traces of their ankles. The only reason these ancient whales, which were in the process of moving into the water and had already withered their hind limbs nearly to oblivion, had such a thing was because they retained it from their artiodactyl ancestors. Like our appendix, it once served a function, and no longer did, but was still hanging around. Good thing, too, because unlike DNA an anklebone is tangible, and it immediately convinced even the most skeptical paleontologists that whales are artiodactyls.

This raises the next question: What type of artiodactyls did whales evolve from? The DNA paternity test gives something of an answer: the closest living relatives of whales are hippos. But hippos and whales are scarcely alike—can you begin to imagine what the common ancestor of a blue whale and a river hippo might have looked like? Plus, the earliest hippos lived in the Miocene, several million years after *Basilosaurus* and *Dorudon* were already gliding through the Eocene seas. This means that hippos are not the ancestors of whales, but first cousins. The actual ancestors

Indohyus

Pakicetus

Ambulocetus

Rodhocetus

Basilosaurus

Sequence of whale evolution. (illustration by Todd Marshall) were the transitional species in the fossil record, the chain that includes *Basilosaurus* and *Dorudon*. Here paleontologists can gloat: it is only the fossils, and not DNA, that reveal the story of how whales moved into the water. It's a tale of how Bambi turned into Moby Dick.

The story begins more than 10 million years before the Egyptian whales, across the Tethyan waves to the east. India was then still an island, but not for much longer. It was rapidly advancing northward through the equatorial seaway, destined for its confrontation with Asia, the first act of the closing of the Tethys. Some 50 to 53 million years ago—give or take—only a narrow stretch of tropical water remained pinched between the two landmasses. This region would soon smash and fold into the Himalayas, the suture between two immovable blocks of crust. For a few million years more, though, it was a quiet backwater, a zone of shallow and sunlit ocean shelf fed by rivers flowing out of India. This unassuming place was the crucible for one of the greatest experiments in evolutionary history.

One of many mammals marooned on the Indian island was a raccoon-size artiodactyl called *Indohyus*. It was a dainty prancer, with a dog's snout on a baby deer's body, which tiptoed through the forest on its stilty limbs, living a humble life of chewing on leaves and hiding from predators. It could outrun most of its pursuers, with a bounding gait orchestrated by its double-pulley ankle. Occasionally, though, it was startled by a more able predator, not on foot but on wing: a raptorial bird. But the little hoofed mammal had a trick: like the African mouse deer of today, it could leap into the streams and lakes and hunker down underwater. An ace swimmer it was not, but it used the water as refuge, and maybe took the opportunity to munch on aquatic plants while waiting for its tormentor to fly away. This is, admittedly, a fun bit of fiction—but it is informed by fossils, which not only illuminate the lifestyle of

Skeletons of fossil whales, showing their transition from land to see, from top to bottom: *Pakicetus*, *Ambulocetus*, *Basilosaurus*, *Dorudon*. (photos by Kevin Guertin, Notafly, and from Voss et al., 2019, *PLoS ONE*) 1 meter scale bar for bottom two images only.

Indohyus, but show that this meager creature, whose skinny build couldn't be further from the titanic proportions of a blue whale, was a whale antecedent.

The first *Indohyus* fossils were collected in Kashmir—the Himalayan border region disputed by India, Pakistan, and China—by the Indian geologist A. Ranga Rao and described in 1971. They weren't much—just a few teeth and part of a jawbone—and when Rao passed away he had little idea what type of animal he had found. His widow, however, was persistent. She kept boxes of rocks from the Kashmir excavation site, many of which were unopened, and passed them to the Dutch-American paleontologist Hans Thewissen. At first Thewissen didn't think much of them, either . . . until his technician accidentally split open a skull embedded in one of the boulders. Thewissen couldn't believe what he saw: the hollow bulla encasing the three middle ear bones was shaped like a conch shell, with a thickened and curved interior wall. One bone nuzzled in the back of the skull might be anonymous to anyone but an anatomist, but it had two huge implications.

First, genealogically, the unusual bulla ties *Indohyus* to whales. Almost all mammals have a delicate bulla, some thin as eggshell, which looks like a bubble. But not whales, with their shell-shaped bullas that are hard as rock. As the double-pulley ankle is for artiodactyls, this bulla is a calling card for whales. Absent DNA evidence, impossible for such an ancient animal as *Indohyus*, it is as close to slam-dunk proof of family heritage as anatomy can provide.

Second, the bulla tells us something about *Indohyus*'s lifestyle. Whales have a strange bulla for a reason. Recall from earlier that, in all mammals, these cavernous bones—one on each side of the head—are noise-canceling headphones, which insulate the middle ear bones as they transmit sound to the cochlea, and then the brain. Whales need to hear underwater, a more challenging medium than air. Hence they need hearing aids, and the thickened rind of the bulla and its curved shelf enhance sound detection. It follows that if *Indohyus* had the same hearing aids, it must have been a keen underwater listener too. This jibes with other oddities of Rao's fossils. The limb bones of *Indohyus* were exceedingly dense, with thick walls and small marrow cavities, a signature of water animals that need ballast to reduce buoyancy and stay submerged. And its body dimensions—small head, robust trunk, twiglike limbs, and long hands and feet—were eerily similar to those of a mouse deer, suggesting that it lived a similar life of foraging on the banks of rivers and streams, and diving into the water when threatened.

Add it all up, and it's clear that *Indohyus* was a land mammal experimenting with the water. In doing so, it made the first step of a long evolutionary journey, which was never preordained. Nature didn't set out to make whales. Evolution doesn't work that way: it doesn't plan ahead, but operates only in the moment, adapting organisms to the immediate challenges they face. When *Indohyus* fled into the water, it was just trying to escape, or find food. It had no idea that its descendants would become sea leviathans. But the Rubicon had been crossed, and now that these petite artiodactyls had dipped their hooves into the water, natural selection could fashion them into more competent swimmers.

The next task was to do something about the arms and legs. The toothpick limbs and hoofed digits of *Indohyus* could provide only a bare minimum of propulsion in the water. If you've ever seen one of those viral videos of a deer that's wound up in a swimming pool, you'll know what I mean. Evolution's answer was *Pakicetus*, the next link in the land-to-sea chain, described by Thewissen's PhD advisor Phil Gingerich, who we met last chapter as an expert on the PETM global warming. Gingerich is equally known as a specialist on early whales, and *Pakicetus* is his greatest prize. It was the size of a large dog, with the long snout and sharp-toothed snarl of a wolf, a signal its diet had changed from vegetarian to carnivore.

What stands out most, though, are its limbs: they are more robust than those of *Indohyus*, and the hands and feet are starting to
look like the foot fins of a scuba diver. *Pakicetus* was still primarily
a walker, in two worlds: it could stroll competently on land with
its double-pulley ankle and could walk on the bottom of shallow
freshwater streams, which fed into the Tethys. It could paddle with
its limbs, too, and wiggle its body a bit, a multipurpose amphibious
vehicle able to get around in many different ways.

Still, though, *Pakicetus* was far from a graceful swimmer, and
it lived a double life shuffling between land and stream. The next
link in the evolutionary chain, *Ambulocetus*, described by Thewissen, ventured off the Indian island, to the salty margins of the
Tethyan coast. The size of a large sea lion, it was clearly more
aquatic than *Pakicetus*, with a longer and more tube-shaped body,
shorter limbs, and broader hands and feet that looked like paddles.
It could power itself through the coastal currents in two ways, one
old and one new: hand and foot paddling retained from its land
ancestors, and up-and-down undulation of the backbone, a talent
devised in the water. It also evolved another hearing aid: a fat pad
in the lower jaw that connected to the bulla bone, which gathered
underwater vibrations and sent them to the ear. *Ambulocetus*—like
today's whales—literally heard through its jaws, an evolutionary
hack that circumvents the ineffectiveness of a land mammal's eardrum underwater, as we can attest when we try to talk to each other
in the pool. An animal with the swimming and sensory prowess of
Ambulocetus had little time for the land. Although it could probably still walk, at least haltingly, it lived more like a crocodile: an
ambush predator that preferred to loiter in the shallows, snatching
fish with its pointy teeth.

At this point, evolution had taken what looked like a miniature
deer and produced a moderately large, flippered, swimming mammal that could hear well underwater, which had broken the bonds

of both land and freshwater and was now paddling, twisting, and ambushing through the coastal ocean shallows fringing the Indian island. There was no going back. Land was in the rearview mirror. Ahead was the ocean—all of it. Not just the margins of one island, but the open swells and dark depths that cover 70 percent of Earth's surface.

The first worldwide whales were the protocetids, exemplified by *Rodhocetus*, another species named by Gingerich. They were still Walking Whales—but barely. Their hands and, especially, their feet were comically oversized—but not because they were runners. No, this was swimming gear, and while protocetids could support their bodies on land, they would have been as awkward as a scuba driver trying to run with foot fins on. They likely lived like seals, spending most of their time in the water, but occasionally hoisting themselves up on the rocks to lounge in the sun, mate, give birth, and nurse their young. Other than that, the land held little use for them. As they all but abandoned their ancestral terrestrial home, they turned to the open ocean. By 40 million years ago—as India was punching into Asia—there were protocetids off the coasts of Asia, Africa, Europe, and North America, and a species called *Peregocetus* as far afield as the Pacific coast of South America. The protocetids, and the afrotherian manatees that were migrating at the same time, followed bats as the second surge of cosmopolitan placental mammals. While bats untethered themselves from geography by flying, protocetids did it by swimming, propelling themselves across the earth with their huge webbed hands and feet.

Walking Whales then stopped walking. With this, they completed their divorce from the land. The Egyptian skeletons of Wadi al-Hitan, particularly *Basilosaurus*, capture this critical moment in whale evolution: when they went all in for an aquatic life and first resembled something we would today recognize as a whale.

Several things set *Basilosaurus* apart from protocetids. For starters, it was gargantuan: it was some fifty feet (seventeen meters) long and weighed more than five tons, an order of magnitude bigger than most protocetids. To move its momentous body through the water, *Basilosaurus* invented a new swimming device: a fluked tail, supported by wide vertebrae, which moved up and down to generate thrust. Such whiplash motion was only possible because the pelvis and hind limb emaciated further, disconnecting themselves from the backbone, giving the tail more flexibility. The legs of *Basilosaurus* were pathetic, smaller than human legs but perched on a torso longer than many yachts. They still peeked out from the trunk, but later whales would lose all external sign of them, retaining a few shrunken pelvic and limb bones internally, which won a reprieve because they anchor genital muscles. The forelimbs of *Basilosaurus*, however, remained prominent, flattened into broad paddles for steering, but they obviously could not support a five-ton body on land. The neck shortened, merging with the body into one seamless torpedo shape, sinuses invaded the bones around the ear to adjust pressure while diving, and the nostrils began moving backward to become blowholes. All of these things helped *Basilosaurus* become a champion swimmer. And not just any swimmer, but one to be feared: it was an apex predator that ate other whales, proven by the *Dorudon* bones found in the stomach of one skeleton.

By the end of the Eocene, all the Walking Whales were gone. The transition was complete: a land animal had become an obligatory swimmer, never able to go back onto the land for anything—to escape a predator, to eat, to give birth, to sleep. From here on out, whales would do *everything* in the water. Evolution, however, was not done tinkering. It never is. Around the Eocene-Oligocene boundary, 34 million years ago, the next phase of whale history began: it was now time to shape these water whales into the best water animals they could be.

At this time, whales diverged into two paths, leading to the two flavors of modern whales: toothed whales and baleen whales, each with their own litany of anatomical features and behaviors exquisitely adapted to life under the sea. Fossils of both groups start appearing around the Eocene-Oligocene boundary. From there, the modern-style whales pushed beyond the limits of protocetids and *Basilosaurus*, by expanding into the highest latitudes and coldest oceans. They now spanned coastal and offshore waters, arctic and tropical temperatures, the shallows and the deep, and even ventured into freshwater, with river dolphins (yes, dolphins are a type of whale) and other species backtracking into the riparian envi-

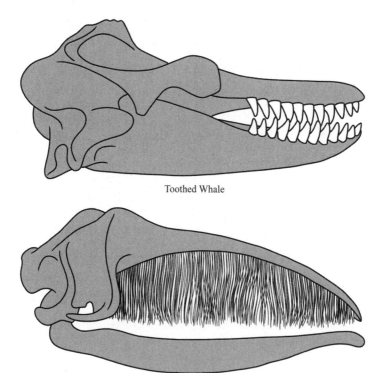

Toothed Whale

Baleen Whale

Skulls of toothed whales and baleen whales. (illustrated by Sarah Shelley)

ronments of *Indohyus* and *Pakicetus*. By all measures, whales had become a global empire.

The toothed whales—formally known as odontocetes—today include sperm whales, killer whales, narwhals, dolphins, and porpoises. They are fierce predators, at the peak of the ocean food web, with three key weapons at their disposal.

First are their sharp teeth, transformed so they no longer seem mammalian. Gone are all the complex cusps and ridges; gone is the lineup of incisors, canines, premolars, and molars; gone is the replacement of baby teeth with adult teeth; gone is the ability to chew. Instead, all the teeth are conical pegs, which simply cut meat off fish or other whales, which the odontocete then swallows. Some odontocetes barely use their teeth in feeding and lazily swallow their prey whole.

Second is something that, astoundingly, is shared with bats, although it obviously evolved independently: echolocation. Odontocetes make high-frequency clicks and whistles by squeezing air through phonic lips, a fleshy constriction in the nasal passages just underneath the blowhole. A blob of fat, called the melon, which bulges out from the forehead acts as an acoustic lens and focuses the sounds, which then reverberate and echo back, detected by a specialized cochlea in the ear. While bats echolocate to snatch insects or find victims to blood-suck, odontocetes use their sonar to locate schools of fish or squid in the dark, turbid murkiness of the deep ocean. It is such a keen sense that these whales no longer need to smell—and in fact, they no longer can.

Finally, odontocetes have stupendously sized brains. The sperm whale has the largest brain of any animal on Earth, perhaps ever. At around twenty pounds (ten kilograms) in weight, it is some five times the heft of a human brain, and bigger than any elephant brain. If you divide its brain size by its body size, to make a rough and relative measure of intelligence, it has the second-highest score of any animal, trailing only humans. Sperm whales are brainy enough to outsmart their prey, and they use tools and can recognize themselves in mirrors.

All these predatory powers evolved long before modern-day sperm whales and kin, beginning around the Eocene-Oligocene boundary. The oldest fossil odontocetes, like *Cotylocara* and *Echovenator* from South Carolina, have conical teeth, big brains, and cranial features related to the production and detection of high-frequency sound, like a bowl-shaped depression for the melon and the muscles that control it, and a cochlea with an expanded base. Odontocetes thus developed oversize brains and echolocation as they diverged from the baleen whales, quickly carving out their own bespoke hunting and cognitive styles. Over the course of the Oligocene and continuing until today, they became ever-better sonar specialists, with bigger facial muscles for making sound and melons for beaming it out.

Some of the ancient odontocetes put the modern ones—even sperm whales—to shame. The Miocene-aged *Livyatan melvillei*, named in homage of Moby Dick, prowled the Pacific off South America around 12 million years ago. It was one of the largest predators, period, in Earth history, with a sixty-foot-long (eighteen-meter) body and a ten-foot-long (three-meter-long) head that could comfortably fit a human inside—although it surely wouldn't have been very comfortable for the human. Its bite was so wide that it could have *easily* engulfed the skull of the largest land predator to ever live: the one and only *T. rex*. And to keep piling on, its foot-long teeth were thicker than railroad spikes, perfect for destroying the bones of its prey—which were other whales, of the baleen variety. As if dreamt by a B-list monster film director, *Livyatan* shared the same waters with the epically famous supershark *Megalodon*—and no doubt the shark would have feared the whale.

As big as *Livyatan* and other fossil odontocetes were, none remotely approached the size of the largest baleen whales. Formally called mysticetes, baleen whales include blue whales, right whales, minkes, and humpbacks. Their skeletons are cartoonish, with engorged heads and slender, naked jaws that bow far outward, as if forming the rim of a garbage can or a basketball hoop. You will not find a trace of a tooth along these strapping jawbones, as mysticetes lost the pegs of their ancestors and replaced them with the thing that gives them their name: baleen, a set of closely spaced, curtainlike plates of keratin—the stuff of our fingernails— suspended from the roof of the mouth. Baleen enables mysticetes to do something no other mammals can: filter feed by straining small prey from the water. The feeding dance of some species is something to behold: they will lower their jaws, open their mouth into a gaping chasm, gulp an unholy volume of seawater, and use their tongues and throat muscles to squeeze the water out of the mouth and through the baleen, all so they can catch thousands (or more) of plankton at a time. It's ironic: the biggest animals ever subsist on minuscule prey—but with the appetite of a glutton. They don't need to echolocate to find their food, or power up their big brains to hunt. All they have to do is meander through the water column and occasionally open their gob.

The fossil record reveals a twist to the evolution of mysticetes. The earliest ones, like the Eocene-aged *Mystacodon* from Peru and *Llanocetus* from Antarctica, still had teeth, some of which looked like the conical pegs of odontocetes, and others elaborate fans of cusps radiating outward from a central peak. They did not have baleen, and they could not filter feed, yet they were already getting much larger than other whales of the time. *Llanocetus*, for example, was at least twenty-five feet (eight meters) long, making it one of the largest whales until other mysticetes and odontocetes like *Livyatan* supersized in the Miocene. Baleen and filter feeding thus evolved after mysticetes diverged from odontocetes, and they were not the secrets to making mysticetes large, at least not initially. Instead, fossils show that the first toothy mysticetes were biters, and then they lost their teeth and became suction feeders, before adding baleen to their naked jaws and unlocking the new skill of filter feeding. As with the transition from walking to swimming in the ancestral whales, the change from teeth to baleen, and from biting to filtering, was a gradual process that proceeded in stages.

Once baleen-based filter feeding evolved, though, it gave mysticetes the means to get bigger and bigger. Unlike odontocetes, which are forever at the mercy of the abundance of the squid, fish, and other whales they eat, baleen whales have a nearly limitless supply of plankton, which they can guzzle without expending much energy. They can passively hang out and gorge on buffets of seafood, particularly during seasonal blooms or in upwelling zones, where nutrients streaming up from the deep ocean nourish swarms of krill. Blue whales—the largest whales that live now, and that have ever lived—are mysticetes. They are the culmination of a long-term trend, beginning around the Eocene-Oligocene boundary and continuing unabated until now, of ever-more-humongous whales. This is unlike the story of land mammal evolution, where as we learned earlier in this chapter, elephants and rhinos reached peak size around the Eocene-Oligocene boundary and have never gotten bigger.

Might this mean that leviathans more massive than blue whales will one day evolve? Could these extreme mammals become even more extreme? It seems like a reasonable bet . . . as long as blue whales and their mysticete cousins can make it through the current whirlwind of climate and environmental change, escape extinction, and find enough plankton to eat in the oceans of the future.

8

MAMMALS *and* CHANGING CLIMATES

Teleoceras

THE AMERICAN SAVANNA, CIRCA 12 million years ago (Miocene)

It was a warm morning, in the early spring. The winter had been long and tedious; not particularly cold or snowy, but dry and dull, more than three months of still air and short days. A few weeks ago, at long last, the rains had returned. The savanna was starting to transform.

As the sun rode up the horizon, it spilled across a land flat and broad, deep in the continental interior, a good thousand miles from the nearest ocean. Grass carpeted the ground. It had died back during the winter, but not quite died away, and was now shooting up from the waterlogged mud like trillions of tiny fingers reaching for the heavens. Spattered about were clumps of trees, too patchy to form a canopy, which from above looked like a speckle of islands in a sea of grass. The trees were breaking out of their dormancy, the tips of their branches bursting open with tender leaves and fragrant flowers.

Last evening's rain was working its way through a tangle of streams, which trickled into a lake. The lake was a waterhole, and it was here where the animals of the American Savanna congregated. To drink, to bathe, to socialize, and to band together in protection from the menace that hid among the trees: a dog the size of a bear, a true hell hound, with bulging jaws and a bite that shattered bone.

The animals loitering near the water that morning formed a motley crew. There were several types of horses, from miniature prancers with three toes on their dainty feet, to stallions whose sprinter's limbs bore but a single hoof. The little ones munched on leafy shrubs poking out of the lakeshore, and the big ones kicked up mud as they raced around the lake's edge, pausing occasionally to crop mouthfuls of grass.

They were joined by deer, and by camels. Some of the camels were standard-issue species with long limbs, goofy smiles, and small humps on their backs. One, though, barely looked like a camel at all. It had the silhouette of a giraffe, with a noodle neck stretched high above its body, some ten feet off the ground, perfect for pinching the tastiest leaves from the taller branches. Normally these giraffe-camels would spend the morning chewing away in the tree islands. But not this morning. Like any dawn after a hard rain, the risk of a hell hound attack was too great, so the giraffe-camels skipped breakfast and sauntered to the waterhole, to seek safety in numbers.

Then there were the rhinos—so, so many rhinos. There must have been at least three hundred of them, maybe more, divided into a few herds. Like the long-necked camels, there was something both familiar and unfamiliar about them. They were rhinos, no doubt, with conical horns sticking up from their noses. But their cavernous bellies, pudgy limbs, swollen heads, and nearly nonexistent necks gave them a hippopotamus vibe. Some of the rhinos were splashing in the shallows, enjoying the feel of cool water on a morning that was becoming sunnier by the minute. Others, however, were not in the mood for indulgence. Their big bellies were rumbling, so they got down to the business of eating. Every day, they would need to devour dozens of pounds of grass to keep themselves going. The earlier they could start feasting, the better.

Although it was a beautiful morning, there was tension in the air. Not only fear of the hell hounds, but something else—something curious was brewing among and within the rhino herds. There were rivalries, scarcely perceptible to all but the rhinos themselves. Each herd had a clear leader: a male, and not just any male, but a beefy one with the largest incisor tusks of the bunch. He surrounded himself with females, so each herd was really more of a harem. Almost all the cows were pregnant, and their due dates were rapidly approaching. Meanwhile, many of them were still caring for last year's babies, which had grown into playful adolescents that still needed occasional comfort, and nourishment, from their mothers.

This family structure left few options for the smaller adult males, of which there were many. Some of them sought company in bachelor herds—gangs governed by testosterone, fueled by the resentment of not fitting in. Every so often, one of these bachelors would work up the courage to confront the dominant bull for control of a harem, but it rarely ended well for the challenger. Sadder still were the loners, the unluckiest males that wandered like zombies on the fringes of the herds. Whenever one of these misfits got too close to the cows, the dominant bull would do whatever necessary to protect his harem. Usually horns and tusks were involved, and the fight didn't last very long.

As one of the bulls eyed a loner approaching his harem—and as rhino calves frolicked in the water and horses snipped grass and giraffe-camels weighed whether it was safe to head into the trees—a boom blasted across the savanna.

Every animal near the waterhole froze.

It was the loudest noise they had ever heard. It was the loudest noise they would ever hear, for it was the clarion call of an apocalypse.

The rhinos and horses and camels craned their necks skyward and saw a cloud disturb the blue vista. The savanna animals wouldn't have known it, but the cloud began as a plume, of what looked like smoke, about a thousand miles (sixteen hundred kilometers) to the northwest. It shot farther and farther into the sky, before fanning into a mushroom shape as it hit the higher reaches of the atmosphere. Whatever it was, the cloud was far away—maybe a few hundred miles distant, at this point. Nothing much to worry about, judging from how the animals reacted. The bull went back to scowling at its wannabe rival, the babies resumed their splashing and the horses their grass-guzzling, and the giraffe-camels decided enough was enough, and no hell hounds could keep them from their leaves.

The cloud, though, did not stop moving. Swept up by the westerly winds, the haze inched across the horizon, as if in slow motion, drifting ever closer to the savanna, getting bigger and bigger.

The sun was rising from the east, the cloud advancing from the west, and these two competing forces collided above the savanna. As the cloud passed between the sun and the land, day turned into night. All went dark.

It was not a calm darkness, but a violent one. To the animals of the savanna, it felt like a strange, intense snowstorm. Something was falling from the sky—the stuff that formed the smoky cloud wasn't just passing through, but was fluttering down. The little particles, most the size of dust flakes, some the size of sand grains, punched holes through the dark. A blizzard was howling across the plains, but this stuff wasn't snow. It wasn't cold, but lukewarm. It wasn't white, but a charred gray. It wasn't wet, but abrasive— tiny bullets of glass scratching the hides of the savanna mammals as it fell, prompting them to itch as they had never itched before. And it stunk. It reeked of sulfur and fire, and turned the air toxic.

Birds started falling from the sky. Their limp carcasses pelted the ground, sometimes ricocheting off the backs of the rhinos. The rhinos couldn't see what was happening, their eyes unable to penetrate the darkness. Plus, their eyes were starting to water—the gritty stuff from the cloud was clogging anything it could find: eyes, ears, noses, mouths. While their ears filled up with filth, the rhinos could still make out the sounds of the savanna. It was chilling. Whooshing winds, a cacophony of coughs from the herds, and dead birds smashing into the grass with the pitter-patter of gunfire. A dead bird hailstorm.

The cloud moved with the wind, and eventually the wind blew

past the savanna. The pitch-black blizzard lasted for a few hours, but to the animals, it seemed like an eternity. As the sun finally broke through the smog, and the cloud continued its journey eastward, the rhinos and horses and camels blinked back to consciousness. What they saw around them—through bloodshot eyes caked with gunk—was otherworldly.

Their savanna was covered in a gray blanket, about half a foot thick. The gray stuff was ash.

It enveloped everything. Not a single blade of grass was visible on the ground, not a single flower or leaf remained clean on the trees. The waterhole was still there, but eddies of ash were swirling across its surface, gradually turning water into paste.

Stupor gave way to panic. The rhinos gathered into one massive mob, harem joining harem, bachelor gangs and loners taking their places alongside breeding cows and dominant bulls. Bodies clanging together like one-ton pinballs, the mob stampeded across the flatlands of ash, trampling the dead birds into pancakes, before running out of energy and whimpering toward the one place they always felt safe: the waterhole. Except it was safe no more, and not a waterhole any more, but a pit of silica ooze.

The next few days were rough. With no water to drink, the rhinos and camels and horses grew parched with thirst, their lips chapping as they tried, to no avail, to suck water out of the sludge of the old waterhole. Some of them ventured too far in and were slurped below, engulfed in a quick-ash slurry that they did not have the energy to escape.

Every one of the animals groaned with hunger. All they could do was try to use their hands and feet and tongues to clear the ash from the grass. But it was of little use: much of the grass had already died, smothered and poisoned by ash that acted like a herbicide. To make matters worse, every time one of the animals tried to clear a patch of ground, they kicked ash into the air—ash that they breathed in.

With each breath, they took in more ash. The particles were small enough to lodge deeply into their lungs, so with every inhale the animals filled their lungs a little further, like sand being added to a sandbag. At first their chests felt a little heavy, but over the next few days, breath by breath, their innards became so petrified that they could barely breathe. So little oxygen flowed through their blood that they got delirious, and their hands and feet started to swell.

One by one, as they stumbled about light-headed on their inflamed limbs, the rhinos and horses and camels gave in to the inevitable. Hungry, tired, thirsty, unable to breathe, they fell, one after the other, into the ash. Bodies piled up around the waterhole and across the plains, like victims on the battlefield of some terrible war.

In the distance, emerging from one of the tree islands, a solitary beast trudged through the ash, leaving handprints and footprints as it navigated the carnage. A hell hound. Mere days ago, it was the greatest fear of every rhino, horse, and camel of the savanna. Now those animals were all dead, and the hell hound almost was, too. Haggard and hungry, it was about to give up before seeing the spread of rotting meat along the lakeshore. It walked up to a rhino and took a few half-hearted bites from its flank, spitting out ash as it gnawed. A last meal, before it lay down beside its prey.

The winds picked up again, an eerie echo across an empty landscape. The ash drifted, like snow, and blew across the waterhole, burying it and all the animals.

AROUND 12 MILLION years ago, during the middle part of the Miocene, a volcano blew its top in what is now Idaho. It was fed by the same system of heat and magma that rests under Yellowstone to-day, which powers Old Faithful and the other famous geysers. Ash was blasted across much of North America, carried eastward by the prevailing winds. About a thousand miles downwind, in what is now Nebraska, a half foot of the nasty, sulfur-smelling, glassy stuff blanketed the ground like snow. Many of the animals living there died right away, particularly the birds, which were caught unaware while in the air and choked on ash as they flew. Others, including many mammals that lived on the ground, endured days or even weeks of hunger, thirst, disease, and exhaustion, before succumbing. Then, in one final insult, they were buried in the very ash that clogged their lungs, polluted their water, and poisoned their food.

This dystopian story is true. The volcanic massacre is captured in stone, the birds and rhinos and camels and horses preserved as they died, crystallized in place, like a prehistoric Pompeii. You can see the skeletons yourself, at one of the world's most breathtaking—and unexpected—museums. From the outside, the "Rhino Barn"—at what is now Ashfall State Historical Park, in rural Antelope County, population circa 6,600—resembles one of those trucking distribution warehouses that dot the open roads of the American midlands. Its low-slung shape seems to merge with the rolling, earth-toned farmland of northeastern Nebraska. You would never guess it, but inside are over one hundred in situ fossil skeletons . . . and counting.

The first bones were discovered here by paleontologist Mike Voorhies in 1971, while prospecting with his wife, Jane, in a ravine on the edge of a cornfield. Voorhies saw the unmistakable glint of teeth, protruding from a soft, gray volcanic ash. Teeth were embedded in jaws, jaws connected to a skull, the skull hitched to a skeleton. Later, a bulldozer removed some sixty-five hundred square feet (six hundred square meters) of topsoil, exposing dozens

The skeleton of the rhino *Teleoceras* (top) and *Teleoceras* and the horse *Cormohipparion* preserved in ash at the Ashfall Fossil Beds (bottom). (photos by Ray Bouknight and Ammodramus, respectively)

more skeletons—mostly rhinos, but also horses, camels, deer, and countless birds, splattered next to rhino hoofprints. Dozens became hundreds as paleontologists continued to excavate, decade by decade. The work continues today, under the flat roof of the Rhino Barn, where visitors can watch scientists and volunteers dust ash off twelve-million-year-old bones in real time.

Fossils are notoriously finicky. So often, all we find are a few bones—if we are lucky. More commonly, as our field crews in New Mexico know all too well, finding fossil mammals usually means finding fossil teeth. Sometimes a few teeth set into a jawbone, but usually single teeth—broken, worn, fragmented. And usually these teeth and bones are not situated where the animal they once belonged to died, but were washed around by rivers, carried by the wind, or scattered by scavengers before getting covered by mud or sand and hardened into a fossil. If paleontologists are detectives, then we are almost always working with crime scenes that have been corrupted by time and circumstance.

Not so at Ashfall Fossil Beds. The Yellowstone supervolcano, in all its destruction, froze into place a snapshot of a Miocene-aged community. It's a tragic portrait, for sure. Birds are at the bottom of the ash pile, a sign they were immediate victims, with mammal skeletons stacked on top, indicating that the rhinos and horses and camels endured at least a few days of torture before dying. Rhino calves cling to their mothers, in one last hopeless cuddle, while last suppers of vegetable mush are suspended within rib cages. On some rhino skeletons, swollen extremities—hands and feet, wrists and ankles—are emblems of asphyxiation. Other skeletons are pockmarked with telltale signs of bone disease caused by lung failure, and gouged by chomps from scavengers—most likely hell hounds, which go by the scientific name of *Epicyon*, a five-foot-long, two-hundred-pound bone-shattering brute. Even the warped mind of Alfred Hitchcock would struggle to create a scene so barbaric.

At the same time, though, this snapshot is exceedingly valuable to paleontologists. For it tells us so much about the lives and behaviors of the Miocene mammals, particularly the rhinos, which belong to a species called *Teleoceras*. Most obviously, if so many rhinos died together, they must have lived together, in social groups. Their skeletons, when studied statistically, divulge the demographics of a prehistoric population. Some of the rhinos still have baby teeth, and they fall into three neat size clusters: one-year-olds, two-year-olds, and three-year-olds. Breeding, it seems, happened once a year. The remaining rhinos, with sets of adult teeth, split into two cliques: those with small incisor tusks at the front of their mouths, and those with large tusks. Fetuses are found inside small-tusked individuals, proving that these are females, and by implication, that the large tuskers are males. This led to an incredible revelation: in the rhino graveyard, there are about five adult females per adult male. This skewed sex ratio, which would make no sense in a human population, characterizes modern mammal species that form harems, where a single alpha male breeds with a group of females.

In their harem structure, these twelve-million-year-old rhinos seem almost modern. Indeed, in so many aspects of their appearance and biology, the Ashfall mammals are instantly recognizable to us. True, the rhinos look a bit like hippos, and some of the camels have long necks like giraffes, but there is no mistaking what these animals are. They are rhinos. They are camels. They are horses. As we learned in the previous two chapters, today's main groups of placental mammals—primates, even and odd-toed hoofed species, carnivorans like dogs and cats, elephants, bats, whales—began to proliferate during the Eocene, about 56 million years ago. With that said, the earliest members of the modern groups were hardly like their descendants that live now. Just look at a Walking Whale, a puppy-size proto-elephant, or one of those wee horses that fell into Lake Messel, and that becomes evident. Ashfall, though, tells a different story. The volcanic victims are mammals that any kindergartner could identify, the same sorts of things that we see in zoos today. By the Miocene, the modern groups had become almost fully modern.

These Ashfall mammals, however, do seem out of place. We expect to see rhinos in Africa and camels in the Middle East, not smack in the middle of the United States of America. Wander the farm country of Nebraska today, and if you think you spot a rhino, you're probably misidentifying an oversized bull of the beef cattle variety. Or hallucinating. Ditto a camel: the only way one of them would end up in Nebraska is if it escaped from a zoo and didn't have the smarts to return.

Why, then, were there rhinos and camels in Miocene Nebraska? Other fossils from Ashfall give us the answer. These fossils are not museum centerpieces like the mammal skeletons, but much humbler specimens: microscopic grass seeds, embedded in the teeth, inside the mouths and throats, and within the rib cages of the rhinos. The seeds would not have grown into the same grasses that rustle in the prairie winds of Middle America today. Quite the contrary, they belonged to subtropical species, similar to those that currently bloom in Central America. Furthermore, other plant fossils from Ashfall hint at small groves of walnut and hackberry trees. Thus, during the Miocene, Nebraska was a savanna: a land shrouded in grass, with occasional clumps of trees, watered by minimal rains. It would have looked a lot like the African savannas today, where lions and elephants and wildebeest romp.

As ludicrous as it sounds, if you were around in the Miocene, you could have gone on safari in Nebraska.

The American savannas were different not only from today's environments, but from what had preceded them, in the Paleocene and Eocene. Recall that the Paleocene world, after the asteroid felled the dinosaurs 66 million years ago, was a hothouse, with jungles covering much of North America and no ice at the poles. At the Paleocene-Eocene boundary, 56 million years ago, the hothouse boiled even further, in a spasm of global warming. Temperatures declined somewhat during the remainder of the Eocene, but it was still a hothouse. The jungles endured, the poles remained free of ice. Then, about 34 million years ago, as the Eocene transitioned to the Oligocene, the world turned. The hothouse flipped to a coolhouse, which would eventually become an icehouse.

The change was sudden, as if the hot water tap suddenly shut off and the cold water tap came on. All told, it took about three hundred thousand years—at most—for global temperatures to lurch downward. Higher latitudes cooled an average of 5 degrees Celsius (9°F), but the effects were even more pronounced inland, deep in the continental interiors, like the areas that would become the American Savanna. Here, temperatures dropped by 8 degrees Celsius (14 degrees Fahrenheit). As land and sea cooled, climates became more seasonal, more variable, and a hell of a lot more unpredictable. It was the most severe, and most permanent, temperature shift since the asteroid hit. Contemporary global warming may eclipse it, in the opposite direction, but that remains to be seen.

The culprit for all this upheaval was a one-two-three punch of coincidences. First, carbon dioxide in the atmosphere was gradually decreasing, meaning there was ever-less greenhouse gas to insulate the earth and keep it warm. Second, summers were becoming cooler than usual, probably due to fluctuations in the earth's orbit around the sun. And third, and perhaps most critical, the continents were still moving. Gondwana, the last remnant of old Pangea, was finally completing its messy divorce.

During the Paleocene and early Eocene hothouse, Antarctica was still connected to both Australia and South America, ever so tenuously. After millions of years of earthquakes, Antarctica snipped free, on both sides, in the late Eocene. Water rushed in to fill the gaps, creating a new frigid-water current that circled the South Pole, preventing warmer waters from reaching the southern ocean. The circumpolar current behaved like an air conditioner, plunging Antarctica into a deep freeze and feeding glaciers that rapidly frosted the polar lands. For the first time in hundreds of millions of years, since the Carboniferous-Permian age of distant mammal ancestors, large ice sheets advanced across a continent. Ice would not yet encrust the north, as it is harder to grow glaciers there because there is not a single Antarctica-style landmass parked over the pole. But that ice would eventually come, much later, bringing with it mammoths and saber-toothed tigers—a story for next chapter.

The Antarctic glaciers were the most glaring consequence of the Eocene-Oligocene cooling, but the effects of the temperature crash were felt globally. Tens of thousands of miles from the new ice fields, the North American inlands were traumatized, too. They became not only cooler, but drier, and plants with shorter life cycles and less wood tissue flourished at the expense of slower-growing trees. The jungles shrank away, replaced first by sparser woodlands, then savannas, then open grasslands. This was a drawn-out process, which took place over the Oligocene (34 to 23 million years ago) and continued into the Miocene (23 to 5 million years ago), the time of the Ashfall mammals, and beyond. With such a wholesale change in temperature, climate, and vegetation, mammals had no choice but to adapt.

MY DAD HATES mowing the lawn. When my parents moved out of my childhood home a few years ago, they claimed it was to be closer to their expanding brood of grandchildren. I suspect it was as much to do with downsizing to a smaller house with a tiny yard. After becoming a homeowner myself, I understood. The grass never stops growing. The winter offers a brief respite, but come springtime, the green blades shoot out of the sod with the ferocity of missiles. You miss one week with the lawnmower, and your house looks like the derelict set of a horror film. But it's never enough. Like a teenager trimming his whiskers, the more you cut your grass, the more it grows back—thicker, heavier, denser.

I really shouldn't complain about grass, because a world without it would be an alien world—and to humans, an inhospitable one. Grasses are more than the decorative carpets on our lawns, parks, and golf courses. There are over eleven thousand species of grasses alive today, and they cover some 40 percent of the earth's land surface: as savannas, prairies, and grasslands, but especially, as areas cultivated by humans. Lawns, yes, but more importantly: cropland. So many of our staple foods are grasses: wheat, corn, and rice, to name a few. For people like me who grew up in farm country, summers spent playing hide-and-seek in the fields and autumn corn mazes made it obvious what these crops—and all grasses—are: a specialized type of flowering plant, which, if allowed to grow tall enough on their straight, skinny, hollowed-out stems, will sprout flowers, and fruit—which can be edible. A wheat kernel is a fruit, albeit a highly peculiar one. So is a corn kernel.

Grass is such an omnipresent part of our world that you might think it's been around forever. Not so. The first 4.43 billion years of Earth history were devoid of grass, but once it evolved, it changed everything. George Gaylord Simpson—the man who made the famous mammal family tree in the 1940s—recognized this. The development of grasslands, he argued, triggered a profound change in the mammals that lived on them, most notably horses. He called it the "Great Transformation." Along with being the preeminent paleontologist of his day, Simpson was a prolific science writer, and he wrote an entire book on the grass-begat-horses story. It is only within the last couple of decades, however, that we've come to understand how the grassy revolution went down.

Caroline Strömberg has written the new history of grasses. She grew up in Lund, Sweden, and as a child collected trilobites on the rocky shores of Gotland, a comma-shaped island in the Baltic Sea to the east of the Swedish mainland. She concurrently studied geology and art, completing a master's thesis on minuscule fossil teeth more than 420 million years old, while serving as apprentice to a scientific illustrator. A PhD scholarship brought her to California, where by chance she attended a lecture on the evolution of horses. She found herself asking: Was the classic Simpsonian story of the Great Transformation correct, or was it more of a tall tale? Only one way to find out: she would need to compile detailed records of both grass *and* mammal fossils over time, to see how they changed together. It would become her PhD thesis, and that thesis would garner her the prestigious Romer Prize for best student talk from the Society of Vertebrate Paleontology, in 2004. This was the first SVP meeting I attended, as a wide-eyed undergraduate, and I've admired Caroline's work ever since.

Grasses are a young phenomenon, in the grand scheme of Earth history. *Brontosaurus* never would have eaten a blade of grass, or even seen one. *Triceratops* might have, but if so, it would have been a fleeting glance. It was only during the latest Cretaceous—the waning days of the dinosaur empire, when mammals were still in the shadows—that grass appeared. The evidence is meager, and a bit disgusting: microscopic silica blobs called phytoliths embedded in the hardened dung of long-necked dinosaurs from India. It was Caroline who identified them. She was shown photos by Indian colleagues, at the same 2004 meeting where she won her prize. Instantly she recognized the Cretaceous blobs as nearly identical to modern-day phytoliths, which grasses deposit in their tissues to provide structural support and protection from overgrazing by animals. She jumped up and down; she knew it was a revolutionary discovery. About a year later, she joined her Indian friends in describing these most modest of fossils. As might be expected from pioneers, these first grasses were small, marginal, forgettable. They were limited to weedy patches, and never got close to forming grasslands. If you ever see an artistic rendering of dinosaurs on grassy ground, it's wrong.

Postasteroid, things mostly stayed the same. The early grasses had to vie for space with the riotous tangle of jungle trees and vines, energized by the blazing temperatures of the Paleocene and Eocene. Grasses diversified to adapt to their surroundings, and some like bamboos became experts at living in the claustrophobic confines of the jungles. Nevertheless, there was little open ground for grasses to fan across, and whatever was available would have been cloaked in ferns and shrubs. Thus, there were no grasslands when the "archaic" placentals were building the first mammal-dominated ecosystems in the Paleocene of New Mexico. Or when the trinity of primates, odd-toed perissodactyls, and even-toed artiodactyls advanced across the northern continents in step with the PETM global warming. Or when the strange mammals of South America and Africa were beginning their evolution in isolation.

Then hothouse switched to coolhouse, as the Eocene ticked into the Oligocene. Climates chilled, precipitation became scarcer, the thirsty jungles dwindled, and landscapes gradually opened. Grasses took advantage, their ability to grow quickly and tolerate harsher conditions allowing them to replace the forests, bit by bit, an army slowly marching across the earth and claiming it as their own. Caroline's doctoral research showed that, in North America, where the fossil record of the transition is most complete, open habitat grasses started to make their presence known in the

Oligocene. They became ever more abundant—engulfing the ever-expanding plots between the ever-evaporating forests—over a course of nearly 10 million years, culminating in full-on grasslands by the Miocene, about 23 million years ago.

This was a big deal. An entirely new ecosystem had been invented, with an all-you-can-eat buffet of greenery that grew, constantly, from the soil. It grew from the bottom up, rather than the top down like the leaves of a tree, and the more it was nibbled, the more it would grow back. Thicker. Heavier. Denser. The mere act of eating grass, therefore, helped grasslands spread. Not only that, but constant grazing ensured that slower-growing plants like bushes and trees had to struggle for any foothold. Continental interiors became oceans of grass, and for horses and other mammals, it was a gift, like the biblical manna sent from heaven to frost the ground, which sustained the Israelites during their desert exodus. Grass, by analogy, sustained the horses—and still does.

Just one problem. Grass is challenging for the palate. Unlike tender leaves, fruits, and flowers, grass is abrasive. Its stringy and fibrous texture is often tougher than that of leaves, but there are two bigger difficulties. First are those phytoliths—the silica blobs that grasses secrete, which are helpful to us because they readily preserve as fossils, but are a nuisance to grazers, as they are in effect little grains of sand within a grass salad. Second is grit. Because grasses grow near to the ground in wide open spaces, they are a magnet for dirt, dust, and other windblown particles. Many grazing mammals today ingest an unholy amount of grit as they nibble. On average, domestic cattle swallow about 4–6 percent dirt, compared to less than 2 percent for leaf-eating browsers. Sheep, which crop grass closer to the ground than cows, have it worse: in New Zealand, they have been observed to consume 33 percent dirt—in other words, one ounce of dirt for every two ounces of grass.

The dirt and phytoliths function as sandpaper, filing down the teeth of grazing mammals as they eat. This is not a trivial concern: grazers today lose about three millimeters of tooth every year, the enamel literally scoured away. That might not sound like much, but consider this. My molar teeth are about a centimeter (ten millimeters) tall above the gum line. If I ate nothing but grass, my teeth would barely last three years. As a mammal, I can't keep growing new teeth, and once each set of baby and adult teeth is gone, then I would need serious dental work. Horses and sheep and cows can't be fitted with dentures, so once their teeth have been ground to dust, the only option is starvation.

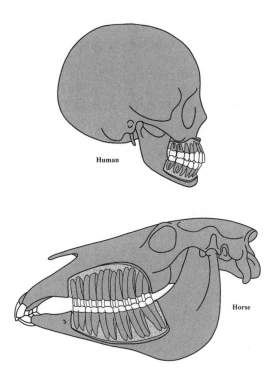

The stretched hypsodont teeth of a horse, with long roots that extend deep into the jaw, compared to the shorter-rooted teeth of humans. Part of tooth exposed above the gumline shown in white. (illustrated by Sarah Shelley)

The Oligocene and Miocene mammals faced a poisoned chalice: so much grassy nutrition, sitting there, waiting to be eaten; the more they could eat, the more it would grow, but eat too much and it would be deadly. Evolution found a solution: hypsodonty. This is just a fancy word for tall teeth, which prolong the period that a mammal can chew. Imagine if my molars were two centimeters (twenty millimeters) tall—then I could eat grass, and its dusting of grit, for twice as long before my teeth would disintegrate. The more I could stretch my molars vertically, the longer window of time I would have to gorge. Now of course, humans have not done this, because our diets don't require it. But many grazing mammals—most famously Simpson's horses—independently happened upon this same simple fix. They evolved molars (and sometimes premolars) that look as if they are made of taffy, stretched out to a comical degree. These teeth became so tall that the tooth crown—the part that is covered by enamel and exposed above the gums—was too big to fit in the mouth, so much of it remained hidden inside the gums and jawbone, gradually erupting over the life of the animal like the lead in a mechanical pencil. In some mammals, the teeth even became ever-growing.

The development of hypsodonty in response to the spread of grasslands is the key narrative arc in Simpson's Great Transformation. It's what supposedly drove the progression of horses, shaping them from anonymous forest-dwelling fruit-and-leaf-eaters into the majestic icons of speed and grace that we cherish today. Caroline's research—and the work of many colleagues—throws shade on such a simple parable of cause and effect. No doubt that grasslands promoted the evolution of horses and many other mammals, but it's a more nuanced—and considerably richer—story than Simpson imagined.

Horses, as it turns out, were late to the game. They were slow to take up their high-crowned teeth, and slow to take full advantage of the new grassy bounty. So slow, in fact, that many other mammals did it before them. As landscapes increasingly opened in the Oligocene, and grasses filled the patchwork between jungles, the smallest mammals adapted first. Rodents and rabbits, with their insane birth rates and fast generation times, were most malleable to the whims of natural selection. They stretched their teeth and began snipping grasses, and coping with grit, at least 10 million years before horses got hypsodont. Then some large hoofed mammals came up with the same trick of elongating their molars—but these were even-toed artiodactyls, among them a multitude of camels that became the most plentiful big grazers of the Oligocene. Meanwhile, Oligocene horses retained the short teeth of their Eocene ancestors, with minimal wear on the enamel, a sign that they were eating soft leaves, while the rabbits, rodents, and camels seized the more open-range, grazing niches.

At the start of the Miocene, 23 million years ago, the jungles were a lost memory and grasslands had taken their place across swaths of North America. It was only then, when there were seas of grass rather than islands between shrinking forests, that horses finally took notice. It's as if they realized their old way of life was gone, and they would reluctantly need to swap leaves for grass. The amount of wear on their teeth suddenly became more extreme, the high-relief cusps of ancestral leaf-browsers turning into blunter mounds decimated by phytoliths and grit. Once the teeth started to wear, they began to get taller and taller, over the generations. Still, hypsodonty came slowly; natural selection was playing catch-up. There was a five-million-year lag between the first early Miocene horses with severe tooth wear and the first middle Miocene horses with genuine high-crowned hypsodont teeth, like those of today's horses. As they became hypsodont, these horses developed other dental tools for dealing with the rigors of grazing: a maze of razor-thin enamel ridges on the chewing surfaces of their teeth, for macerating and cutting, which sharpen as they are worn down by grass and grit.

They took their sweet time, but by the end of the Miocene, 5 million years ago, horses had perfected the art of grazing, becoming some of the most skilled grass guzzlers in the history of life. They were not alone: at least seventeen groups of hoofed mammals independently went hypsodont, including those potbellied rhinos that were buried en masse in Nebraska, when ash exterminated their grass supply. The animals of the American Savanna adapted in other ways, too. The horses, camels, and other hoofed species became champion runners, lengthening and straightening their limbs into stilts, so they could gallop across the open rangelands. Horses simplified their limbs down to a single finger or toe, rendering them into levers whose only job was to race, all the better for life in the fast lane. No longer tied to the forests, the rodents and rabbits experimented with new ways of moving, like hopping and bounding, and new ways of protecting themselves, like burrowing into the soil, where they would be concealed by the grass above.

All these grass-fed herbivores grazing in the open was an invitation for predators. They, too, had an all-you-can-eat smorgasbord, if they could catch it. The Oligocene and Miocene was the stage for an arms race, a tango of predators diversifying alongside their prey, each side striving to outlast the other. Many new meat-eating mammals terrorized the American Savanna: bears, cats, and dogs, of innumerable shapes and sizes, some with saber teeth for slicing flesh, others with piston molars for cracking bone. Some were ambush predators, able to hide in the tall grass or remaining groves of trees before shocking their victim with a burst of violence. Others lengthened their limbs so they could pursue their prey over short distances before pouncing in for the kill.

Looking at this menagerie of murderers, it's hard to decide which of the predators were most terrifying. Maybe the borophagines,

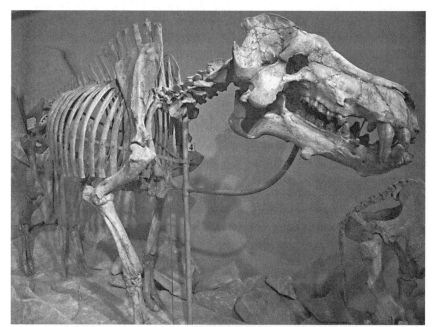

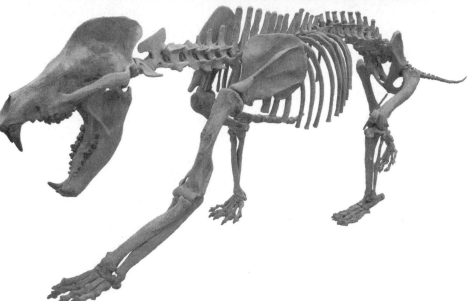

Predators of the American Savanna: the "hell pig" *Daeodon* (top) and the "bear dog" *Amphicyon* (bottom). (photos by James St. John and Clemens v. Vogelsang, respectively)

a thankfully extinct group of dogs—including the hell hound *Epicyon*—that acted like hyenas in wolves' clothing, chasing down prey, then dismembering it with bone-crushing bites. Or maybe the amphicyonids, another group of dead dog relatives nicknamed the "bear dogs" because they seem like a nightmarish mashup of these two creatures blended together. The group's namesake, *Amphicyon*, was about eight feet (two and a half meters) long and weighed thirteen hundred pounds (six hundred kilograms), making it one of the largest carnivores to ever live in North America since the extinction of *T. rex* 66 million years ago.

My money, however, is on the entelodonts. These so-called hell pigs could have easily dispatched a hell hound, if they needed to. They were hideous thugs, with enormous heads, fat bodies, humped backs, and, yet, slender legs for running, tipped with hooves—a dreadful combination of strength and speed. The biggest of them, *Daeodon*, stood 7 feet (2.1 meters) tall at the shoulder, and weighed nearly 1,000 pounds (450 kilograms). Their canine tusks and vise-like jaws could handle pretty much anything, whether leaves or roots, dead carcasses, or live prey. For the animals of the American Savanna, the only solace was that the hell pigs probably spent more time ripping into one another than hunting other species. Their heads were scabbed with gnarly knobs and blisters of bone, a hideous mask to make rivals think twice before starting a fight. Many of their fossil skulls are slashed with wounds and bite marks, battle scars earned in clashes over mates or territory.

When you survey the American Savanna, Simpson's Great Transformation looks equally like a Great Diversification. The changeover from forests to grasslands introduced a new set of mammalian characters: grazers like the horses and rhinos, hyper-carnivores like the hell hounds and hell pigs, speed demons, jumpers, hoppers. The grasslands spawned new niches, to add to those already present in the forests. Cooler temperatures, open spaces, and grasses worked in tandem to turn a limited cast of mammals into a larger, more diverse, more specialized, and more interesting cavalcade of species compared to the jungle dwellers. It's like contrasting the original 1989 cast of *The Simpsons* to the hundreds of characters in today's show, each with their own recurring roles, their own catchphrases.

The more characters in your story, the more convoluted it can be. Alas, there were still many plot twists to come in Simpson's Great Transformation.

The evolution of horses might seem like a tidy narrative, of leaf-browsers becoming grass-grazers as the savannas expanded, and others like rhinos following suit, and then predators growing fiercer to keep up with their prey. This, however, was only one story line of many tangled tales playing out on the American Savanna. When horses evolved high-crowned teeth to mow through the grasslands, not all of them went along on the journey. Many of the Miocene horses were perfectly fine eating leaves, and they stayed in the forests, which of course didn't disappear, but retreated, at first to bubbles within the grass oceans, and then to the warmer tropics and subtropics, in tune with the changing climate. Remember that it wasn't only the singled-toed, grass-eating stallions congregating at that Nebraska waterhole before the ash fell. There were smaller, three-toed horses, too, less adapted for running, and with stubby teeth that couldn't handle much grit, but which were perfect for grinding leaves.

More than an orderly march toward grazing, the Miocene was one big horse dance party. It was their heyday, a time of astounding diversity. Browsers thrived at the same time as grazers, dividing the forests and grasslands into many niches, so that up to a dozen species at a time were able to coexist. Two families thrived in unison, for nearly 20 million years: the grass specialists of the equine group, and leaf-browsers called anchitheriines. Transport yourself back to the Miocene, and it was equally likely you would see a horse eating grass as one eating leaves.

THESE GLORY DAYS, which persisted until just a few million years ago, were the climax of the horse saga. From that point, it became a coming-of-age drama, of loss, and the most unexpected redemption.

In the Pliocene, the time interval after the Miocene, the coolhouse became an icehouse. Glaciers crept across the northern continents, and drier, open grasslands spread farther. The browsing horses all went extinct, not only in North America but around the globe, leaving only the grazing equines. These became the horses of today—the genus *Equus*, which originated about 4 to 5 million years ago in North America. Then *Equus* declined even further, going extinct in North America about 10,000 years ago, a victim of climate change and overhunting, by a terrifying new predator that stood on two legs, more cunning and deadly than any hell hound or hell pig. By dint of circumstance, some *Equus* escaped to the Old World, where they were domesticated by a band of those hominin hunters around 6,000 years ago, in Asia. The Asian horses found their way to Europe, and then the Spanish brought some back to North America during their brutal conquest, a few hundred years ago. Any time you see a pack of "wild" horses on the American plains today, these are not natives with an unbroken pedigree tracing to the Miocene grazers who transformed their teeth and bodies on the American Savanna. They are feral descendants of the Spanish horses.

This goes to show that focusing only on North America can obscure the bigger picture. After all, the entire world was cooling in the Oligocene and Miocene (albeit with a few heat spikes thrown in here and there). Grasslands went global, but at different paces in different places. East Asia seems to have transformed at the same time as North America, starting in the Oligocene and ramping up in the Miocene. In western Asia and Europe, savanna ensembles of horses, rhinos, giraffes, and antelopes were widespread, from the Balkans to Afghanistan, by the end of the Miocene. There were grasslands in Africa by the close of the Miocene, too, the progenitors of today's safari ecosystems of lions, wildebeest, and zebras. South America, cut off below the equator, took it more slowly. Some of its mammals, including Darwin's ungulates, invented hypsodont teeth long before their distant North American cousins. But as Caroline's research showed, this was to deal with ash from the rising Andean volcanoes, not grass, which spread across the nontropical parts of the continent only at the very, very end of the Miocene.

And then there is Australia, which after cleaving from Antarctica around the Eocene-Oligocene boundary, became secluded near the bottom of the world. Outback grasslands would come much, much later here, after a Miocene spectacle unlike anything happening elsewhere.

EACH YEAR, FOR about a week, Mike Archer blows up small parts of the Australian Outback. He usually arrives by helicopter: sometimes military choppers, more often ones rented from cattle wranglers or hunters who shoot deer from the air. Once, he was chauffeured by a lunatic (or in Aussie English, a larrikin), who got buzzed role-playing that famous scene in *Apocalypse Now*, when the squadron of American gunners ambushes a Vietnamese village. Each evening, this pilot—who Mike thinks may have served in Vietnam, although he's not quite sure—would come streaking down the Gregory River, a trickle of moisture in this parched nowhere of northwest Queensland. He would hug close to the water, about a meter above the surface, and keep his rotors as quiet as possible, until he spotted an unsuspecting boater. Then he would blare "Ride of the Valkyries" on loudspeaker, ramp his chopper vertically into the sky, and laugh at the chaos down below.

"Years later, I do find myself wondering if that pilot is still alive," Mike said, chuckling as we navigated the eleven-hour time difference between Sydney and Edinburgh during a pandemic chat.

No matter who's flying him, Mike will take a seat and cruise above the scrublands, scouting for patches of chalky gray limestone peeking out from the khaki-colored grass. He'll look for places where the ground has recently been scarred by fire, exposing the rocky scalp normally overgrown with sharp stalks of *Spinifex* grass and the occasional tree. When he spots stone, he'll instruct the pilot to drop down and let him off. A regiment of students and colleagues will follow him out of the hatch—a scientific army, which, like a real one, is armed with plenty of firepower. Their weapon of choice is detonating cord, thin plastic tubes that resemble computer cables, filled with a mild explosive. When sparked, it slices the rock into pieces, like a knife through cake. As it cuts, it booms—but with a subdued rumble. It wasn't always this way, though.

The first time he tried to liberate fossils with explosives, Mike used a quarter stick of secondhand gelignite, a substance like dynamite that was a favorite of the Irish Republican Army. "When the fuse went off, the rock just evaporated," he told me. Well, almost. "There was one larger piece, about the size of a computer, that launched straight up in the air and nearly took out one of our vehicles." Recounting those wild times in the 1970s, when few people were looking for Australian mammal fossils, and even fewer knew how to collect them, Mike just shrugged. "We were immortal at that point, I guess." But it was lesson learned. If you wanted to study fossils, skulls and skeletons would be much more informative than a pile of rubble.

Mike could have been that mad pilot and not a wisecracking

professor, had circumstances turned out differently. Born in Sydney, but raised in the United States, he won a one-year Fulbright Scholarship, which he took back to Australia on the advice of one of his undergraduate professors, who told him there were fossil mammals to be found in the Outback if only somebody would look for them. This was 1967, as the US was getting hopelessly mired in Vietnam. Mike told his local draft board in New York that he was leaving the country unless ordered otherwise. Befitting the incompetence of the war effort, they ignored him. That is, until two months after he arrived in Australia, when they sent him a letter summoning him back to take his physical exam. The Fulbright commission refused to pay for Mike's flight, so he was spared . . . but only for a few months, as his one-year scholarship was fast ending, and the fund was on the verge of bankruptcy. If the legend is true—and Mike thinks it is—the scholarship administrator took the dregs of remaining money, bet it all on a horse race, won big, and suddenly there was another year of funding for Mike to continue with his PhD on fossil and modern Australian carnivores. Not long after, the lottery let Mike off the hook and the war ended. But by then Mike was totally hooked on Australia. He sent a letter to his dad saying it had been fun but he wasn't coming back from Australia, and he's been there ever since.

In the mid-1970s, Mike found himself at Riversleigh Station, in the empty spaces of northwestern Queensland. Mammal fossils had been reported from the area, but few seemed to care—at least not enough to weather the extreme heat and loneliness of working in such an outpost. Mike found fossils, too—and he kept finding more, as he returned year after year. His greatest haul came in 1983, at a place he nicknamed the Gag Site. "I looked down at my feet and there were skulls and jaws just bristling out of this rock and it was just awesome, everything we had been dreaming about finding in Australia, here it was," he recalled. Only one problem: the

Collecting fossils at Riversleigh, Australia: team members prying out limestone blocks with mammal fossils (top), Mike Archer sitting on a box of used explosives (left), a helicopter delivering supplies (opposite). (photos courtesy of Mike Archer)

rock was hard as concrete, so the fossils could not be removed the traditional way, with hammers, brushes, and dental tools. Mike's team would need to use their boom-cords, carve the rock into more manageable chunks, and bring them back to the lab, where the limestone could be slowly dissolved by acetic acid—basically a dilute vinegar—leaving the bones behind. Many of the skulls are so pristine, with glistening teeth and bleached white bone, that they look fresher than your average modern-day roadkill.

The Riversleigh fossils span, with some gaps, from the latest Oligocene, about 25 million years ago, to the early Pleistocene, an interval of around 24 million years. When Mike started collecting in the 1970s, everyone thought that Australia was swathed in grasslands from the middle of the Miocene, like North America and much of the rest of the world. But the more fossils Mike found, the more he began to doubt it. "It was painfully obvious that not one of the herbivores I was looking at could survive in grasslands," Mike told me, noting that their tooth enamel was paper-thin and they had low-crowned molars, not the obscenely stretched hypsodont teeth of the horses, rhinos, and other grazers of the American Savanna. Instead, their teeth seemed suited for feeding on softer, lusher vegetation, like leaves, flowers, and fruits of the rain forest. It was also painfully obvious that Mike's mammals were not horses, or rhinos, or camels, or hell hounds or hell pigs, or any of the Miocene savanna denizens of North America. Nor were they like anything known from the Oligocene and Miocene of Europe, Asia, Africa, or South America.

Every one of Mike's Oligocene and Miocene mammals, bar a few egg-laying monotremes and heaps of bats, was a marsupial. Some skeletons even had joeys preserved in their pouch regions, mother and baby fossilized together—in a final tragic embrace— during that helpless stage of the infant's life when it was permanently attached to its mother's teat.

It's not a surprise, really, because Australia is overrun with marsupials today. Many of the most charismatic mammals from Down Under are marsupials: koalas, kangaroos, wallabies, bandicoots, possums, and Tasmanian devils, among around 250 total species. There are two exceptions. First, a handful of monotremes like the platypus and echidna—Cretaceous holdovers that we learned about a few chapters ago. Second, some limited placentals: bats, which first flew to Australia during their worldwide diaspora in the Eocene and then diversified into many species, rodents that rafted southward from New Guinea and Indonesia just a few million years ago, and invasive pests like dingoes and rabbits, brought much more recently by the most invasive placental of all time: *Homo sapiens*.

How did marsupials get to Australia? This question has gnawed at Robin Beck, one of the many students Mike has trained over the years. I became Friday pub buddies with Rob when he was a post-doctoral researcher at the American Museum of Natural History, when I was doing my PhD. Before Rob came to New York, he took a similar globetrotting path as his mentor Mike, albeit without the threat of war in the background. Rob is from northern England, won an international scholarship to do a PhD, and chose Australia after Mike tempted him with a project on "weird fossil marsupials," including one he called "Thingodonta." Moving thousands of miles away to study extinct pouched weirdos was a leap of faith, for Rob and his family. The night before he departed for Sydney, his mother looked at his suitcase and sighed. "So you're really going, then?"

The same could be said for the marsupials. As Rob's research shows, Australia was their final stop on a multimillion-year, round-the-world adventure. Recall that during the Cretaceous, the metatherian ancestors of marsupials were prolific across the northern continents. The asteroid put an end to their dreams of dominance, as they nearly went the way of *T. rex* and *Triceratops*. Some were able to flee to South America, where they mixed with Darwin's Ungulates and xenarthrans (like sloths and armadillos) to form an island continent ensemble. Not content to stay marooned in one place, the metatherians continued their trek, using as superhighways the tendrils of land connecting South America to Antarctica to Australia. For whatever reason, the South American placentals made incursions into Antarctica and at least one group appears to have reached Australia, but they didn't establish a lasting foothold there—leaving the metatherians alone to mingle with the monotreme natives. As Australia broke free in the Eocene, it became a laboratory where marsupials could do as they pleased. Many converged on placentals; there are marsupial versions of anteaters, moles, wolves, lions, and groundhogs. Others, as we shall see, did their own things.

The Australian marsupial radiation had begun by the earliest Eocene, about 55 million years ago—the age of the oldest Aussie metatherian fossils. But it really kicked off in the Oligocene, as recorded by the first fossils of modern lineages like kangaroos and koalas at Riversleigh, and reached a crescendo in the Miocene. Although climate and temperature fluctuated over the Oligocene and Miocene, and forests oscillated between denser jungles and more open woodlands, the overall environment remained the same. This was a rain forest realm, not a grassy one. It would have been saturated with humidity, vibrant with the giant leaves and colorful fruits of soaring trees, perfumed with the smell of ripe plumlike fruits—which have actually been found fossilized at Riversleigh. Vastly more common, though, are the fossils of mammals—the jewels of Mike's dreams—which became encased in limestone as they fell into midforest lakes or plummeted into caves.

One of the best examples is *Nimbadon*. More than two dozen of these wombat cousins were found jumbled together in an old cave, having tumbled through what was probably a concealed pit trap on the forest floor. Modern-day wombats are cute, stocky furballs that clamber slowly on all fours, frequently stopping to deposit one of their trippy cube-shaped poos. *Nimbadon* looked nothing like this. With muscular arms longer than its legs, and big hooked claws, it was an ace climber, at home in the trees. Maybe, Mike thinks, they could have hung upside down on their crampon claws, like a sloth. Adults weighed about 150 pounds (70 kilograms), making them the largest tree-living mammals in Australia at this, or any other, time. The leaves and buds of the canopy fueled their plump bodies, and they probably traveled together in mobs.

Surely a *Nimbadon* was a tasty meal—if you could snatch one from the branches. Two types of predators were primed to do just that. The first were the thylacines, dog-mimic hunters that are now extinct, but held on until 1936, when the last stripe-backed "mar-

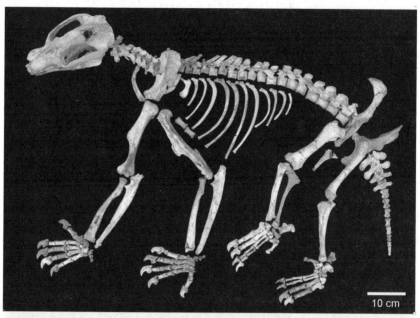

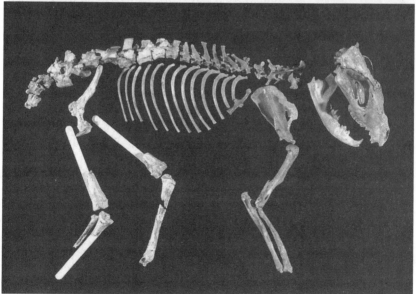

Riversleigh marsupial fossils: the wombat cousin *Nimbadon* (top) and the "marsupial wolf" cousin *Nimbacinus* (bottom). (photos from Black et al., 2012, *PLoS ONE* and Mike Archer, respectively)

supial wolf" breathed its last in a Tasmanian zoo. Many different kinds of fossil thylacines have been found at Riversleigh, including one fox-size species called *Nimbacinus* entombed in the same cave as all those *Nimbadon*s. Its muscular skull and fierce bite could have torn apart prey much larger than itself—like, perhaps, a burly *Nimbadon*. Not all thylacines lusted after blood, though. Across their evolutionary history, they ranged from standard carnivores to bone-crushing specialists, from omnivores to insect-eaters.

The second predators were even more ferocious. These were the thylacoleonids, with a name confusingly similar to the thylacines, but with a distinctive—and stranger—heritage. They were members of the wombat and koala group, otherwise languid planteaters, which went rogue by evolving a hypercarnivorous lifestyle. Nicknamed the "marsupial lions," they became some of the most blood-curdling killers in mammal history, by refashioning their final premolars into huge razors, which would slice flesh like two opposing guillotines as the upper and lower jaws snapped shut. If you were a *Nimbadon*, escaping up the trees was futile: the marsupial lions could climb, using their flexible forelimbs and shoulders. These monsters are now extinct, but the first Indigenous Australians would have encountered the biggest and scariest one of all: *Thylacoleo*, with the dimensions of a real (placental) lioness, up to 350 pounds (160 kilograms) in weight, and with premolar guillotines so powerful that they doubled as bolt cutters, for crushing bone as well as slicing flesh.

Rounding out the roster of Riversleigh marsupials are thousands of other fossils, belonging to dozens of species. Some of these were forerunners of today's marsupials, reminders that almost everything now considered a distinctive "Australian mammal" has its roots in the long-gone rain forests. There were kangaroos, many of which galloped rather than hopped, and a closely related critter

The "marsupial lion" *Thylacoleo*. (photo by Karora)

called *Ekaltadeta* which marshaled its large, shearing premolars for eating both plants and small game—earning it the nickname "killer kangaroo." There were koalas, of different types, living together in the treetops, probably lazy and loud like the one species still around today, but generally smaller, and more like monkeys in some of their behaviors. Shockingly, Mike's team even found the delicate bones of a marsupial mole, which burrowed through leaf litter and moss on the rain forest floor. As contemporary marsupial moles swim through desert sand, this fossil is a reminder that the Australia we know today—which is generally a flat and arid land—was populated by a bunch of rain forest residents who adapted as their environments changed.

Then there are the Riversleigh marsupials that were just wacky:

totally unlike any surviving species, vestiges of a time when pouched mammals were even more diverse Down Under. *Malleo-dectes*, an ever-so-distant cousin of the Tasmanian devil, turned one of its upper premolars into a bloated hammer-tooth, resembling a bowling ball cut in half, which it used to crush the shells of snails to extract their juicy insides. And what about "Thingodonta," formally *Yalkaparidon*, the puzzle Mike used to lure Rob to Australia? Rob spent years poring over its teeth, and as he finished his PhD, he proposed a radical interpretation. Its enormous, ever-growing incisors and small molars with boomerang-shaped crests were a tag team for gouging holes in wood to collect and chew soft bug larvae—making it a marsupial woodpecker, a pouched mammal convergent on a bird!

Such remarkable animals, living in such a verdant habitat, could not last forever. Eventually, grasslands did come to Australia, during the Pliocene, around 5 million years ago. Grass advanced, and wombats and kangaroos became grazers, evolving stretched hypsodont teeth to deal with the phytoliths and grit. Forests shriveled, forcing the koalas into restricted groves where they truncated to a single species hyperspecialized on a single type of tree suited to the new, drier conditions: eucalyptus.

Once again, the cause of all this disruption was climate change. The same climate change that doomed the horses and rhinos in North America and ended the reign of the American Savanna. The relatively stable coolhouse of the Oligocene and Miocene tottered into an icehouse. *The Ice Age was coming.* Glaciers invaded from both the north and south, and much of the earth went into a deep freeze during the Pliocene and ensuing Pleistocene—frigid, dry, windy. Mammals responded, as they always have. This time, some of them became huge, some got woolly, and others leapt down from the trees, began walking on two legs, and inflated their brains.

9

ICE AGE
MAMMALS

Megalonyx

WHO DISCOVERED THE FIRST MAMMAL fossil? It's a simple question with an unknowable answer. People have encountered fossils for millennia, but until recently, few have kept detailed records of what they found, and when. Then there is the issue of what is meant, exactly, by discovery. Who should get the credit: The first person to spot a fossil? The first to collect one? Or the first to correctly identify it and understand that it belonged to a particular type of animal that used to live in a distant era?

Here's something we do know. The first people in North America to find and accurately identify mammal fossils, and have their impressions recorded in writing, were African slaves. The entire enterprise of American vertebrate paleontology can be traced to a group of forced laborers, their names lost to history, kidnapped from their homes in modern-day Angola or the Congo and made to toil in the malarial swamps of coastal South Carolina.

Their discovery happened sometime around 1725, at a plantation called Stono, on the outskirts of Charleston. A little over a decade later, the name Stono would become infamous, the site of the bloodiest slave rebellion in the American colonies, which claimed over fifty lives and led to a brutal crackdown on the already-meager rights of assembly and education granted to the Africans. Later, a skirmish broke out here during the American Revolutionary War—an embarrassing Yankee loss that claimed the life of future president Andrew Jackson's brother. History being what it is in these parts, Stono again saw action during the Civil War. The Confederates captured a Union steamship on the Stono River, part of a string of victories before the war turned and the slaves were freed.

Long before the battles to come, a group of Stono slaves were digging in a swamp, perhaps planting cotton or rice. As they reached into the mucky ground, mosquitoes abuzz in the humidity, they felt something solid. Then more of them. Each was the size of a brick and covered in glistening enamel, with corrugated ridges aligned in parallel on one of the surfaces. To us, the pattern would recall the sole of a running shoe. The slaves, however, needed no analogy: they knew exactly what these things were.

They showed the plantation owners what they had found. Their masters, dumbstruck, realized that these things belonged to a large creature. But what? They defaulted to the explanation favored by so many people of the time, confronted with strange curiosities pulled from the ground: they must have been body parts of biblical beasts, which perished in Noah's Flood. You could imagine the slaves rolling their eyes. No, they insisted, they could explain.

The objects were teeth. Elephant teeth.

More precisely, they were molar teeth, the so-called grinders that elephants use to mill grass and leaves. The slaves were familiar with elephants, having lived alongside them back home in Africa. There were no elephants in the Carolina swamps, though, or anywhere else on the American continent, as far as anybody knew at the time. These were exotic, foreign animals. One can imagine the disbelief on the faces of the plantation owners. Preposterous! The slaves must be wrong.

But the slaves were correct—and everyone would soon recognize it. More "grinders" began to turn up across the northern and eastern reaches of the American colonies, and they were often associated with long, curving ivory tusks. It became evident that there were two varieties of molars: ones with corrugated ridges like the Stono specimens, and others with rows of pointy, pyramid-shaped cusps. For a while, all these fossils were grouped together as "mammoths," in reference to the similar elephantine brutes whose bones were thawing out of the Siberian permafrost. Later, anatomists realized that there were two different American elephants: true mammoths, with ridged teeth for grazing grass, and mastodons, with cusped teeth for snipping and grinding leaves.

One American colonist became obsessed with mammoths. During the late 1700s, there was a lot on Thomas Jefferson's mind—writing the Declaration of Independence, winning a Revolutionary War, preventing his new country from falling apart, running two of the most contentious presidential campaigns in American history, and raising (or at least producing) two families. Through it all, he kept thinking about mammoths. And writing about mammoths, begging people to send him mammoth bones, ordering generals to procure mammoth skeletons. In part, this was escapism. Jefferson loved nature, and, in his words, preferred "the tranquil pursuits of science" to the pugilism of politics. But he had grander reasons, too. In a bestselling book, the noble French naturalist Comte de Buffon had presented his "Theory of American Degeneracy," which held that the cool and wet climates of North America caused its animals to be "feeble" and its people to be "cold," compared to the grandeur of the Old World. Hyper with patriotism, Jefferson saw the mammoth—an elephant larger than those of Africa and Asia!—as his ultimate comeback. It was proof that America was not a backwater, but a land of vitality, with a bright and industrious future.

In his quest to prove Buffon wrong, Jefferson became enamored with other giant bones, too. On March 10, 1797, Jefferson rose to address the American Philosophical Society in Philadelphia. Just six days earlier, he had been sworn in as the second vice president of the United States, after a nail-biting loss to John Adams in the race to succeed George Washington as president. Yet here he was, talking about a bunch of limb bones found on the floor of a Virginia cave. Three of them were claws—and they were big, sharp, and scary. With gifted rhetorical flourish, Jefferson identified them as belonging to "an animal of the lion kind, but of most exaggerated size." Here, he thought, was a titanic American lion, three times as big as those supposedly superior cats of the Old World. A beast so fierce deserved a fitting name, so Jefferson called it *Megalonyx*: the "great claw." It turned out to be even weirder than he thought. A few months later, Jefferson—ever the voracious reader—came across an obscure report from Paraguay, of an "enormous animal of the clawed kind." It had the same claws as his *Megalonyx*, but the rest of its skeleton matched that of a sloth—of monstrous proportions. Later naturalists concurred, and *Megalonyx* was formally named as a species of giant ground sloth, with the species epithet *Megalonyx jeffersoni* in his honor.

One thing Jefferson couldn't accept was that mammoths and giant lions (or sloths) were extinct. Extinction, to him, was impossible: it upset the natural order; by removing a link in the chain of being, it would throw all creation into disarray. In prominently arguing this view, he attracted another rival, once again a Frenchman: Georges Cuvier, the upstart virtuoso anatomist who formally recognized the distinction between mammoths and mastodons. Cuvier contended that both were so different from the well-known African and Indian elephants that they must be distinct species. But nobody had ever seen a mammoth or mastodon in the flesh; all they had were bones and teeth, and the occasional frozen carcass. To Cuvier, the simplest explanation was that these "megafauna" mammals were once alive, but no longer. Jefferson, though, would have none of it. "In the present interior of our continent there is surely space enough for elephants and lions, and for mammoths and megalonyxes," he said in his 1797 address, holding out hope that once the American West became better explored, living megafauna would emerge.

A few years later, Jefferson could finally do something about it. He ran for president again in 1800, a rematch against his now bitter adversary John Adams. Technicalities in the Electoral College instead narrowed the competition to Jefferson and the man who was supposed to be his vice presidential running mate, Aaron Burr

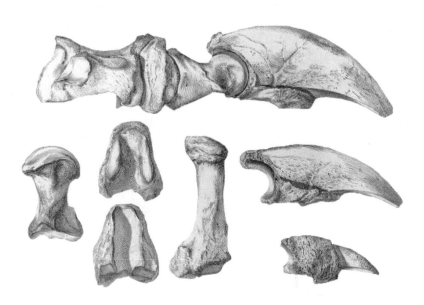

Thomas Jefferson's ground sloth *Megalonyx*. The opening lines from Jefferson's 1797 research paper, an early illustration of the bones, and a modern rendition of the skeleton. (skeleton photo by MCDinosaurhunter).

N°. XXX.

A Memoir on the Diſcovery of certain Bones of a Quadru-ped of the Clawed Kind in the Weſtern Parts of Virginia. By THOMAS JEFFERSON, *Eſq.*

Read March 10, 1797. IN a letter of July 3d, I informed our late moſt worthy preſident that ſome bones of a very large animal of the clawed kind had been recently diſ-covered within this ſtate, and promiſed a communica-tion on the ſubject as ſoon as we could recover what were ſtill recoverable of them.

(who would later kill Alexander Hamilton in a duel). In February 1801, as the capital was dripping with intrigue and his political future hung in the balance, Jefferson was corresponding with a doctor, trying to acquire even more mammoth bones, this time from New York. Meanwhile, the election was thrown to Congress, and on their thirty-sixth ballot, Jefferson was finally elected president. With the powers of the national purse, he bought a vast spread of western North America—the Louisiana Purchase—from the French in 1803, and commissioned a politician and a lieutenant—Meriwether Lewis and William Clark—to explore it. Their tasks were many, but there was one thing Jefferson personally asked them to do: find animals "deemed rare or extinct," to prove Cuvier wrong.

Alas, on their journey from the Mississippi River to the Pacific Ocean, Lewis and Clark didn't spy any living mammoths, giant sloths or lions, or other megafauna. Jefferson remained undaunted. Not long after he returned from his western expedition, Clark was given another presidential commission: to gather megafauna bones from a place called Big Bone Lick, near the Ohio River in northern Kentucky. Here, Native Americans had found countless skeletons, which they considered as giant buffaloes and other tasty prey, killed by "tremendous animals" that were consequently smited by lightning bolts from the "Great Man" in the sky. Digging fossils was an easier mission than finding a living mammoth, and Clark was wildly successful. He returned with over three hundred bones, which Jefferson scattered across the floor of the East Room of the White House. As a break from matters of state, the president would dip into the bone room and connect the thighbones and shinbones and backbones into skeletons, like he was working on a huge puzzle.

As his second presidential term came to an end, Jefferson surely sensed the inevitable. The American West was becoming ever-better explored, and nobody had spotted a real-life mammoth or mastodon yet. Each day the odds became longer, Cuvier the more likely victor. When Jefferson left the White House, he took part of his bone collection to his self-designed home, Monticello, where some specimens remain today. Others went to the Academy of Natural Sciences in Philadelphia, where on a research trip to study dinosaurs as a PhD student, I was treated to one of the most surreal moments of my life: donning a pair of dainty white gloves—like those once worn by Michael Jackson—to open a cabinet and pick up the claws of *Megalonyx*. The very fossils Jefferson himself handled, scrutinized, identified, and announced as a feisty American lion that day some 225 years ago.

In 1823, an aging Jefferson wrote a letter to his old foe John Adams, now once again a friend. Amid a reflective passage, Jefferson grudgingly admitted that "certain races of animals (have) become extinct." With his concession, the debate was over. The implications were enormous—and came to be widely accepted.

North America was once inhabited by a multitude of mammalian giants. Some were larger versions of species still around today, like beavers. Others were variants of animals still alive, but no longer present in North America, like sloths and elephants. Still others were bizarre creatures with only the most tenuous links to modern mammals. And the giants lived elsewhere in the world, too, and very recently, until just a few tens of thousands of years ago. Many of them held on until about 10,000 years ago—the same time humans were building temples and cities in the Middle East, domesticating cattle, and cultivating grains. The megafauna mammals might be dead now, but our ancestors would have known them.

SPEND ENOUGH TIME in northern Illinois, where I grew up, and you might start believing the conspiracy theories. It. Is. Flat. The type of flatness that dulls the senses, and makes you forget that we live on a spherical, three-dimensional planet. It's a place where corn and bean fields sweep toward infinity, Chicago somewhere in the distance. Where roads are razor straight for dozens of miles, and lonely grain silos stand as the only monoliths against the monotony. It's a flatness that seems almost unnatural, as if somebody took a gigantic iron and smoothed out the landscape. And in a sense, that is what happened.

My surroundings did not inspire me as a child. I wanted to dig up dinosaurs in the badlands, traverse deserts, climb mountains. Then, as a sophomore in high school, I took a geology class. My outlook changed. Mr. Jakupcak—that special teacher who made it all click—taught me not only to appreciate the geography around me, but to read it and recognize its subtleties. Yes, northern Illinois is flat: because a few tens of thousands of years ago it was covered by glaciers, which crept down from the Arctic during a cold snap, in that time of global deep freeze that we call *The* Ice Age. The ice sheets were up to a mile thick, and they moved, flowing like molasses, expanding and contracting with the heartbeat of rising and falling temperatures. As they advanced, the glaciers scoured the land, ripping up rocks and dirt, filling in valleys, sanding down hills. Behaving more like a Brillo pad than an iron, they ravaged the topography and made it horizontal.

Mostly horizontal, that is. After all, these continental ice sheets were not smooth and featureless, like some perfect layer of frosting draped atop a cake. They were slashed with crevasses and hollowed by tunnels, where streams of liquid water glided as blood through veins. They were shattered by faults that rumbled like tectonic plates as the ice moved across uneven ground. And they were very, very dirty. The glacial front would have been a sloping, soupy mess of soft ice mixed with sand, gravel, and dust scraped off the land, a scaled-up version of the stuff pushed to the side of the road by a snowplow. All this complexity left marks on the land-scape. As the glaciers melted, they left behind a crime scene of vandalism—largely flat, but pockmarked and scabbed in places, subtle wounds that, to the knowing eyes of geologists, are calling cards of an ice age.

On class field trips, and summer jaunts in his Buick to collect fos-sils, Mr. Jakupcak trained me to spot the clues. As my eyes became better tuned, signs of ice popped out from the farmlands, as invis-ible ink held to the light, transforming the seemingly boring ter-rain into a tapestry of glacial features. About ten miles south of my hometown of Ottawa—on the edge of a valley where floodwaters from a melting glacier surged 19,000 years ago—there is a long, curving mound of earth, elevated some two hundred feet above the plains. It is one of several concentric moraines, which on a map appear to emanate outward from Lake Michigan like ripples from a stone dropped into water. In actuality, they are waves of retreat, directed inward: each marks the place where the glacier stopped in its tracks, dumped its sedimentary load, and withdrew ever farther to the northeast when temperatures warmed. Other landforms are even more inconspicuous. Tiny ponds, which I had thought were farm runoff, are actually "kettle" lakes formed from melted ice chunks orphaned by the retreating glaciers. Small sinuous "esker" ridges are the sandy bottoms of the streams that coursed within the glaciers, and cone-shaped "kame" mounds are gravel and sand that accumulated in depressions on the surface of the ice sheet.

Many of the moraines, ridges, and mounds in the Ottawa area are mined for stone, to put into the concrete of those razor-straight roads linking the farms. Growing up, we called the quarries gravel pits, but they contain more than gravel. Inside you can find any-thing that happened to be frozen inside the glaciers and shed when the ice melted. There's finer material like sand and the windblown dust that makes Illinois's rich agricultural soil, bigger rubble like cobbles and boulders, and sometimes . . . fossils. For entombed within the chaotic mélange of glacial garbage are bones and teeth of all of Thomas Jefferson's favorites: mammoths, mastodons, and giant ground sloths, along with a host of other oddities: huge beavers, bison, musk ox, and stag moose. These were the animals that endured the Ice Age. They would have gawked at cliffs of ice taller than skyscrapers, shivered while foraging for food in the snow, and felt the bone-chilling breezes whipping off the glacial front.

It's astounding to think about it: a little more than 10,000 years ago, about half of North America was a frozen wasteland. What is now Chicago, New York, Detroit, Toronto, and Montreal were roofed by ice, many thousands of feet thick. It was not simply an American phenomenon: much of northern Eurasia was icebound, too, with glaciers covering Dublin, Berlin, Stockholm, and my current home of Edinburgh. South of the equator, the Antarctic ice sheet wasn't able to do the reverse, and creep northward onto the continents, but that's only because the big landmasses were too far away to reach. Glaciers, however, did grow over the Andes, spilling onto a portion of Patagonia. And although much of the rest of the south was ice-free, many regions became cool and dry, a strange type of desert.

What's even more astounding is that this global freeze—which began about 130,000 years ago, was at its coldest peak about 26,000 years ago, and ended about 11,000 years ago—was not *the* Ice Age. It was but one phase of the Ice Age, one of dozens of cycles of glacial advances and retreats over the past 2.7 million years (mostly during a time called the Pleistocene) that, added together, make up what we call the Ice Age. Rather than a long stretch of frozen hell, the Ice Age has been a roller coaster of cold snaps when glaciers spread from the poles far onto the continents (called glacials) and warm spells when the ice melted back (called interglacials). And a harrowing roller coaster it has been, with extreme climate variability and violent shifts between hot and cold. Within the last 130,000 years alone, there have been times that Britain was buried by mile-deep ice, and others when it was warm enough for lions to hunt deer and hippos to bathe in the Thames. Changes between these bipolar states have happened fast, often within a few decades or centuries. Sometimes, indeed, within the span of a single human lifetime.

Most astounding of all is one simple fact. *We are still in the Ice Age*. We're just in an interglacial period, and the ice sheets are on hiatus. Before too long, we're due to lurch back into a glacial phase, when ice will smother Chicago and Edinburgh yet again. However, all the greenhouse gases we're pumping into the atmosphere will probably suppress it: one positive, and perhaps unexpected, side effect of global warming.

Only a handful of times in Earth history qualify as true ice ages, when glaciers capping the poles become overambitious and expand to lower latitudes, far into the continental interiors. It takes remarkable changes to the climate system to make it happen. The roots of our current Ice Age go back to the Eocene-Oligocene transition, 34 million years ago, which we learned about in the last chapter, when Antarctica became stranded over the South Pole and frosted with glaciers, crossing a climatic threshold that turned a hothouse world into a coolhouse. This provided a baseline of cool global temperatures, which got even colder in the Pliocene, between 3.3 and 2.7 million years ago. Another threshold was crossed, and a big ice cap nucleated at the North Pole. With ice sheets on both poles, itching to grow and move, the earth officially entered an ice age.

What caused temperatures to get colder in the Pliocene? There seem to be two major factors. First was something we hinted at in the last chapter: a long-term decline of a key thermostat controlling global temperature—the amount of carbon dioxide in the atmosphere. This was probably due, in a roundabout way, to the rise of the Himalayas, Andes, and Rockies over the last few tens of millions of years. As mountains grow taller, they are inevitably mowed down by erosion. When rocks are eroded, they dissolve and react with carbon dioxide, forming new minerals—effectively locking up carbon dioxide and preventing it from warming the atmosphere. Taller mountains mean more erosion, and thus more carbon dioxide sequestered from the atmosphere, weakening the greenhouse effect and cooling the earth.

Then second, on top of this longer-term trend came an unexpected event of geography, a chance meeting of strangers that would set the world on a new course. Since before the dinosaur-killing asteroid South America had stood alone, an island continent with its own peculiar fauna, rich in marsupials but lacking in placentals, save for homegrown oddballs like sloths and Darwin's Ungulates, and the descendants of those rodents and primates that transatlantically rafted from Africa. About 2.7 million years ago, South America's bachelorhood ended. The Isthmus of Panama formed, and North and South America came together, spindles of crust approaching each other like the hands of God and Adam on Michelangelo's Sistine Chapel painting. As North and South gently touched, the new Central American land bridge dammed the cross-ocean current that had flowed through the Gulf of Mexico, linking the Pacific to the Atlantic. Flow was rerouted and more Atlantic water streamed northward, bringing more moisture to the North Pole. More moisture meant more raw material to freeze into ice, and so the glaciers swelled.

The new marriage between North and South had other, more immediate consequences for mammals. Two realms separated for over 100 million years were now connected by a migration super-highway, and mammals streamed across in both directions, mixing and mingling like easterners and westerners in those frenzied moments after the fall of the Berlin Wall. It's such an important event in mammalian history that it's given a grand name: the Great American Interchange. For the South American species, long ensconced on their island continent, it was a prison break. Like many escaped convicts, however, they didn't do so well. Only a few gained a foothold up north, including armadillos, sloths, and opossums—the first marsupials to raise their pouched babies on North American shores since their northern wipeout millions of years earlier. Darwin's Ungulates tried to make the move, and for a brief while one species called *Mixotoxodon* lived in Texas, but they couldn't stick—eventually dying out entirely on both sides of the land bridge at the end of the Ice Age.

The story was very different for North American mammals. For them, it was an opportunity for a hostile takeover, and they pushed their way into a new neighborhood, seizing the South American rain forests and grasslands and displacing the natives. Many hoofed mammals swarmed southward, including camels, tapirs, deer, and horses. Pressure from these invaders—whose descendants include llamas and alpacas, two of the most iconic South American mammals today—is probably among the reasons Darwin's Ungulates met their end. Many predators also moved south, the ancestors of today's jaguars and cougars, and southern wolves and bears. They may have delivered the death knell to the incumbent apex predator mammals, sparassodonts like the "marsupial saber-tooth" *Thylacosmilus*, although these once dominant pouched hunters seemed to be on their way out anyway.

Meanwhile, as the mammals were marching and mixing, the ocean currents kept delivering more moisture northward, and the ice caps grew. Then the ice started to pulsate: expanding and contracting as temperatures went up and down. Playing pacemaker were the celestial cycles, small and repetitive variations in the earth's movements that control the amount of sunlight our planet receives. Earth's orbit is not perfectly circular, but elliptical, and the shape of the ellipse stretches and compresses over time. The earth rotates on a tilted axis, and the degree of tilt changes over time, too. As does the amount of wobble on the axis, if you think about the earth as a spinning top. In all three cycles, there are times when the earth, or portions of it, will be closer to the sun, and thus showered in more sunlight, and other times farther away, receiving less sunlight.

Often the three phases are out of tune, three musicians in a bad high school garage band, and they swamp one another out. Sometimes, though, they line up in harmony. Like the bass line joining the drums and guitar to make beautiful music, there were times when the cold phases of the cycles coincided and conspired to make temperatures colder. Colder temperatures, more ice, the ice sheets moved onto the continents. These were the glacials. Then, when the cycles got out of tune, temperature warmed and the ice melted back: the interglacials. These cycles are always operating—even as we speak—but it is only when they have enough ice to play with that they can cause havoc and conduct an ice age.

When a mile-deep pane of ice moves onto a continent, the effects are hyperbolic. This is best illustrated by focusing on the most recent glacial advance, the one that peaked 26,000 years ago and littered the northern Illinois countryside with moraines and mammoth bones as it withdrew. As the celestial cycles synchronized, global temperatures crashed more than 12 degrees Celsius (24 degrees Fahrenheit) from their interglacial high. Cold air holds less moisture, so the world became not only cooler, but drier. The northern ice cap bloated, sucking water from the oceans to make into more ice, dropping sea levels by over 100 meters (330 feet). Large areas of continental shelf were exposed, connecting hitherto detached landmasses—like Asia and North America over the Bering Strait, or Australia and New Guinea. A vast belt to the south

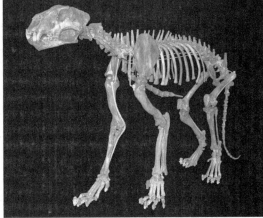

Montage of charismatic Ice Age megafauna: Irish elk (above left), cave lion (above right), dire wolf (opposite, top right), glyptodont (opposite, top left), woolly rhino (opposite, bottom). (photos by Franco Atirador, Tommy from Arad, Mariomassone & Momotarou2012, Ryan Somma, and Didier Descouens, respectively)

of the ice sheets—stretching from Britain and Spain, across the entire Asian continent, and jumping the Bering Land Bridge to continue across North America—was transformed into a new ecosystem. This so-called Mammoth Steppe was a dehydrated prairie, chilled by frosty winds sweeping off the glaciers, where only the heartiest grasses, low bushes, wildflowers, and small herbs could grow. Trees, if present, were restricted to the edges of frigid rivers that bled off the glacial front. Average winter temperatures were minus 30 degrees Celsius (-22 degrees Fahrenheit)—and perhaps even colder.

The Mammoth Steppe was the single most expansive biome on Ice Age Earth, and not an easy place to settle. As always, some mammals found a way. Many of them were huge, and outrageously furry. These were the quintessential megafauna, among them the giants that so inspired Thomas Jefferson. Woolly mammoths, foremost and obviously, as they give their name to their ecosys-

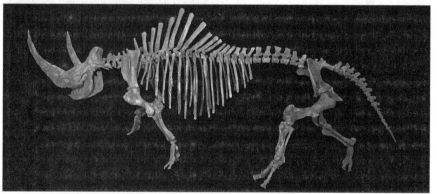

tem. They lived alongside woolly rhinos, two-ton brutes with tremendous nose horns and thick shaggy coats, so unlike the nearly naked and almost reptilian skin of today's rhinos. Then there were monstrous bison, and the last American horses, fed on by a cavalcade of predators, like cave lions larger and more muscular than any modern lion and bone-eating hyenas that hid in caverns. It's easy to understand how bushy fur and bulky bellies kept these megafaunal mammals insulated from the chill.

More temperate environments, farther south of the glaciers, sported an equally remarkable fauna. There were short-faced bears that stood twelve feet tall and weighed a ton—the biggest and baddest bears to ever live—and chisel-toothed beavers larger than humans. There were ground sloths, like Jefferson's *Megalonyx*, that could reach more than a story into the canopy when rearing up on their hind legs, and could have dunked a basketball without jumping. There were cats with saber teeth, an American version of a cheetah, dire wolves, and the misnamed "Irish elk": a deer with antlers so ridiculously oversized as if designed by a mad plastic surgeon.

Elsewhere, far from the wintry touch of the glaciers, Southern Hemisphere mammals had to nonetheless survive the cold, and the dryness. Many of them were enormous, too. Australia was populated by marsupials of the most preposterous dimensions. Wombats that weighed three tons, like *Diprotodon*, the biggest pouched mammal of all time. Pug-faced kangaroos whose 500-pound (225 kilogram) bodies were too fat to hop. And those "marsupial lions" we met last chapter, as big as today's proper placental lions, with premolar bolt cutters to kill and process their prey like some type of terrifying lion-hyena hybrid.

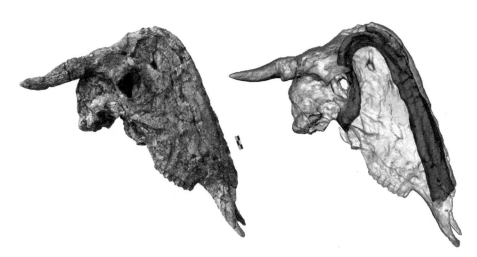

The African wildebeest *Rusingoryx*: photo of skull and CAT scan showing its internal hollow structure and looping nasal passages. (images by Haley O'Brien)

In South America there were armadillos called glyptodonts, the size—and shape—of Volkswagen Beetles, living with ground-dwelling sloths bigger than Jefferson's *Megalonyx* on the other side of the Panamanian land bridge. Plus, the last-surviving and plumpest of Darwin's Ungulates, like the one-and-a-half-ton *Toxodon*, whose fortuitously preserved proteins, as we saw a few chapters ago, proved that these southern hoofed mammals, which so flummoxed Darwin, were relatives of horses and rhinos.

Africa boasted its share of stunners too. There were two-ton cattle called *Pelorovis*, the beefiest cows that have ever existed, with arching horns that resembled a handlebar mustache. Nothing, though, was weirder than *Rusingoryx*: a wildebeest with a bulging, dome-shape snout, hollow inside. Haley O'Brien, who CAT-scanned one of its skulls to see into the dome and understand its function, described it to me as a "wildebeest that could make fart noises with its face." To communicate with each other, of course.

During the Ice Age, no matter where you were, no matter how close you were to the glaciers, there were strange, stupendous, shaggy, and, most of all, supersize mammals. It was a time of mammalian magnificence, and on the clock of Earth history, it was only a few ticks ago.

OF ALL THE Ice Age megafauna, two get most of the attention. Mammoths and saber-toothed tigers. They are the superstars—the most popular museum exhibits, the ones that have become metaphor (it was Jefferson himself who popularized the term "mammoth" as a synonym for big), and the literal Hollywood stars. No surprise, it's Manny the Mammoth and Diego the Saber-Toothed Tiger headlining the never-ending *Ice Age* movie franchise. As with any celebrities, misconceptions about mammoths

Columbian mammoths. (Todd Marshall)

and sabertooths abound. So let's indulge in a proper biography of these two Ice Age icons.

Mammoths are probably the best-understood extinct animal, period. Unlike *T. rex* or *Brontosaurus*, or nearly every prehistoric mammal throughout this book, we actually know what mammoths looked like. That's because we—our *Homo sapiens* ancestors and Neanderthal cousins—saw them alive, and drew them, constantly, on cave walls. Grottoes in France and Spain are plastered with mammoth images, the earliest forms of human graffiti. Our ancestors, it appears, were as obsessed with mammoths as Jefferson.

The artwork is surprisingly accurate. We know, because we can cross-check with real mammoth carcasses, from Siberia and Alaska. In deep freeze for tens of thousands of years, these are not mere skeletons, but ice mummies, covered in hair, muscle still on bone, hearts and lungs and other internal organs inside, eyeballs staring out from eye sockets, genitals exposed, last meals lodged in the viscera. They're not as rare as you might think. The Arctic holds tens of thousands of mammoths, maybe more, and as the

Mammoth and ibex drawn by Paleolithic humans at Rouffignac Cave in France, around 13,000-10,000 years ago.

permafrost thaws, they fall out, like walnuts in a puddle of melting Rocky Road ice cream. Another unexpected side effect of global warming—and a profitable one, for those who can source mammoth tusks from the tundra, to sell on the ivory black market.

As you read this, legions of Russian "mammoth hunters" are on the trail, grizzled men like the gold miners of yore, warmed by pints of vodka, looking for that one big score to save them from a lifetime of post-Communist poverty. It's a messy job. The hunters use firefighting equipment to slurp water from rivers and blast away the permafrost, collapsing riverbanks and polluting the environment. It's also a dangerous job: hunters tunnel for tusks, digging makeshift caves deep in the permafrost that are liable to collapse at any moment. Injured? Good luck—there might be a hospital several hundred miles, and many days' journey, away. And, just to be clear, it is illegal.

The cave paintings and Arctic mummies present an evocative image of the real-life woolly mammoth. It was about the size of a modern-day African elephant: males were around three meters (ten feet) tall at the shoulder and weighed up to six tons, and females were slightly smaller. Tall and compact, with a domed head, humped shoulder, sloping back, bloated belly, and sturdy limbs, there would have been no mistaking a mammoth on the steppe. They were front-loaded: two long, curving tusks lashed forward from the face, but the tail was a small, noodlelike appendage, kept tiny to spare it from frostbite. For the same reason, the ears were small too—much smaller than the satellite dishes of today's elephants, which are oversized for the opposite effect: to shed heat on the boiling savanna.

The most palpable difference between our elephants and woollies, clearly, is the hair. Woolly mammoths were cloaked in it: a scruffy layer of outer guard hairs, each one up to three feet (ninety-one centimeters) long, swaddled a shorter, fluffier undercoat. The hair was associated with extralarge sebaceous glands, which secreted fluids to repel water and improve insulation. Although they are often depicted in films and books as uniformly brown or orange, mammoths had a diversity of hair colors just like humans. Hair ranged from blond to orange to brown to black; some hairs were so light as to be essentially transparent, whereas others were bicolored, a mishmash of multiple hues on the same strand. Light and dark hairs coexisted on the same animal, with light or colorless hairs especially common in the undercoat and darker hairs in the outercoat. Some mammoths may have had a mottled, salt-and-pepper look; others, a more dappled coat of clashing color patches. Overall, light-furred mammoths were much rarer than darker ones, which may seem peculiar, as many Arctic mammals today—like polar bears—have bleached white fur to blend in with this snow. Really, though, this is not a surprise: it's not like mammoths were living on the glaciers, but on the steppe, which was probably snowy only in winter.

You might be wondering how we know all this. In part, it is because the mummies preserve hair, and we can physically observe it. But there is another clue to hair color, and so many other aspects of mammoth biology: their genes. The mummies are so well preserved that they contain genetic material: DNA from a dead species! In one of the most remarkable achievements of modern science, geneticists finished mapping the entire woolly mammoth genome in 2015—an inventory of more than three billion base pairs: the A, C, T, and G letters that wrote the nuclear code that built, operated, and perpetuated the mammoth species. Shocking but true: we know more about the DNA of mammoths than most living mammals.

The mammoth genome reveals many secrets. When subjected to the DNA paternity test and used to build a family tree, it confirms that mammoths are indeed elephants. In fact, they are highly advanced elephants, nested deep in the family album, more closely related to today's Indian elephant than African elephants. Mastodons, that other group of famous Ice Age pachyderms that Jefferson lumped together with mammoths, are a distantly related lineage, not quite true elephants but archaic cousins. One study found that woolly mammoths shared an astonishing 98.55 percent of their DNA with African elephants. Many of the differences have to do with specific features that allowed mammoths to survive in the cold. Geneticists have identified the genes that made mammoth ears and tails small and sebaceous glands large, their hair luxurious, and their bodies more insulated with fat. There were genes that modified their circadian rhythms so they could thrive in the long winters of high-latitude darkness, tweaked their temperature sensors so that they wouldn't freeze, and altered their hemoglobin so their blood could carry enough oxygen in the frigid temperatures.

These genetic mutations were critical adaptations, because mammoths evolved from elephants that migrated from warmer climates. The woolly mammoth—technically known by the species name *Mammuthus primigenius*—is not the only mammoth, but the last of a once populous family, with insatiable wanderlust. Mammoths originated in Africa during the Pliocene, about 5 million years ago, before the North Pole ice cap began to crawl onto the continents. A couple million years later, they leapt north and fanned across Europe and Asia, spinning out new species as they pushed farther into new territory. Around 1.5 million years ago, one of these mammoth species traversed the Bering Land Bridge during a glacially induced drop in sea level and found itself in North America, where it became the Columbian mammoth. Then, a little over a million years later, that same ancestral Asian stock wandered into North America again during another sea level fall. These became the woolly mammoths—one of the *last* megafauna immigrants into North America. Woolly mammoths met the incumbent Columbian mammoths and the two reached an understanding: the woollies kept mostly to the steppes on the fringes of the glaciers, whereas the Columbians preferred the warmer grasslands to the south. Occasionally they would intermingle in the North American midlands, and they still shared enough DNA that they could successfully interbreed—as shown by preserved genetic material of both species.

Far from their hotter African homes, with big bodies and fast metabolisms and a need to stay warm, woolly mammoths developed a voracious appetite. Modern elephants can eat up to 300 pounds (135 kilograms) of plants per day, and mammoths must have eaten at least as much, if not more. They were opportunistic steppe-grazers, and as the gut contents of ice mummies show, preferred grass and flowers (like buttercups), especially in the long summer growing season, supplemented with leaves, twigs, and bark in the winter.

They had several tools for procuring and devouring their favorite foodstuffs. First were their tusks, which despite their menacing appearance, were used less as weapons but more as snow shovels to clear patches of grass and spades for uprooting tubers and roots. The tusks—the modified first incisors of the upper jaw—were gargantuan: up to 14 feet (4.2 meters) long, curved outward and upward in a stylish curlicue, and they grew a few inches per year throughout the life of the mammoth. In some ways, they were akin to our hands: they varied tremendously in size and shape across mammoth populations, and they were asymmetrical, perhaps indicating left- or right-handedness.

Once the tusks gathered food, it was up to the molars to mill it down. The molars were massive—the largest about a foot long and weighing around four pounds (two kilograms)—and complex, with a corrugated surface of parallel enamel ridges for grinding vegetation. It was this unmistakable morphology—the size, the shape, the ridges—that the Stono slaves immediately recognized as elephantine, when they pulled those mammoth "grinder" teeth from the swamp and made the first recorded successful identification of a North American fossil mammal. Like today's elephants, there were only four molars in the mouth—one on each jaw— at a time, and new ones erupted from the back, in conveyor belt fashion. There's something else intriguing about mammoth molars: they were tall. There is a word for this condition, which we saw last chapter: hypsodont. Mammoths, thus, did the same thing as horses on the American Savanna in the Miocene: make their teeth extra deep so they could wear down slowly, an adaptation for eating abrasive vegetation like grasses.

Mammoths were social animals, and they spent at least some time grazing their grass suppers in groups. One site in Alberta, Canada, preserves the dinner-plate-size handprints and footprints of a mammoth herd, walking on sand dunes fashioned from dust howling off the glacial front. Tracks of large adults, intermediate-size subadults, and smaller juveniles are intertwined, in nearly equal proportions. There is reason to believe that most of these tracks belonged to females. Today's elephants are matriarchal, organized into small pods of mothers and their offspring. Males are part of the herd while young, but peel off and become independent, or cohabit in bachelor herds, in their teenage years. Cave paintings depict groups of smaller-size mammoths, with classic female features, congregated together—one of the earliest instances of animal social life recorded by humans.

Mammoth motherhood, or childhood, would not have been easy. This we know, from the stunningly preserved ice mummy of a one-month-old calf discovered on the Yamal Peninsula, a frozen finger of Siberia protruding into the Arctic Ocean, in 2007. The cadaver—the size of a large dog—was spotted by a Nenets reindeer herder named Yuri Khudi, briefly stolen and traded for two snowmobiles, gnawed by wild dogs, rescued, put in a museum, and named Lyuba after Yuri's wife. Some 41,800 years ago, the calf perished while crossing a river, hopelessly mired in the bank and asphyxiated by inhaling cold, wet mud. A short and hard life, a casualty of the relentlessness of the Mammoth Steppe. But in dying, the infant became a time capsule of information on mammoth growth and development—which, ironically, may help resurrect its entire species, should mammoths ever be cloned.

Lyuba is small: a St. Bernard–size baby, whose birth weight would have been around two hundred pounds (ninety kilograms). Barring her mishap, she would have grown into a four-ton adult over a sixty-year life span—a number determined from the count of growth rings in adult mammoth bones and tusks, which were deposited once a year like bands in a tree trunk. Mammoth pregnancies would have lasted well over a year, probably about twenty-one

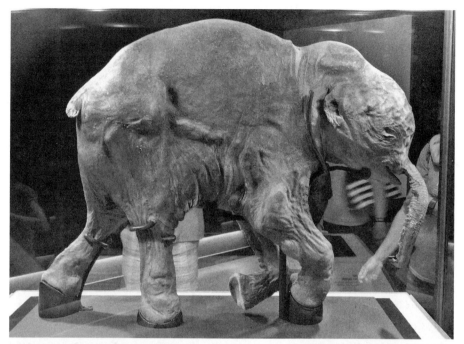

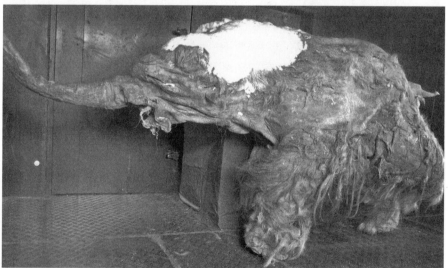

Woolly mammoth mummies found frozen in Siberian permafrost: Lyuba (top) and Yuka (bottom). (photos by Ruth Hartnup and Cyclonaut, respectively)

to twenty-two months like in modern elephants. Mating season was probably in summer or autumn, with births in spring or summer, indicated by the time of year that Lyuba had died, just weeks after she came into the world. Her belly was full—but not with copious amounts of grasses as would the stomach of an adult mammoth, as Lyuba was still nursing. She probably would have relied on her mother's milk for another few years and started eating plants in her second or third year of life. Other juvenile mammoth fossils record this moment of weaning as a shift in the isotope composition of their bones and teeth. Today's elephants wean earlier; delayed weaning in mammoths was likely a consequence of decreased food quality and quantity in their colder, darker habitats. Lyuba didn't ingest only milk, though. Also found in her gut were remnants of fecal material: she probably ate her mother's droppings. Disgusting as it sounds, this is a normal thing for many mammals, to ensure the youngsters develop their gut bacteria.

Nature got to Lyuba, but if she hadn't gotten stuck in that riverbank, predators might have eventually caught up to her. The Mammoth Steppe abounded in carnivores—cave lions, hyenas, wolves, bears—and mammoth would have been a tasty dinner. Healthy adult mammoths were so large as to be unobtainable to even the fiercest of predators, which would have instead targeted the young and infirm. Except maybe, just maybe, there was one meat-lusting monster capable of hunting full-size mammoths.

WHEN YOU THINK about fossils, and where to find them, your mind probably goes straight to the Discovery Channel stereotype. Badlands, in some anonymous nowhere; a guy who looks like Indiana Jones brushing sand off bones, pausing every so often to wipe the sweat from his brow. You probably don't think central

The saber-toothed *Smilodon*. (Todd Marshall)

Los Angeles. Yet, a quick drive south of Hollywood, and just east of Beverly Hills, is one of the world's most incredible fossil tombs.

During the Ice Age, beasts stalked the Valley and the Hills. With apologies to Tom Petty, mammals that looked like vampires would have walked through the Valley, and probably moved west (and presumably east) along what is now Ventura Boulevard, and glided—or at least ambushed their prey—over the rugged terrain of Mulholland Drive. They were big cats, with knives for canine teeth jutting below their jaws in an ominous overbite.

Saber-toothed tigers.

During the last glacial advance, between 40,000 and 10,000

years ago, many of them fell into a trap. Asphalt seeped to the surface, in what is now the La Brea neighborhood, and like flypaper trapped scores of mammoths, bison, camels, and Jefferson's giant sloths. Lured by what they thought was a free lunch, many sabertooths got stuck in the tar, too: at least two thousand of them went "Free Fallin" right on in, judging from the number of skeletons already collected from the La Brea Tar Pits, now a popular tourist site. The asphalt embalmed their bones, making them resistant to decay, providing an unparalleled record of this most famous Ice Age predator.

Saber-toothed tigers. The name strikes fear, a certain anxiety in our subconscious, probably inherited from our ancestors who, each time they went out to hunt or gather berries or partake in some Ice Age gossip, knew these machete-mouthed demons were lurking. Although undoubtedly evocative, the name is half true, half false. Obviously, saber-toothed tigers had sabers for teeth: each upper canine was nearly a foot long. The La Brea sabertooth's formal name, *Smilodon*, means "scalpel tooth," and for good reason: these canines were big, and sharp, and also quite thin, like a surgical blade. But *Smilodon* was no tiger. DNA extracted from La Brea bones, and other fossils across *Smilodon*'s North and South American range, confirms that the sabertooth is a cat, but part of an archaic family that branched off the family tree more than 15 million years ago, only distantly related to today's Siberian and Bengal tigers.

Smilodon was the last of its kind. The sabertooth family was a dynasty, with roots in the Miocene, and over their long evolutionary reign they sired dozens of species. Early on, their domain was Europe and Asia, but during the Pliocene—before the ice sheets began to dance—one of them made it to North America. This migrant got bigger and thirstier for new territory, and as the ice caps grew, it split into two species: *Smilodon fatalis*, the menace of La

Smilodon skeleton and skull. (photos by Ninjatacoshell and Bone Clones)

Brea, and an even larger one called *Smilodon populator* that took part in the Great American Interchange, marching across the Panamanian land bridge into South America around a million years ago. About the dimensions of a modern African lion, but bulkier and up to 880 pounds (400 kilograms) in weight, *Smilodon populator* was one of the largest cats of all time. It had a worthy adversary: *Smilodon fatalis*. The La Brea sabertooth, too, made the journey across Central America. It was a double invasion, which put the sloths, armadillos, Darwin's Ungulates, and marsupials of South America on the defensive. The two sabertooths reached an uneasy truce: *fatalis* prowled the Pacific coast west of the Andes, and *populator* the eastern portions of South America. They would sometimes cross paths, and you could imagine the consequences. Up north, however, *Smilodon fatalis* had no such concerns: it had North America to itself.

Smilodon fatalis was a terrifying animal, worthy of the hyperbole. Yes, it was smaller than its South American cousin *Smilodon populator*, but not by much. At around 620 pounds (280 kilograms), *fatalis* was about the same weight as a modern Siberian tiger, although its bones were more robust, its body bulkier and more muscular, its limbs more buff. Think tiger on steroids, and that's *Smilodon fatalis*. When it sauntered, its high shoulder would have protruded above its back, a warning symbol, like the fin of a great white shark projecting from the water. Its fur color is anyone's guess. Was it striped like a tiger, spotted like a leopard, uniformly drab like a lion? Unlike with woolly mammoths, we do not—yet—have any *Smilodon* ice mummies with lusciously preserved hair. Finding one will be difficult: *Smilodon* was not a denizen of the Mammoth Steppe, where it could be flash frozen if it fell into an icy river or snowbank. Its range was farther south of the glaciers, in the warmer and more pleasant grasslands and forests.

What this means, perhaps disappointingly, is that saber-toothed tigers and woolly mammoths were not adversaries. They might have been casual acquaintances, which occasionally met on the fringes of their ranges, where the Mammoth Steppe gave way to more temperate biomes. They were not, however, Batman and the Joker, or Sherlock Holmes and Moriarty, or *T. rex* and *Triceratops*. Any battles between them would have been far outside the norm. At least some sabertooths did eat at least some mammoths, though. There is a cave in Texas, once the den of *Smilodon*'s saber-toothed relative *Homotherium*, chock-full of juvenile Columbian mammoth bones, gouged with pits and scratches—prehistoric bite marks. As the Columbian mammoth was larger than the woolly mammoth, and *Homotherium* smaller than *Smilodon*, this fight would have been all the more impressive.

Smilodon was a big-game hunter. It preferred forest-dwelling prey, like deer, tapirs, and woodland bison. While it wouldn't have turned down a horse or camel if on offer, these faster-running grassland-grazers were more regularly eaten by the other carnivore found in abundance in the La Brea Tar Pits: dire wolves, of *Game of Thrones* fame—untamed dogs that were slightly larger than today's wolves, with much stronger bites. Dire wolves were true pursuit predators: they chased prey over long distances, with limbs optimized for speed but unable to grab or pounce. To snatch and kill, it was up to their jaws alone. Predators with such stamina came late in Earth history, only during the Pliocene. Even on the Miocene grasslands of the American Savanna there were no true pursuit predators of the wolf variety—the most recent invention of mammalian carnivory.

Smilodon could surely run for short distances, but it was not a pursuit predator. It was an ambusher, which would lie in wait— perhaps camouflaged among the trees and grasses—for an unsuspecting victim. When it pounced, it deployed its saber teeth—very carefully. Something sticking so far from the jaw could easily break, and as a mammal, *Smilodon* could not regrow a cracked canine. Hence, it could not behave like a knife-wielding maniac, slashing with abandon. Nor was it a specialist in blunt-force trauma, or able to latch onto the throat and suffocate struggling prey, like a lion. No, *Smilodon* was a precision killer. It would have stormed out from its lair, subdued its prey with its muscular arms, opened its mouth wide, and delivered a perfectly aimed puncture to the throat with its sabers, then stood back to watch its victim bleed to death.

The saber canines, thus, were not so much knives as ice picks. And they merely delivered the final blow—although it was less of a blow than a poke. This might sound fanciful, but it is backed up by computer simulations of *Smilodon* skulls, which show that not only was this type of predatory strike plausible, but necessary. For all of its assets, *Smilodon* had a weak bite along the postcanine part of its tooth row, so it packed its punch into the sabers. It was, however, blessed with slacked jaws that could open outlandishly wide, to neck-stab prey much larger than itself, like bison. Perhaps, even, Columbian mammoths—especially if one was snared in a tar pit, a siren beckoning with the poisoned promise of an easy meal.

Being sucked into tar was not a pleasant way to die. But before their traumatic last moments, the La Brea sabertooths lived hard-scrabble lives. Their skeletons are riddled with injuries. Paleontologists cataloging the La Brea fossils have found some five thousand *Smilodon* bones with wounds, breaks, and other signs of hurt. They have nearly double the number of pathologies as their rivals, the dire wolves. In part, this is probably because the ambush lifestyle was more perilous than pursuit hunts, which is supported by the high incidence of injuries on *Smilodon* bones involved in leaping onto and wrestling prey, like the shoulders and back vertebrae.

There may be another reason for all the scars: *Smilodon* was likely a social animal, more so than many of today's big cats, and it might have engaged in fierce conflicts over mates and territory. The evidence for *Smilodon* sociality is limited, but captivating. First, it would be strange for so many individuals of a solitary species to wind up in the same tar pits. Second, the hyoid bones of the *Smilodon* voice box, which anchor the muscles and ligaments of the throat, have the characteristic shape of modern-day cats that roar. Roaring—in addition to being petrifying—is how social cats communicate: to project power and warn their prides of danger. If a single *Smilodon* ambushing and knifing its prey isn't horrifying enough, think about a pack of them killing and feeding together.

It wasn't all crime and grime, and there were moments of tenderness in the tough lives of *Smilodon*. Sabertooths were caring parents, as revealed by a fossil site in Ecuador where the bones of a mother and two cubs were found together. This family of southern *Smilodon fatalis*, descendants of that second wave of migrants that crossed the Panamanian land bridge, was swept into a coastal channel, interred with seashells and shark teeth. The youngsters were at least two years old, indicating that they stayed with their mother long after birth. In this, they were like lions, which become independent at about three years of age, and not tigers, which leave their parents much earlier, by eighteen months postbirth. In growth rate, however, *Smilodon* babies matured rapidly like tigers, not slowly like lions. Thus, they developed in a unique way, combining the fast growth of a tiger with the extended childhood of a lion.

Why the long adolescence attached to their mothers? *Smilodon* cubs were born with the robust and muscular proportions of adults, so they didn't need to grow into their bodies. But they were not born with saber teeth, and in fact, it took a good year until their baby-tooth canines were fully erupted. Once those shed, they were replaced by a second and final set of canines—but these adult teeth were not properly in place until about three years after birth, sometimes longer. Without full-on canines, perhaps the young

*Smilodon*s could not perform the hunting ritual of quick ambush and delicate throat-piercing. And maybe, it was such a specialized hunting style that they needed a long apprenticeship before trying it themselves.

Regardless, once they hit three years old, or four at the very most, the kitties were all grown up. Their sabers were footlong weapons, craving the flesh of bison, deer, maybe Columbian mammoths, and other Ice Age megafauna. The hunt was on.

A LITTLE OVER 10,000 years ago, woolly mammoths and saber-toothed tigers rapidly became rare. Some woollies managed to escape to Wrangel Island, a speck of frozen ground a little smaller than Jamaica, above the Arctic Circle north of Siberia. There, they did what mammals often do on islands, since at least the time of dinosaurs: they got smaller. The Wrangel woollies managed to hold on for several thousand years, but they were not well. The island could support a population of only a few hundred, maybe a thousand at most. They accumulated genetic defects, which spread like wildfire through such a small community. Their sense of smell atrophied, their hair lost its luster and became a boring satin. Around 4,000 years ago—as the pharaohs were building pyramids and irrigating cropland along the Nile—the last of the Wrangel mutants perished. And with that, the woolly mammoth was extinct. It joined the sabertooths, which had died out a few thousand years earlier.

Before this pathetic end to the megafauna, during the depths of the Ice Age, all around the world, you would have seen giant mammals like woolly mammoths and saber-toothed tigers. I intentionally use the word "you" here, and I don't mean it in some sort of imaginary, editorial way. We are products of the Ice Age, and members of our species, *Homo sapiens*, would have seen, encountered, hid from, and engaged with many of these megafauna mammals. Today, mere tens of thousands of years later, the giants are almost entirely gone. Only a few land mammals that can genuinely be called megafauna remain: rhinos, bison, and moose among them, along with the closest remaining relatives of mammoths and sabertooths, the endangered elephants and big cats, which if current trends continue, may be the last gasps of their once proud, once diverse, once globe-conquering groups.

If the world seems a little empty now, that's because it is. The megafauna should still be here, and food webs haven't yet fully adjusted to their absence. There is a woolly-mammoth-size hole in the tundra, a saber-toothed-shaped hole in Los Angeles. Their ghosts linger. Why are they no more?

10

HUMAN MAMMALS

Ardipithecus

THE GLACIAL MARGIN, WISCONSIN, 12,500 years ago.

Winter had come with a vengeance. A lone mammoth stood on the crest of a moraine, his seven-foot tusks projecting upward in a pose of defiance. Wind thrashed from the northeast, blasting his woolly orange coat with snow. The sun was low, the evening darkness approaching. Undeterred, the bull raised his hairy trunk to the sky and let out a trumpeting roar, as if tempting the elements. Instantly, his breath turned to smoke.

It was a tough time to be a woolly mammoth bachelor. While the all-female pods hunkered together to keep their babies warm, the males had little to do for the next six months, until the buttercup blossoms signaled the start of the mating season. Some gathered into their own makeshift herds, but this mammoth preferred to stay alone. At thirty-six years old, he was in the prime of his life. If all went well, he would survive another twenty years, maybe more. He always kept to himself, and it served him well. Dozens of the mammoth calves on the steppes and in the spruce forests near Lake Chicago were his.

Ice no longer physically covered this area. It was in retreat, shrinking back toward the Arctic Circle, but it was not going away gently. A short day's walk north, the glaciers lurked. They still controlled the seasons; breezes streaming off the glacial front kept summers cool and made winter hell, preventing the emerging spruce forests from completely reclaiming the grassy tundra. The glaciers controlled the topography, too: the moraine the mammoth was standing on was a new addition to the landscape, a garbage dump of gravel, dirt, wood, and bones dropped by the ice as it fled.

As the moon rose through the dusk, the winds grew fiercer. This was no normal blizzard. Dust devils howled through the chill, the ice sheet exhaling the fine particles of rock it had once pulverized on its march south. The dust mixed with the snow to make a dirty, sludgy mess. The glacier seemed angry, and the highest point amid the flatness was no place to linger. The mammoth looked to the north: a no-go zone of mile-thick ice. The east was no better: the refrigerated waters of Lake Chicago, where icebergs calved off the glaciers in the summer, now a frozen pane. The south was more promising, but once the bull turned to face that direction, he felt the brunt of the frost, focused and following the contours of the lake as a wind tunnel.

Westward ho, the only choice.

Carefully, so as not to fall tusks-first into the snow, the mammoth eased itself down the moraine. One furry foot in front of the other. Each step brought more danger, the blackened snow barely able to hold the mammoth's six tons. Until it wasn't. One foot gave way, then the others, and the mammoth capsized onto his back and tumbled down the ridge.

A graceful sledding performance it was not, but thankfully the moraine wasn't too high, and the stumble was more embarrassing than debilitating. As the mammoth propped himself up and brushed the snow from his face with his trunk, he inhaled deeply with bruised ribs. Time to pause, and wait out the weather. He would have to spend the night here.

There were worse places to pass an Ice Age evening. The mammoth's slide had taken him to the edge of a pond, crusted over with ice. It was poised in a valley, held in place by two moraines, which provided shelter from the wind and dust devils. A spattering of spruces emerged from the snowbanks, their green needles blanketing the ground, inviting the mammoth to lie down and slumber through the storm.

Something, though, didn't seem right.

The mammoth felt a chill, but not from the cold. No, it was the type of instinct that perked up when a dire wolf or cave lion was nearby. A predator was hiding amid the trees, camouflaged by the leaves and the dusty snow and the darkness. The mammoth heard a rustle of branches, and then a curious sound. It was not a roar or a bark, a growl or a snarl. It was much more complex, high pitched at first, rising and falling in a melody. It was a song—one fierce and urgent. Then another mimicked it, and soon it crescendoed into a chorus: many monsters shouting in unison, communicating, taunting their quarry. They were marching off to war.

Moonlight filtered through the spruce grove, and the mammoth finally saw it. Charging ahead was the silhouette of a hunter. It did not run on four legs like a wolf or lion, but only on two. It did not have a long snout of gnarly teeth, but a giant, round head perched on proud shoulders. Its arms were not capped with claws but held a spear. It was not alone, but part of a pack, whooping and hollering as they attacked.

The lead hunter launched its spear, and during the few seconds the weapon hung in the air, the eyes of predator met the eyes of prey. Mammoths, with their big brains and social skills, were long accustomed to being the smartest animals on the steppe. Wolves and lions were vicious, but they were small, and they weren't very bright—at least compared to a mammoth. As the spear lodged into his neck, and the bull gasped for air, a final thought crossed his mind. This new predator was different. Keener, shrewder, deadlier.

The hunters gathered around their prize. After checking that the mammoth was dead, they went to work. Out from their mammoth-skin tunics they pulled a variety of tools. Blades, scrapers, pounders—all made from shiny, hard chert rocks, which they had sourced from the glacial gravel dumps. Racing against the night and the storm, they butchered with purpose and precision. They cut off the feet and legs first, with cleaving blows, and then went for the meat along the shoulders and flank. By the time they were done, a jumbled pile of giant bones was all that remained. They hoisted the mammoth meat on their shoulders and gathered up their stone implements. Why waste a good blade? But in their haste they forgot two of their tools, hidden under the mammoth's pelvis in the bone pile.

With a victory yelp that echoed through the blizzard, the hunting party moved on, over the moraines, on their way to their encampment near Lake Chicago. The dust and snow subsided, the sky brightened, and the summer mating season came—minus one alpha male mammoth. The glaciers continued to melt, their waters trickling into the pond between the moraines, covering the mammoth bones and the stone tools.

THIS IS NOT a book about humans. We, *Homo sapiens*, are but one of more than six thousand species of mammals alive today. Viewed from the longer perspective of mammal evolution, we are a single point, among millions of species over more than 200 million years.

It's not all about us. The tale of mammal history told in these pages—extending back to those scaly critters in the coal swamps, enduring mass extinctions and dinosaurs and cruel climates—was not some mere backstory to our inevitable coronation. Submarine-size whales, woolly mammoths and sabertooths, ocean-rafting monkeys, and echolocating bats are all special in their own right. As, of course, are we: a big-brained, nimble-handed, two-legged primate whose intelligence and capacity for destruction are unparalleled among mammals. That, and we are the only ones that can contemplate our own origins.

One of the most fascinating humans I ever knew was a man I would regularly see walking around the Hyde Park neighborhood of Chicago, when I was an undergraduate. I'll say this in the most delicate but honest way I can: he looked like a drifter. He moved ponderously as if he had nowhere to go, back hunched and a floppy red fishing hat draped over his pockmarked face. Usually he was mumbling something to himself, the words muffled by his long, white beard of the Gandalf style—or was it more like the beard of God? A pocket protector, crammed with pens and notecards covered in the tiniest handwriting, nestled in the breast pocket of his faded flannel shirt. Sometimes our glances would meet, and for a brief second I would peer into his eyes—gentle and a little sad, but with the glint of genius, concealed behind glasses with lenses so big and frames so thick that not even the trendiest hipster would attempt to make them cool.

Leigh Van Valen was his name. Despite appearances (or some might say, befitting of them), he was a professor, ever so slightly on the right side of the razor separating brilliance from madness. He was nominally an evolutionary biologist, but he dabbled in mathematics and philosophy. When he wasn't wandering the campus, he was holed up in his office, said to contain thirty thousand books. To enter was to risk your physical safety: stacks of photocopied papers teetered like Jenga towers, taller than Leigh himself. Some of the graduate students who I befriended worried that one of the stacks would fall and crush him. It was a genuine concern; his students adored him.

"He was a 'character' in the truest sense, as opposed to just a 'jerk,' which seems to be the case for many others in academia who garner that description," Christian Kammerer reminisced to me in an email chat. We met Christian earlier as an expert on the origin of the mammal line. As part of his PhD apprenticeship at the University of Chicago, he took Leigh's rambunctious Evolutionary Theory class, whose required reading included Leigh's original poetry—like one salacious piece on dinosaur copulation, which was quoted in Leigh's *New York Times* obituary when he died in 2010.

Stories about Leigh were legendary. His lunch, rumor had it, was a quart of milk and a rotten banana. He had received a skin graft from his butt to replace a cancerous lesion on his face. Many times, I witnessed him hijack a seminar by asking the visiting

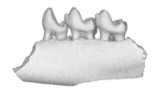

2mm

Leigh Van Valen in his book-strewn office in Chicago (left), photocopied images of *Purgatorius* teeth printed in his self-published journal (top right), and a modern-day CAT scan rendering of *Purgatorius* teeth. (from Wilson Mantilla et al., 2021, Royal Society Open Science)

speaker a meandering question that took at least three minutes to complete, punctuated by high-pitched wheezes the grad students called "gronks." By the end of the question the speaker had usually forgotten what was asked, but did not dare request a repeat. Leigh had even gone into the publishing business: annoyed with journals that regularly rejected his theoretical papers—which were far over the heads of his mortal colleagues—he started his own journal, which he typeset and printed himself. "Substance over form" was its motto, his way of admitting it looked ugly.

Of all his interests and obsessions, Leigh's true specialty was fossil mammals. His greatest discovery—which he published in *Science* in 1965, about a decade before he began printing his homemade journal—was small in stature, but could hardly be grander in its implications. It revealed our deepest origins.

One year earlier, a field crew led by the University of Minnesota's Robert Sloan collected six teeth, from the ominous badlands of Purgatory Hill in Montana. It's remarkable they ever saw them: they were tiny, the largest molar barely three millimeters (about a tenth of an inch) from root to crown. And they were also very old, coming from rocks deposited in the early Paleocene, which we now know was just a few hundred thousand years after the end-Cretaceous asteroid killed the dinosaurs. Most small mammal teeth from this time, and the earlier Cretaceous, had tall and sharp cusps for shearing insects—the easiest way for a featherweight mammal to make a living. To Leigh, however, these six teeth were subtly different. Underneath his microscope, he saw gentler and more bulbous cusps, whose size, shape, and position were not primed for cutting insect exoskeletons, but more suited for eating softer vegetation, like fruits. He made the connection: these wee teeth, which he named *Purgatorius* after the Hill, were a link between insectivorous ancestors and today's primates. The title of his 1965 paper with Sloan—which otherwise is full of opaque jargon about cusps and enamel ridges—says it all: "The Earliest Primates."

For my colleague Mary Silcox of the University of Toronto, who I consider today's leading expert on primate origins, Van Valen's realization was revolutionary. "Understanding that *Purgatorius* was a primate was a thing of genius! It doesn't look like a primate, but we now know it was the starting point," she told me during the Q and A session of an online seminar she gave to my students in Edinburgh, during the pandemic. It was in response to a simple question of mine, much shorter than anything I ever heard Leigh ask a speaker: What is the one primate discovery that I *must* profile in my book? To Mary it was a no-brainer. Although she was inspired to become a paleoprimatologist by reading Donald

Johanson's books on Lucy and other fossil humans, she got swept up by *Purgatorius* and focused her research on the very oldest primates instead.

Purgatorius is a plesiadapiform, a tongue-twister name for the ancestral stock from which primates evolved. Some scientists like Mary follow Leigh in calling them primates, others prefer to call them "stem primates" and restrict the primate title to the crown group consisting of today's species and all descendants of their most recent common ancestor (these are the "true primates" we discussed earlier, which migrated widely during the global warming pulse at the Paleocene-Eocene boundary). Whatever. It's nomenclature, a human exercise in organization and not important. What is important is that *Purgatorius* is the oldest-known animal on the primate bloodline, after it diverged from the other major groups of mammals, and the first to show key changes to diet and behavior indicative of a new lifestyle.

Not only were these first primates eating more vegetation with their modified molars, but they had moved from the ground into the trees—a refuge from the chaos and climate swings and postdinosaur predators. Fifty years after Leigh's original paper, Stephen Chester described the ankle bones of *Purgatorius*, each smaller than a fingernail clipping. They have highly mobile joint surfaces, which would have permitted a wide range of motion for clinging and climbing, as in arboreal mammals today. A few years later Stephen confirmed his inference by describing a more complete fossil: the oldest good skeleton of a plesiadapiform, called *Torrejonia*, collected by Tom Williamson in 62-million-year-old rocks in New Mexico. Not only were its ankles highly mobile, but its shoulder and hips were, too, allowing its arms and legs to glide and rotate in many directions.

These changes in diet and habitat were happening very fast, right after the dinosaur extinction. "These primates seem to rise out of the ground like Zeus," as Mary describes it. The oldest *Purgatorius* fossils, for the moment, are from Montana, found in the narrow band of rock immediately postdating the extinction, no more than a couple hundred thousand years after the asteroid. There are rumors of older ones. Leigh himself described a seventh tooth from Montana, found alongside a *Triceratops* skeleton, which he thought was Cretaceous in age. It has since been interpreted as a Paleocene tooth tangled together with Cretaceous dinosaurs, by an ancient river out to trick paleontologists. Still, DNA family trees of mammals imply that primates originated in the Cretaceous, so maybe our primate ancestors actually survived the asteroid instead of just taking advantage of it.

As the Paleocene unfolded, and eventually gave way to the Eocene, plesiadapiforms diversified from their *Purgatorius* origins. They became a hugely successful group of more than 150 known species, which lived across North America, Europe, and Asia. They made it as far north as Ellesmere Island, above the Arctic Circle: the most poleward primates until humans. With their astounding diversity of size, diet, and behavior, they were one of the first great radiations of a placental mammal subgroup that endures today. The smallest were barely heavier than a grape—the meekest primates ever—while others got as large as an organ-grinder's capuchin monkey. Among their foods were fruits, gums, seeds, pollen, leaves, and some of the insects favored by their pre-*Purgatorius* ancestors. All of them were arboreal, but with different skill levels. The bigger ones probably moved slowly and stuck to tree trunks, others bounded through the canopy, and one called *Carpolestes* had long fingers and opposable toes for grasping firmly onto branches.

As you trace the evolution of plesiadapiforms during the Paleocene, some trends emerge. Animals like *Purgatorius* went into the trees first, before developing solely vegetarian or frugivorous diets, and before their brains became huge. Even once they started eating fruit more exclusively, their brains remained small—as Mary showed by CAT scanning the skulls, and digitally reconstructing the brains, of various plesiadapiforms. Their visual sense was limited, too, as the optic regions of the brain were trivial and their eyes faced to the sides, precluding sophisticated 3-D vision and depth perception. Thus, big brains and high intelligence were not necessary to launch primates into the trees, nor to start eating fruit. But the arboreal habitat and fruit-rich diets may have been prerequisites for making a huge brain, by providing high-calorie meals inaccessible to most other mammals.

What does seem to be linked, however, is the evolution of grasping and fruit-eating. Features of the hands and feet that permitted better branch-grabbing evolved alongside dental features enabling

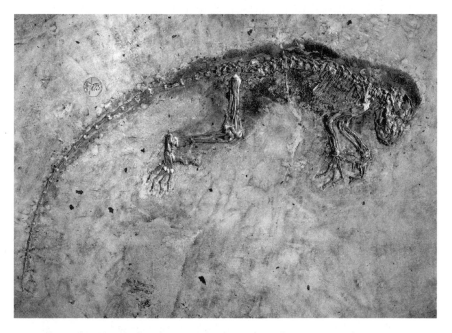

The early primate fossil *Darwinius* from Messel, Germany, showing its gracile fingers and toes and grasping thumbs and big toes. (photo from Franzen et al., 2009, *PLoS ONE*)

better fruit-chewing. To Mary, this suggests that early primate evolution was driven by "terminal branch feeding": primates inching to the edges of branches to pluck fruit and choice leaves. Some of the fundamental components of our human blueprint—like our long-fingered hands, flexible wrists and ankles, and gently cusped molars—are vestiges of this time when our ancestors lunged for food in the trees.

As the Paleocene turned to the Eocene, and temperatures boiled at the PETM global warming spike 56 million years ago, some plesiadapiforms elevated their climbing and sensory repertoires to a higher level. In doing so, they became the true primates—the crown group, the ones everyone agrees deserve the title. *Teilhardina*, that globe-trotter we met earlier that colonized the Bighorn Basin of Wyoming and the rest of the northern continents in rapid succession, is the archetype of these first canonical primates.

They were different from their *Purgatorius*-type ancestors in two main ways. First, they became even better arborealists, able to leap through the branches. To do it, they turned their claws into flat nails, added an opposable thumb and long toes to the opposable big toe and long fingers of their ancestors, and fashioned a more constrained ankle that still could move in many directions but was now stable enough for jumping and sticking the landing. Second, they became much smarter and sharper eyed. Their brains not only got bigger but were reorganized, with a larger neocortex for sensory integration, and larger visual regions that developed as the olfactory regions truncated, reflecting a trade of smell for eyesight. Their bulbous, forward-facing eyes could see in 3-D, and some in technicolor. Bigger eyes and brain plus atrophied nose meant a flatter face, with less of a snout. Many other aspects of our human blueprint—our bug-eyed and pug-nosed faces, our capacity to give a thumbs-up sign and grab objects with our hands, our fingernails and toenails, our ability to see the beautiful hues of a rainbow or the colors of a traffic light, and the roots of our intelligence—are vestiges of this time when our ancestors leapt through the trees as the world roasted.

With these new adaptations, primates spread and diversified with an intensity that matched the extreme Eocene temperatures. Quickly they hopped southward to Arabia and Africa. Their family tree swelled, spinning out lemurs—with lower incisors and canines modified into a comb to groom their fur—and then New World monkeys, which made their ridiculous cross-ocean journey from Africa to South America. At least once, and maybe twice, lemurs also rode the waves, hitching an easterly ride on currents streaming off Mozambique and landing on an island that would become their playground, and later their only sanctuary: Madagascar. Around one hundred lemur species live on Madagascar today, but until just a few hundred years ago there were many more, and they were magnificent. There were koala lemurs as tall as humans, sloth lemurs that hung upside down from the trees on their spindly limbs, and most hallucinogenic of all: the giant lemur *Archaeoindris*, the size of a gorilla, a 440-pound (200-kilogram) silverback that nibbled leaves while trudging across the ground. They all went extinct around the same time, in a cascade of die-offs on islands around the world, following the extermination of woolly mammoths, sabertooths, and other continental Ice Age megafauna.

If the hothouse of the Eocene promoted primate diversification, the shift to cooler temperatures in the Oligocene did the opposite. As Antarctica was orphaned over the South Pole and southern glaciers nucleated, primates were decimated in Europe and eradicated in North America. For unclear reasons, they were not replaced by any of the South American New World monkeys, which after their improbable transatlantic odyssey suddenly grew tired of wandering, making it as far north as Central America before going no farther. For this reason, there are no primates—other than humans and zoo animals and some feral monkeys in Florida—in the United States or Canada today. Things were better in the warmer parts of the Oligocene world, however. While Asian primates did suffer, there were tropical pockets where some—mostly lemurlike forms and smaller ones related to monkeys—persisted.

The real success story, though, was Africa. As the higher latitudes became cooler and drier, the sun still shone brightly on the rain forests of Africa and the Middle East. The epicenter of primate evolution shifted there. During the Oligocene, African primates prospered, recorded by a surplus of fossils from the Fayum of Egypt, above those Eocene rocks bursting with whale skeletons. Some of these primates became the Old World Monkeys, others apes. And from the apes . . .

GADA HAMED WAS an Afar tribesman, who lived in the Awash region of Ethiopia, near the corner of eastern Africa where the Red Sea meets the Gulf of Aden. He went by the nickname Gadi, but the American paleoanthropologists called him Zipperman. They first crossed paths in the early 1990s. As members of the University of California team were searching for fossils, they looked up to see a short, hunchbacked man scowling down at them, an ammo belt across his chest, rifle in hand. His front incisors were filed into sharp fangs, and around his neck was a chain of zippers—kill trophies from when the tribes fought the Communist military junta during a decade-long civil war.

This was not a man to be messed with; the Ethiopian members of the expedition knew it. For they had been scarred by the war, too: one of them, Berhane Asfaw, had been hung upside down and tortured by the Communists, before somehow escaping death, getting a degree in geology, and becoming Ethiopia's leading expert on human origins. He and the other Ethiopian team members

pleaded with the expedition's leader—the eminent American paleoanthropologist Tim White—to pack up camp. White, a pug-
naceous perfectionist and one of the most tenacious sleuths on the
trail of human origins, was not normally one to quit. But this once,
he had to.

The next year White's team returned, and again they met the
Zipperman. This time they brokered a deal: Gadi would join the
team. He soon became a key cog in their operation, multitasking
as a guard, guide, enforcer, laborer, and fossil collector. White always made a point of working collaboratively with his hosts, but
he forged a special bond with Gadi. The Zipperman became a regular companion in White's vehicle, silently brandishing his firearm
as his boss sped through the desert on the fossil hunt. One day in
late December 1993, White had a hunch and stopped near some
rocky outcrops. The odd couple got out for a look.

"Doctor Tim," Gadi yelled out, "come here."

Among a clutter of stones, he had spotted a small bleached
tooth: the molar of a hominin, an early member of the human lineage. Another human trophy for the Zipperman, but one of decid-
edly older vintage, some 4.4 million years old.

White called for his crew to assemble, and all the PhDs scoured
the area where the Afar warrior, trained in combat and camel
herding but never formally schooled in science, had made his discovery. The team kept finding more fossils: a canine first, then
others, comprising a set of ten teeth from one individual. White,
Asfaw, and their Japanese colleague Gen Suwa later described it
as a new species: *Ardipithecus ramidus*—"ardi" meaning "ground"
and "ramidus" "root" in the Afar language. No doubt it was on
the human line: it had small diamond-shaped upper canines, a
hallmark of humans, and not the daggers of chimps and gorillas,
which sharpen as they rub against the lower premolars. Almost
everything else about it was uncertain. Did it walk upright on its

hind legs like humans or climb trees like apes? What did it use its hands for? Did it have a massive brain like ours or a smaller one as in chimps? Those questions would have to wait, for more complete fossils.

They would come, next field season. In November of 1994 White was back in Ethiopia, and to kick off the expedition, brought his team back to Gadi's discovery site. They didn't have much hope; they thought they picked it clean the previous year. But to their surprise, more bones had surfaced. Yohannes Haile-Selassie, another standout Ethiopian paleoanthropologist trained by White, found a couple of hand bones about 50 meters (175 feet) north of Gadi's teeth. With that, it was game on. The team didn't just comb

The joint Ethiopian-American team, of Tim White and colleagues, searching for fossil hominins at Aramis, Ethiopia. (photo by Kermit Pattison)

the surface, but sieved sediment for smaller fossils, and dug into the ground. Over the rest of this field season and the next, they collected over 100 bones from a skeleton of a female, which would have weighed about 110 pounds (50 kilograms) and stood around 4 feet (1.2 meters) tall.

As the team worked, Gadi stood guard atop a ridge overlooking the excavation site, rifle slung over his shoulders. Sadly he wouldn't be there to see the project completed. In 1998, he got into a shootout with a rival tribe. He took a bullet to the leg, the wound got infected, and he died. White paid his respects with a framed photo of the Zipperman on his desk back in Berkeley. That, and by announcing the discovery to international fanfare in 2009, after fifteen years of meticulous study. The so-called Ardi skeleton got its own special issue of the journal *Science*, and a documentary on the Discovery Channel. If any fossil earned its hype, it was this: an animal that seemed part human and part ape, which could walk bipedally on its hind limbs—that signature superpower of humanity—but retained an opposable toe and enormous arms and hands for climbing in the trees. Its face was small like a human, its brain small like a chimp.

Ardipithecus is a hominin—just ever so slightly on our side of the family tree, after we split from our ape relatives. Since the time of Darwin it has been recognized that humans are closely related to chimpanzees, gorillas, and orangutans, a fact confirmed more recently by the DNA paternity test. Chimps are our nearest cousins: we share some 98 percent of our genes. We can place the start of humankind at that evolutionary fork in the road on the family tree: hominins going one way, chimps the other. The split was recent, and messy. It began by at least 5 to 7 million years ago, in the latest Miocene, but was less a quick separation and more a long goodbye: it seems like genes kept flowing between the human and chimp branches until around 4 million years ago, in the early

Pliocene. That was the time of *Ardipithecus*. But don't think that *Ardipithecus* evolved from chimps, or even an ancestor that looked like one. Today's chimps are highly specialized animals, the product of more than 4 million years of independent evolution from humans. Just like today's humans are highly specialized after 4 million years of independence from the chimp line.

These evolutionary changes were happening against a backdrop of climate and environmental change. Before their split, chimps and humans had more distant ape ancestors that thrived in the Miocene. It was a real Planet of the Apes during this time, at least in the Old World. Apes ranged across Africa and Asia and reinvaded Europe after primates were nearly extinguished there during the Oligocene cooling. Although grasslands were spreading in the Miocene, there were still swaths of forests, and the apes adapted to a tree-swinging life by developing long arms and shoulders with a wide range of motion, and losing their cumbersome tails. During this Miocene heyday, some apes in Africa became gorillas, and one population reached Southeast Asia and became orangutans. As apes diversified, they experimented with new ways of moving in the trees. Some walked along branches on their feet, supporting themselves by clinging to overhead branches with their hands. They might have looked like a toddler—learning to walk but not quite there—grabbing on to furniture while shuffling sideways on their feet. Still, it was a start.

Then, in the Pliocene, the cooling and drying trend continued. Now, what had been happening in fits around the world properly hit Africa: the tropical forests withered into pockets, replaced in many areas by open grassland. The African apes adapted, a story told by an incredible sequence of fossils from the Great Rift Valley, a scar stretching southward from Gadi's tribal lands in Ethiopia through Kenya and Tanzania, which has been dropping as Africa slowly rips apart, accumulating sediments and fossils from the Miocene onward. Some apes stayed in the shrinking forests, like gorillas and chimps. But the hominins took another path: out of the forests and onto the open range. To do so, they had to learn to walk, properly.

Standing and walking on two legs is not easy. When you think about it, it's exceptionally rare in the animal kingdom. There's a reason Plato defined humans as "featherless bipeds": if you set aside birds (and their dinosaur antecedents), we are some of the only animals that habitually move on two feet, freeing our hands to do other things. Plato could have also mentioned other human trademarks: our enormous brains and intelligence, our dexterous hands that can grab and hold in so many different grips, our use of tools and ability to tame fire and band together in cultural groups. This is the package deal that makes a human a human, and not a chimp or gorilla or monkey. There has long been debate about how these components came together: Did they evolve in one fell swoop, or more gradually, in a piecemeal way? Did one trigger the evolution of the others? We now know that some parts of our blueprint are relics inherited from those earliest primates studied by Leigh Van Valen, but many others were forged only after the hominin line split from the chimps.

Of all the things that combine to make us human, bipedal and upright walking looks to be the keystone. Gadi's *Ardipithecus* is our earliest good glimpse at how it evolved. Ardi could walk on its hind limbs: it had prominent muscle attachments on the hips to anchor the powerful quadriceps muscles of bipeds, and a strong and broad foot that could push off the ground in heel-off and toe-off motions. Yet it was not exclusively a walker: its opposable big toe and gangly arms were climbing gear. Ardi shows that early hominins did not exchange tree clambering for ground walking in one brief flourish, but went through a stage where they could walk and climb and spent time in both the canopy and on the grasslands.

They were generalists—but clearly exploring new terrain. Why they were moving into the open is unclear: Were they fleeing predators, hungering for new types of food, or just trying to survive as the forests shrank? What we do know is that these early hominins started walking on two legs *before* sprouting big brains and learning to shape tools from stone. Walking upright seems to have enabled these other human innovations—probably by liberating the hands from locomotor tasks and enabling these hominins to eat new calorie-rich foods that could be transformed into brain tissue.

With a little more time and the guiding hand of natural selection, these half-walkers and half-climbers became habitual bipeds that moved, and lived, almost exclusively on the ground. To complete the transformation, these hominins underwent a whole-body renovation. The head was repositioned to sit atop the neck instead of sticking out in front of it. The backbone, which is horizontally oriented and perpendicular to the hind limbs in horses and mice and whales and basically all other mammals, rotated to become parallel to the legs, and took on a curved shape. If you suffer from neck or lower back pain, you can blame it on these anatomical reconfigurations of our ancestors.

This new human profile—tall, dignified, legs and back and neck and head aligned and balanced on two arched feet—is seen in *Australopithecus*, another type of early hominin that lived slightly after *Ardipithecus*. *Australopithecus* is the most famous human antecedent of all, because it is represented by one of the most famous fossils ever found: the "Lucy" skeleton, discovered in Ethiopia in 1974 and named for the Beatles song "Lucy in the Sky with Diamonds," which played on a loop as her bones were excavated. Her co-discoverer, Donald Johanson, made a career out of popularizing Lucy in books and documentaries, after penning an initial scientific description of the skeleton with a young Tim White, long before he made the acquaintance of Gadi, and *Ardipithecus*.

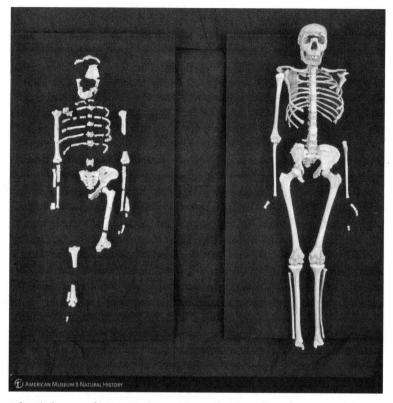

The skeletons of *Australopithecus* (Lucy) and Turkana Boy (an early member of our genus, *Homo*). (photo by AMNH Library)

Australopithecus was a walker, of this we are sure. It is not purely conjecture based on the shape of its bones, because around 3.6 million years ago a group or two or three of these hominins (or perhaps a very close relative) left their footprints in a layer of volcanic ash that had blanketed the ground like snow, then turned to a wet cement. The trackways look like ones we might make on a beach: footprints only and no handprints, deep heel and toe impressions with the subtle imprint of the arch in between, the sign of a confident strider. The trackmaker wasn't very smart, though: *Australopithecus* had a small brain that was organized like the brain

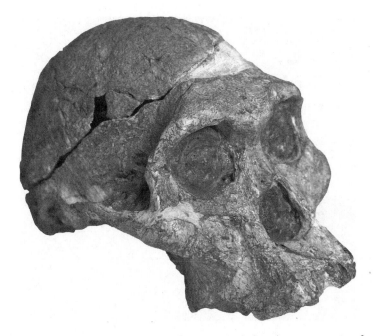

Skull of *Australopithecus*, an early hominin and close human cousin of ours. (photo by José Braga and Didier Descouens)

of chimps. It did, however, develop slowly over a long period of growth—in other words, Lucy and her kin enjoyed long child-hoods like we do, another human trademark.

From *Ardipithecus* and *Australopithecus*, a rich family tree of hominins blossomed. It wasn't a simple, ladderlike tree of *Ardipithecus* evolving into *Australopithecus*, which begot modern humans, in a tidy sequence of grandmother-daughter-granddaughter. Our family tree is more like a bush, with a lush and thorny tangle of ancestors and cousins. The bush was firmly rooted in Africa— the homeland of humanity—for the first few million years of our history. Initially humans were an endemic group, restricted to Africa, where all our great inventions of bipedalism, braininess, and tool use took place. We made Africa our home, and became part of its fabric, as much a savanna native as lions, elephants, and gazelles. By at least the middle Pliocene, about 3.5 million years ago, numerous hominin species were living together across eastern and southern Africa. It's logical, really: just as there are many types of dogs and cats, there were many types of humans. Such diversity was normal for humankind and would persist until very recently. Our single modern human species, *Homo sapiens*, is the lowest ebb of our diversity, and very much an exception to the historical norm.

Two *Australopithecus* leave their footprints in Tanzania, ca. 3.7 million years ago. (photo by AMNH Library)

All this incredible early human diversity was underpinned by diet. Different hominin species specialized on different foods and procured and processed food in different ways. The earliest hominins were probably similar to other apes and ate a smorgasbord of fruits, leaves, and insects. As they moved from the forests into patchier habitats, and then grasslands, some of them became hard-object feeders, with deep jaws studded with huge premolars and molars for grinding tougher foods like roots and tubers. They were, in a sense, a human version of those hypsodont horses on the American Savanna. Others made an even more profound dietary shift: they started eating meat. These first human carnivores left their calling card: butchered animal bones with cut marks from stone tools first appeared about 3.4 million years ago, followed shortly thereafter by the tools themselves. This is the beginning of the archaeological record. No longer was our presence documented only by our fossil bones and teeth and footprints, but now too by the things we made.

Eating meat was a game-changer. Meat has a lot more calories than leaves and bugs, and those calories fueled larger brains. Gluttonous, energy-rich meals also meant that these humans could spend less time looking for food, and less time and energy grinding nutrition out of roots and foliage. Their teeth and chewing muscles grew smaller, bestowing our welcoming smiles and lean faces. More idle time meant more opportunities for socializing, communicating, teaching, learning—the origins of our culture. Meanwhile, toolmaking was a game-changer, too. For the first time, humans didn't need to wait for natural selection to produce new utensils—teeth, or claws, or whatever—for obtaining new foods. We could make the cutters, scrapers, and pounders ourselves. This diversity of weaponry, alongside our flexible and ever-expanding appetites, helped us become supremely adaptable. Around two million years ago, humans were living in a kaleido-

Tools made by various species of humans: core and chopper likely made by primitive *Homo* species ca. 2 million years ago in Tanzania (left), tools likely made by Neanderthals ca. 42,000 years ago in Iran (middle), tool and engraved ochre fragment made by Middle Stone Age *Homo sapiens* in Africa (right). (photos from Mercader et al., 2021, *Nature Communications*; Heydari-Guran et al., 2021, *PLoS ONE*; Scerri et al., 2018, *Trends in Ecology & Evolution*, respectively)

scope of African environments: meadows, woodlands, lakesides, grasslands, dry steppes, and fire-charred landscapes.

Out from the tangle of ancestral branches on the family tree came a new type of human, which first shows up in the fossil record about 2.8 million years ago. *Homo*, our genus. It—or, I should say, we—made our appearance as the local climate lurched into an even drier, more variable state, with even more grass and open territory. There were many species of early *Homo*, and the classification of our closest relatives is an utter mess. From the morass arose *Homo erectus*: taller, standing more upright on longer legs, with shorter arms that signaled a complete divorce from the trees, a flatter face, and a much bigger brain than the hominins that had come before. It was a runner, which chased prey over long distances. Freed from the shade of the forests and exposed to the sun, it probably was among the first humans to lose its mammalian fur. It seemingly was social, and probably harnessed fire to cook its food, and it was particularly violent. It made beautiful tools: pear-shaped hand axes, intricately fashioned by the expert hand of a stonemason, so advanced beyond the simple flaked rocks of the first human toolmakers.

Homo erectus was also the first hominin to move widely, and as far as we currently know, the first to leave Africa.

The long, convoluted, and ever-more-ambitious saga of human migration began with *Homo erectus*, first within Africa. By about 2 million years ago, it had expanded from the rift valleys of northeast Africa to the southern tip of the continent, where it mingled with the last-remaining *Australopithecus*, pushed to the bottom of their world some half a million years after they went extinct up north. *Homo erectus* would not stop there. Its southward journey held in check by the ocean, it ventured northward instead, streaming out of Africa, into the Middle East, and then Europe and Asia. My prose here implies a heroic crusade, but in reality, these *Homo erectus* groups were just tracking food and suitable habitat as the Ice Age climates waxed and waned. They arrived in Asia around two million years ago, leaving behind stone tools embedded in cliffs of glacial dust. Among the Asian *Homo erectus* was Peking Man: a population that lived around 750,000 years ago, whose skeletons are found in caves on the outskirts of Beijing.

As *Homo erectus* fanned across Asia, they hit a wall once again. Upon reaching the fringes of Southeast Asia, they would have looked out at a restless blue sea. Perched anew at the end of a continent, they now went farther. They would cross the water. This, indeed, *would* have been a heroic journey; for them, it was like us flying to the moon. It was much, much more than animals diffusing along migration corridors in step with climate shifts. This trip would require planning, forethought, and teamwork. These *Homo erectus* tribes would have needed to build watercraft and navigate the unknown waves. They probably required some type of language to pull it off. However they did it, they did it. And multiple times.

Different *Homo* populations reached different islands, becoming at least two new species: *Homo luzonensis* on Luzon (now part of Philippines) and *Homo floresiensis* on Flores (now Indonesia). When they settled the islands, these humans did what mammals often do when marooned: they got smaller. The Flores dwarf is nicknamed the Hobbit, and for good reason. Barely three-and-a-half feet (one-meter) tall as an adult, it weighed a scant 55 pounds (25 kilograms) and had a minuscule brain that reverted to the size of a chimp's. But it was well suited to its environment, living for many hundreds of thousands of years in isolation, going extinct only about 50,000 years ago, around the same time that another species of wandering human passed through Southeast Asia on its way to Australia.

While *Homo erectus* was off exploring, the genus *Homo* continued to evolve in Africa, too. Some of these *Homo* populations meandered around the Mediterranean coasts, into the Middle East, the Caucasus, the Balkans, and across Europe. Others stayed in Africa, at least for a while. About 300,000 years ago, in what is now Morocco, the first crumbling bones of *Homo sapiens* entered the fossil record. If you were able to look them in the face, you would see yourself staring back. Put a suit and tie on one of them, stick them on the New York subway, and nobody would notice. Yet these first *sapiens* were not exactly like us: they had our flat and small face, but lacked our globe-shaped cranium, which houses our inflated brain—the largest of any human species (with one possible exception, below).

Over the next few hundred thousand years, other *Homo sapiens* were fossilized, all around Africa, and they were a variable bunch. Some had a flat face, others didn't; some had a projecting chin, others a small one; some had prominent brow ridges, others tiny rims; some had a swollen head that was filled with a globular brain, others not. It seems like *sapiens* did not evolve in one corner of Africa, after a clean break from the other *Homo* species, but had a more mosaic, pan-African genesis. There was a constellation of early *sapiens* populations across Africa, and they mated and migrated within a continent-wide petri dish, mixing their combinations of anatomical features until, sometime between 100,000 and 40,000 years ago, our classic modern human body plan became fixed. Small and flat face, pointed chin, small brow ridges above our left and right eyes separated above our nose, and that stupendously sized brain engorged within our cranium like a balloon, added to the myriad other features we inherited from our hominin and primate and mammal ancestors.

A couple of remarkable things happened around the time that our modern-style *sapiens* bodies took shape.

First, human population sizes increased, and technological and cognitive innovations spread. Like changes to our bodies, this didn't happen all at once, but coalesced over time, as different *sapiens* populations developed new tools and new ways of thinking, and then came into contact with each other. By around 50,000 years ago, tools and other artifacts had become more elaborate. Humans had begun to produce a diversity of ornaments and artworks, perform more complex burials of their dead, manufacture entire tool kits of implements with specific purposes—from projectile points to drills, and engraving tools to knife blades—and build dwellings and other structures complex and resistant enough to survive as archaeological materials today. More or less, this appears to be when humans became modern—not just in the way they looked,

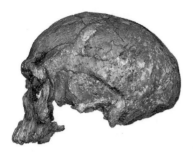
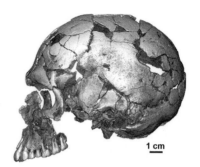

Evolution of the *Homo sapiens* brain region, from a flatter condition in a ca. 300,000-year-old skull from North Africa, to a more globular shape in a ca. 95,000-year-old fossil from the Levant. (photos from Scerri et al., 2018, *Trends in Ecology & Evolution*)

but the way they thought, communicated, worshipped, and sought meaning in the world around them. These people were us.

Second, these humans—*Homo sapiens*, us—were on the move once again. In actuality, we had probably been in motion all along, zigzagging around the Mediterranean, and into Asia, but unable to get too far north because of the ice sheets. An approximately 210,000-year-old *sapiens* skull, described in 2019 from a Greek cave, might be proof of one of these earlier migrations. Then around 50,000 to 60,000 years ago, those occasional forays became a tsunami, as *sapiens* poured out of Africa. They pushed farther than any *Homo sapiens*, or *Homo erectus*, before them, arriving in Australia by 50,000 years ago, crossing the emerged Bering Land Bridge into North America during a glacial interval 30,000 to 15,000 years ago, then rapidly coursing into South America, and later colonizing even the most far-flung Pacific islands and setting foot on ice fields of Antarctica, and then the moon, in 1969.

When the wave of *sapiens* departed Africa, and first entered Europe and Asia, they did not find themselves on virgin territory, unclaimed by other humans. No, they would have encountered at least two other species, both close relatives and also classified in the genus *Homo*: the Neanderthals in Europe, and the Denisovans in Asia. These humans were offshoots of those earlier *Homo* wanderlings, which were ambling around before *Homo sapiens* became a fixed species, with our classic body plan.

Put yourself in the mammoth-skin moccasins of our distant *sapiens* compatriots. Having left the warm confines of Africa to settle the steppes of Europe, where woolly mammals roam and mile-thick slabs of ice encroach from the north, you take shelter from the frozen winds in a cave. A flicker of light dances across the cavern walls, illuminating the figures of deer and horses, painted in red ochre. The light gets closer, and the shadow of an animal leaps across the artwork. You cower with fear, as you're confronted by something that looks like you, but a little stockier, with shorter limbs, and a bulging nose, and wilder hair.

Sapiens, meet Neanderthals. When our ancestors came into contact with these native European humans, they would not have met some dim-witted, drooling, mouth-breathing cavemen of stereotype. Neanderthals were, in so many senses, like us. Their brains were about the same size as ours. They were cultured. They were social. They probably painted, and buried their dead, and cared for the sick by making medicines from plants; they wore jewelry, and maybe makeup; they brandished fire, they performed rituals, they constructed structures—perhaps spiritual shrines—out of stalactites and stalagmites. They could probably talk.

To the east, in Asia, we know far less about the Denisovans. In fact, we didn't know anything about them, at all, until a lone finger bone from a young female was found in a Siberian cave in 2008. Its genome was sequenced in 2010, and to the shock of paleoanthropologists everywhere, was found to have a highly unusual genetic code, divergent from *sapiens* and Neanderthals, indicative of a new species. A ghostly species, still known only from a scatter of

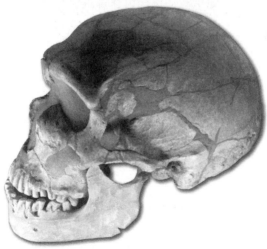

Neanderthals: reconstruction of a possible burial site at Chapelle-aux-Saints, France (top) and a skull from La Ferrassie, France (bottom). (photos by Wikipedia 120 & V. Mourre and Wikipedia 120, respectively)

bones, our knowledge of its DNA far eclipsing our understanding of its appearance and behaviors. We do know they ranged across much of Asia, beginning at least 195,000 years ago—the age of the oldest Siberian cave fossil. They even made it to Tibet, tweaking their blood hemoglobin so they could take in more oxygen to survive in the punishing altitudes, on the Roof of the World. But what did Denisovans look like? Would we recognize their faces? How big were their brains? Did they have culture, make art, practice religion? The skeleton of a Denisovan is—to me—the greatest outstanding prize of paleoanthropology, and it is sure to become as famous as Ardi and Lucy.

So what happened when our African *sapiens* ancestors met the Neanderthals of Europe and Denisovans of Asia? We all bred with one another—in what seems like a mad, transcontinental bacchanalian carnival that lasted for a few tens of thousands of years.

Although three different species, we were related closely enough for genetic exchange to occur—and occur it did. Neanderthals and Denisovans mated and could produce viable offspring, as proven by another improbable discovery of a single bone in that same Siberian cave, announced in 2018. Usually it's skeletons, not a lone shard of limb bone, that earn catchy nicknames, but Denny deserves her fame. The thirteen-year-old girl to which the bone fragment belonged was a first-generation hybrid, with a Neanderthal mother and Denisovan father. Our *sapiens* forebears, too, mated with both Neanderthals and Denisovans, and all this ancient promiscuity contributed to our genomes: East Asian and Oceanian people today share between 0.3 and 5.6 percent of their genes with Denisovans, and all non-African people—myself included—are 1.5–2.8 percent Neanderthal. (Modern-day Africans have, mostly, descended from the *sapiens* populations that remained in Africa, so their ancestors would not have met the European Neanderthals.) And so, our family tree isn't a ladder, and it isn't really even a bush.

It's more like a hedgerow, of many bushes, twisted and mangled and growing together.

When *Homo sapiens* expanded beyond Africa, and into the wider world, nothing would be the same again. As we marched, we hunted. We burned. We brought other animals with us, traveling mini-ecosystems of invasive species, ourselves the most invasive of all. We conquered, and colonized, and killed. Many Neanderthals and Denisovans probably met their fate at the end of a *sapiens* spear. The rest of them were, basically, absorbed into us—living on in our genomes, and maybe in some of the advanced behaviors and rituals and habits that *they* taught *us*. By the end of it all, around 40,000 years ago, the Neanderthals and Denisovans were no more, and those last remaining island remnants of *Homo erectus*—like the Flores Hobbit—were also gone.

The sprawling family tree of humanity had been pruned down to one species. Us, alone, to ponder where we came from.

WHEN *HOMO SAPIENS* left Africa and dispersed across the world, we would have encountered not only other types of humans, but many unfamiliar animals. Woolly mammoths, saber-toothed tigers, and Thomas Jefferson's ground sloths in North America. Other gargantuan sloths, car-size armadillos, and Charles Darwin's strange Ungulates in South America. Monstrous marsupials—multiton wombats and pug-faced kangaroos—in Australia. We met the Ice Age megafauna, and then much of that megafauna died.

The dying happened over the last 50,000 years, during what is called the Near Time Extinction. It was not an equal-opportunity extinction: the deaths happened almost wholly on land and not in the oceans, and the victims were disproportionately large, and many were mammals.

During the brunt of the Ice Age in North America, there were about forty mammal species larger than one hundred pounds (44 kilograms); now about a dozen are left. Everything bigger than about a ton died, as did about half of all mammals between one ton and seventy pounds (32 kilograms). Among the fatalities were the last North American elephants (mammoths and mastodons) and odd-toed perissodactyls (horses). It was worse in South America: more than fifty large mammals were extinguished—more than 80 percent of the megafauna—including every last one of Darwin's Ungulates and all the big sloths and armadillos, leaving only their meager modern-day cousins. Meanwhile, in Australia, everything larger than about a hundred pounds disappeared, such as the giant wombats and "marsupial lions." Things were better in the Old World: about 35 percent of the large mammals died in Europe and northern Asia, whereas Southeast Asia and Africa barely felt the pang of loss. Today, it is only here—on the African savannas and in the Asian rain forests—where "supermegafauna" over two tons remain: elephants and rhinos.

All this death was not normal. The Near Time Extinction was one of the largest mass mortality events since the asteroid killed the dinosaurs 66 million years ago. It far eclipsed the killing power of both the PETM global warming spike and the lunge into the Ice Age, and it is the only mammal extinction over this time that focused intensely on large-bodied species. Logically, it is straightforward to understand why larger mammals might have died. Bigger animals have lower reproduction rates, and produce fewer offspring that take longer to develop, compared to smaller species. Any forces that disrupt population structure and heighten juvenile mortality can thus topple these big-bodied slow breeders. What, then, might have done it? This question has fostered intense debate, which still rages.

The most obvious answer is us: humans. There are a few inconvenient clues that point the finger in our direction. Almost all the charismatic megafauna of North America, South America, and Australia—landmasses untouched by humans of any kind until the recent *Homo sapiens* migrations—went extinct *after* our arrival. We also know that humans can wipe out entire species: as we hopped from island to island over the last millennium, dodos on Mauritius, foxes on the Falklands, the "marsupial wolf" on Tasmania, lemurs on Madagascar, and so much else died in our wake. If we exterminated islands in historical time, maybe we exterminated whole continents beforehand. This led paleontologist Paul Martin to propose his "blitzkrieg" hypothesis for the Ice Age megafauna extinctions: as *Homo sapiens* entered new continents, we surged across the land as a tidal wave of murderers, literally hunting and killing the large mammals until none of them were left.

It's a provocative—if gratuitously violent and overwhelmingly sad—idea. But the blitzkrieg theory has its holes. Most problematic: Where are all the bodies? If we hunted dozens of large mammals to death, then the Near Time fossil record should be riddled with butchered mammoths and sabertooths with stab wounds. There are a few examples: the story that opens this chapter is based on two mutilated mammoths from Wisconsin, found with the stone tools that dismembered them. These are rare, though, and far outnumbered by tool-marked bones of another large North American mammal: bison, a survivor! Similarly, butchered ground sloths and giant armadillos are known from South America, but far fewer than you would expect from an all-out assault. And then there's Australia: so far, not a single convincing case of a megafaunal marsupial killed by humans has emerged.

The lack of crime scenes may point to a different, more deceptive killer. For many paleontologists, the other obvious suspect is climate change. The megafauna was living in an Ice Age, after all, and during the last 50,000 years climate, temperature, and precipitation have changed dramatically. A measly 26,000 years ago much of North America was roofed by ice, then by about 11,000 years ago the glaciers had retreated. Australia was not icebound, but experienced extreme cooling and drying. Some of these very climate changes probably allowed humans to migrate, but perhaps they had the opposite effect on the megafauna. The whiplash from cold to hot and dry to wet was too much to handle, and they died. But there is a glaring problem with this hypothesis, too: the last glacial advance and retreat were but one of dozens during the Ice Age. The megafauna had dealt perfectly fine with the roller coaster of glacials and interglacials before. Why would the Near Time— only the last 50,000 years out of more than 2 million years—be any different?

Humans, that's why. The last glacial-interglacial cycle is the only one, of the entire Ice Age, in which *Homo sapiens* were living around the world.

As debate about the megafauna extinction continues, there is growing realization that it was probably humans and climate, working in tandem, which spelled doom. While we certainly did hunt some of the megafauna, we probably didn't destroy them through blitzkrieg. We can kill in so many other ways, which don't leave physical marks on dead bodies strewn across a battlefield: overexploitation of prey species over generations rather than outright massacring them, introduction of exotic species that disrupt ecosystems, and environmental destruction, like using fire to clear landscapes. Add these to climate change, and it can be a potent cocktail. A harrowing prospect for our modern world.

How exactly did humans and climate conspire to kill the megafauna? There is still much to learn, but an intriguing study of North American and northern Eurasian megafauna reveals one potential mechanism. Over the last 50,000 years, extinctions of large mammals were concentrated during quick warming episodes, which shocked the climate out of a stable state. DNA extracted from their bones suggests that megafauna population sizes got smaller during these warming intervals but would then normally replenish over time, through dispersals that mixed different populations. All this can be explained by climate. But then, if humans were around to interrupt those connections between populations, declines in small groups of megafauna scattered here and there could coalesce across an entire continent, turning climate-caused local shrinkages into human-caused complete extinctions.

Ultimately, this exact domino-sequence of climate and human impact may not explain all the megafauna extinctions. There is sure to be some variability: different mammals might have died out for different reasons, and the pace probably proceeded differently in different places. It might have been lightning fast in the Americas, with most of the big mammals gone within a few thousand years of human arrival, but slower in Australia, where humans arrived earlier, and where the cool and dry climate may have brought different pains than the glacial-interglacial shift in North America. There's also the matter of the African and Southeast Asian extinctions—or lack thereof—to grapple with, but maybe it's as simple as these mammals, having lived alongside humans (*Homo sapiens* and many other species) for millions of years, coevolved with us and adapted to our tricks.

It comes down to this: if our human species had not spread around the world, then a lot of the megafauna would still be here. Maybe not all of them, but probably most. Dinosaurs like *T. rex* and *Triceratops* were felled by an asteroid. For mammoths and saber-tooths, *we* were the asteroid.

WHILE WE WERE killing the megafauna, humans started to do something that no mammals had done before. We started to make our own mammals.

Sometime around 23,000 years ago in Siberia, during the bitter depths of the most recent glacial maximum, both *Homo sapiens* and wolves found themselves isolated by the geography of the ice sheets. We interacted, probably fleetingly at first. Wolves would have been fierce predators for us to avoid, but some started to hang around our campfires, the smell of a mammoth barbecue too enticing to ignore, and there was little we could do about it. The wolves came back more often, began to stick around, scavenged the scraps, and maybe we threw them a mammoth bone or two every once in a while. They learned to live alongside us, to help us hunt, to become indispensable companions on the steppe. Not only did they live with us, but we took care of them, controlled their breeding, modifying their genes to make them more docile. These wolves became dogs: the first domesticated mammal species. When the Siberian *sapiens* crossed the Bering Land Bridge into the new frontier of North America, they brought their dogs with them. They took them to Europe, too, and across the world. Your dachshund, or pug, or golden retriever are all descended from these first domesticated Ice Age wolves—as are the nearly one billion dogs alive today.

We didn't stop with wolves. As we traveled, we scouted for other animals to mold to our needs, in exchange for food and security—a mutual partnership. We sought those that could become our food, or help us procure food, or provide transportation, or companionship. More than twenty-five times we've taken a wild mammal and tamed it into something we ultimately control through selective breeding: a human-powered version of natural selection, under our auspices rather than the vagaries of random mutation. From these breeding experiments came the billions of pigs, sheep, cows, and so many other mammals that are integral parts of our world and comprise an enormous part of the total biomass on Earth today—some fourteen times as much pure weight of muscle and skin and bone as all wild mammals combined.

Much of this domestication occurred in a frenzy beginning around 12,000 to 10,000 years ago. This was the Neolithic, the last part of the Stone Age, and the next big step in human history after the development of our *sapiens* body plan and advanced cognition by some 40,000 years earlier. During the Neolithic, we learned how to domesticate plants, too, ushering in the Agricultural Revolution. Our lifestyles quickly changed from nomadic hunting and gathering to a more sedentary existence, with many people settling into towns and villages. Our fields and irrigation canals and buildings transformed the landscape, leading to more environmental devastation, far beyond what we did to the megafauna. With a ready source of ever-available food, our population expanded exponentially. And because only some people had to spend time actually producing that food, our blossoming brood was free to take up new tasks, divide labor, and create societies. Some of us could become doctors, priests, architects, couriers, builders, teachers, politicians.

And scientists. A little over a quarter century ago, here at the University of Edinburgh where I now teach and write, a scientist did something that was long thought impossible. Ian Wilmut took an adult sheep—one of the countless descendants of an agricultural species domesticated during the Neolithic—and made a copy. Dolly, the first mammal clone. Cloning has now become somewhat commonplace, to the point where people can pay many tens of thousands of dollars to make duplicates of their dearly beloved pet cats and dogs—both domesticated mammals, of course.

Cloning technology is progressing rapidly, leading to the inevitable conundrum of whether humans can—or should—clone ourselves. There is also constant chatter about using cloning to bring back something that is completely dead: woolly mammoths. Those mammoth hunters in Siberia do find frozen mummies with genetic material, and the mammoth genome is now fully sequenced, and mammoths are extremely closely related to modern Indian elephants, which could be a potential "mother" species for a mammoth clone. Making a mammoth certainly won't be easy, and we can debate whether it is ethically and morally acceptable (with all the carbon dioxide we are pumping into the atmosphere, it will soon be much warmer than any mammoth ever experienced), but I do think it will happen, and there will be a Nobel Prize in it for somebody.

And if it does happen, it will give us that rarest of opportunities: a chance to make amends for our sins, and resurrect something that would still be alive, if it wasn't for us.

EPILOGUE
Future Mammals

African lion

IT'S LATE JULY and I'm in Chicago. My fieldwork digging up mammal fossils in New Mexico has finished, so it's the rare few weeks I can spend with my parents and brothers before flying back to Scotland. On a day like today—summer heat radiating from the sidewalks, the air sticky with humidity as a thundercloud gathers—it is impossible to believe that a little over 10,000 years ago this area was underneath an ice cap. The ice is gone, but left part of itself behind. Lake Michigan, from where the Windy City's breezes blow, is a giant puddle of glacial meltwater.

We're at the Lincoln Park Zoo, about five hundred feet west of the lake. Mammoth hunters once camped here, not too long after they crossed the Bering Land Bridge and entered a New World. Their inheritors, many successive tribes of Native Americans, settled permanently. When Europeans rudely arrived in the 1600s, and confronted the Potawatomi, the lakeshore was a soggy swamp, wafting with the stench of wild onions—a plant some of the natives called *shikaakwa*, which the French wrote down as Chicago. A handful of European loners stuck around, and the US military put up a fort that was sacked by the Potawatomi (now aligned with the British) during the War of 1812, but it was not until the 1830s—less than 200 years ago!—that Chicago became a town.

From the first couple hundred residents, things moved fast. Skyscrapers were invented here, and now surge from the lakefront as man-made mountains. Tentacles of rail lines radiated from downtown, creeping across the United States, connecting a continent. Later came the horrors of O'Hare, one of the world's busiest airports, as *Homo sapiens* became a globe-trotting species, able to hopscotch continents in a single day. For a time, Chicago grew faster than any city in the world, and so heaved with people that they reclaimed habitable land from the lake, by dumping sand—once moved by glaciers—to fill in the mess the ice left behind. As I peer over the zoo walls, traffic is backed up on Lake Shore Drive.

Each car silently belches out carbon dioxide, invisibly percolating into the atmosphere, one molecule at a time.

A roar pierces the afternoon stillness. Everyone around us stops and looks over to the lion house.

An alpha male, mane in full bloom, is strutting upon a rock cliff like a dictator on his balcony, snarling at the plebs below. Chicago is a Bears town—although it's odd that my beloved (and always exasperating) football team was named after animals the European settlers pushed out of Illinois by the 1870s. Today, however, the lion is staking his claim. A savanna cat, once widespread across Africa and southern Europe and Asia, his kind is now reduced to a few tens of thousands in the wild, in shrinking bubbles of grassland. For now, their fate is suspended, as many thousands pace away in zoo purgatory. Big cats seem out of place in a midwestern metropolis, but during the Ice Age American lions would have lived here, and saber-toothed tigers, too.

This ornery lion is one of many mammals at the Lincoln Park Zoo. These furry creatures are—in the overall scheme of 4.5 billion years of Earth history and the many millions of species that have ever existed—our very dearest cousins. Each one carries the hallmarks of mammalness that evolved, piece by piece, over the last 325 million years, starting when those scaly critters in the coal swamps broke away from the reptile line. Hair, big brains, awesome senses of smell and hearing, incisors and canines and premolars and molars, fast-paced and warm-blooded metabolism, babies raised with milk. All these mammals—including us—had ancestors that endured mass extinctions, hunkered in the shadows of dinosaurs, and faced both asteroid and ice.

As my wife, Anne, and I stroll past the pens and cages, the story of mammal evolution unfolds in front of us—past, present, and future. Most of the mammals here are placentals, but there is a red kangaroo on loan from Australia, a marsupial carrying its baby in a pouch. Giraffes and zebras—two of many hoofed mammals that proliferated as grasslands spread—tempt the lions from a distant enclosure. An afrothere (an aardvark) and a xenarthran (a sloth) harken to those times that Africa and South America were island continents, incubating their own weird mammals in isolation. There are bats: mammals that reinvented themselves by turning arms to wings and taking to the air. Thankfully, there aren't any whales in this small urban zoo, but there are seals: another type of mammal that refashioned its land-living body into the physique of a swimmer. Much less exotic are the cows, pigs, goats, and rabbits: mammals we know well, the ones we eat and keep as pets, the ones we domesticated as we built cities and civilizations.

Not all these mammals are well. Many, like the lions, are taking refuge from an unpleasant present, holding out for an uncertain future. At the northern edge of the zoo are the polar bears. No animal is so symbolic of loss as these bleached carnivores—the biggest meat-eaters remaining on land today—that are losing their hunting grounds as the ice melts. During the Ice Age, the Arctic ice cap was right here, 42 degrees north latitude; in the coming centuries, there might be no ice cap at all. A short walk away are some camels—they're far from vulnerable, but we can't forget that they originated in North America, lived here for tens of millions of years, and then disappeared with the megafauna. Thus, there are Asian camels in this zoo, but no North American camels, and of course, no woolly mammoths or giant sloths. There *is* a black rhinoceros: one of the last "supermegafauna" survivors, but how long does it have left? As I'm pondering this thought, I hear the chatter of something more human, but not quite. Chimps and gorillas, our very closest ape relatives, both very endangered.

Now is not a good time to be a mammal. Our family has not been this threatened since the end-Cretaceous asteroid nearly wiped us out, back when our ancestors were mousey vermin trying to avoid the stomp of a *T. rex*.

More than 350 mammals have gone extinct since *Homo sapiens* started meandering late in the Ice Age. Of these, about 80 have died over the last 500 years. That means roughly 1.5 percent of all mammal species have been extinguished in the short time humans have been keeping records. It may not sound like a lot, but this rate of extinction is more than twenty times greater than the background rate in the prehuman era. If this blazing pace continues, we're on track for about 550 mammal extinctions before 2100: nearly 10 percent of mammalian diversity. And if all mammals currently at risk ultimately go extinct, we'll be left with only half the number of species relative to a mere 125,000 years ago. Even if everything was frozen in place right now, and extinctions were stopped and mammals given the opportunity to rebound, it would take many thousands of years to recover the lost diversity.

It's not only about total numbers of species dying, though, just as a country's economy is about more than raw GDP. Mammal distributions are shifting fast, and populations are in a state of unrest. The largest mammals went extinct when the megafauna died, and are still more likely to disappear today, but now species of all sizes are dying. If the trends persist, in a few hundred years the rhinos and elephants will be gone, and the largest mammals may be domestic cows. Mammal communities are not only downsizing, but becoming more homogenized, and the near future might be one without great apes and lions, but one overrun with rodents. Meanwhile, mammals are migrating with the desperation of refugees. Larger mammals—like the once-real Chicago Bears—are being exiled from climate zones dominated by agriculture and cities, and pushed into colder and drier regions. Smaller mammals are, for the moment, taking their place and moving into the farmlands and suburbs, but it's not clear how long they can last.

I hate to be the bearer of bad news, but this is all happening because of us.

As our population multiplies, we require ever-more resources, and transform the earth ever more into our own playground and pantry, with less and less room for other mammals. We clear rain forests and plow savannas into farmland. We pollute. We burn. We hunt and poach. But more than anything, we change the climate.

Temperatures are rising—that's a fact. As we've seen throughout this book, temperatures have risen before, but what's different now is the speed. Warming spikes that took many tens of thousands of years at the end-Permian and end-Triassic—which, lest we forget, were mass extinctions!—are now taking place within a few human generations. Very soon—next century at the latest—we are projected to reach a climate state like that of the Pliocene, before the Ice Age. If we keep emitting greenhouse gases, then we will reach Eocene climates within a few centuries. Recall that this was a hothouse world, when rain forests cloaked the Arctic and crocodiles lounged where there is now—at least for a little bit longer—an ice cap. Put another way, a few centuries of human activity might rewind the clock by 50 million years, reversing the cooling trend that has played out across most of the time mammals have been dominant.

I don't know exactly what the future holds. I'm not going to get too speculative, because with climates changing so fast, we're entering unprecedented territory. What I do think, however, is that our human species is in trouble. The newer, higher temperatures will be outside the range in which we evolved. We'll be ripped out of our cozy interglacial, the backdrop to the entire history of global *Homo sapiens*, where temperatures are pleasant, the polar ice conducts ocean currents that bring warm water to higher latitudes, and crops are easy to grow in ground blanketed in glacial soil. Sea levels are rising, as they often do, but this will be the first time they encroach on our cities, so many of which are perched so perfectly where land meets sea today. We could go extinct, or we could adapt. That's the choice, and because we are sentient beings with big brains and tools and technology and global reach, it really is a choice.

And what about other mammals? Extinctions will surely continue, in one form or another. There is much talk about a "Sixth Extinction," a human-caused mass die-off on par with the "big five" extinctions of Earth history, three of which we've discussed in this book (the end-Permian, end-Triassic, and end-Cretaceous). Might our current predicament level up to these annihilations of prehistory, or worse? So far, the number of modern extinctions is far below the apocalyptic death tolls of past mass extinctions, so we shouldn't be too alarmist. Plus we have a remedy: we can choose to conserve the endangered species still with us! Yet the fear is that if today's extinctions continue, they could domino, and ecosystems could fall like houses of cards, and global communities could collapse like a cascading blackout on a failing power grid. If so, to a paleontologist of the future, the death of the Ice Age megafauna, and the marsupial wolves and Falkland foxes in recent history, and lions and gorillas in the future would all blend into one. Condensed to a thin line in the rocks. Many mammals below, few (if any?) mammals above, as stark as the line that divides the Age of Dinosaurs from the Age of Mammals.

It leaves me uneasy, and as I stare past the polar bears to the darkening skies over Lake Michigan, my glum thoughts are broken by a scream.

The howler monkeys are howling.

I turn to my wife and smile. These monkeys descended from those Eocene rafters that, somehow, tolerated a transatlantic journey on the waves, reluctantly leaving Africa but adapting to the new landscapes and climates of South America. Long before, those rafters had ancestors that made it through three mass extinctions. The monkeys, the mammals, are resilient.

And I also know this. Evolution endowed us—a single but splendid mammalian species, *Homo sapiens*—with big brains and the ability to cooperate in groups. We know what we are doing to our planet, and we can work together to come up with solutions. The mammoths and sabertooths and millions of other extinct mammalian cousins never had the same powers, either to alter the world or to ameliorate it. We do.

I don't know what lies ahead for the human dynasty and our mammalian family, but I'm hopeful the reign of mammals will continue.

NOTES ON SOURCES

The following notes mention supplementary material and sources that I used, and which I direct you to for more information on the subjects covered in the chapters.

In general, I relied extensively on a handful of excellent books, including Tom Kemp's *The Origin and Evolution of Mammals* (Oxford University Press, 2005), Liam Drew's *I, Mammal* (Bloomsbury, 2017), David Rains Wallace's *Beasts of Eden* (University of California Press, 2004), Donald Prothero's *Princeton Field Guide to Prehistoric Mammals* (Princeton University Press, 2017), Donald Prothero and Robert Schoch's *Horns, Tusks, and Flippers* (The Johns Hopkins University Press, 2002), Kenneth Rose's *The Beginning of the Age of Mammals* (The Johns Hopkins University Press, 2006), Zofia Kielan-Jaworowska's *In Pursuit of Early Mammals* (Indiana University Press, 2012), and Ross MacPhee's *End of the Megafauna* (W.W. Norton & Company, 2019). When describing the paleogeography of ancient Earth, I used the industry-leading maps of Ron Blakey (https://deeptimemaps.com/).

CHAPTER 1: MAMMAL ANCESTORS

The story of the "scaly critters" is set in a Carboniferous-aged coal forest. In some parts of the world, the Carboniferous is regarded as a single geological period spanning from about 359 to 299 million years ago; in other parts of the world, notably North America, it is split into separate Mississippian (359–323 million years ago) and Pennsylvanian (323–299 million years ago) periods. To bring the coal swamp to life, I relied on descriptions of the Mazon Creek fossil site in Illinois, most importantly the review of Clements et al. (*Journal of the Geological Society*, 2019, 176: 1–11), Shabica and Hay's influential book *Richardson's Guide to the Fossil Fauna of Mazon Creek* (Northeastern Illinois University Press, 1997), and Jack Whitry's books *The Mazon Creek Fossil Fauna* (2012) and *The Mazon Creek Fossil Flora* (2006), both published by the Earth Science Club of Northern Illinois. I supplemented this with information from another similar coal swamp fossil site, Joggins (Nova Scotia), as summarized by Falcon-Lang et al. (*Journal of the Geological Society*, 2006, 163: 561–76).

My description of the fictional "scaly critters," which represent the most recent common ancestor of synapsids and diapsids, is based on *Hylonomus* (the oldest well-known unequivocal diapsid in the fossil record) and *Archaeothyris* (the oldest well-known unequivocal synapsid with reasonably complete fossils). The DNA molecular clock estimate for the synapsid-diapsid split is a mean of 326 million years ago (range of 354 to 311 million years ago), from Blair and Hedges (*Molecular Biology and Evolution*, 2005, 22: 2275–84). A very similar divergence estimate (mean 324.51, range 331–319 million years ago) was reported by Ford and Benson (*Nature Ecology & Evolution*, 2020, 4: 57–65), using morphological clocks on a phylogeny of extinct amniotes. The phylogenetic relationships reported in the Ford and Benson study are the framework I use to discuss the relationships of amniotes. This exciting new study, based on a comprehensive data set and analyzed with a variety of methods, finds some novel relationships compared to the long-held consensus view of early amniote genealogy. Most notably, the varanopids—long considered early synapsids—group with the diapsids. This is why I do not include varanopids in my narrative on early synapsid evolution, whereas older publications often do so.

For more information on the climate of the coal swamp world, and during the Carboniferous Rainforest Collapse, there are two excellent papers by Isabel Montañez and colleagues (Montañez et al., *Science*, 2007, 315: 87–91; Montañez and Poulsen, *Annual Review of Earth and Planetary Sciences*, 2013, 41: 629–56). For information on how oxygen has changed over Earth history, and how geologists calculate past oxygen levels, please consult David Beerling's 2007 book *The Emerald Planet* (Oxford University Press) and Berner (*Geochimica et Cosmochimica Acta*, 2006, 70: 5653–64).

There is a huge literature on the origin of tetrapods and the early evolution of amniotes. There is no better source than the masterful book *Gaining Ground* (Indiana University Press, 2012) by Jennifer Clack, the world's expert on the fish-tetrapod transition, who sadly passed away in the spring of 2020, as I was writing this book. There are also two outstanding pop-science books on the subject, by two of the finest science writers I know: *Your Inner Fish* by Neil Shubin (Pantheon, 2008) and *At the Water's Edge* by Carl Zimmer (Free Press, 1998).

The Florence synapsids *Archaeothyris* and *Echinerpeton* were described by Robert Reisz in a 1972 paper (*Bulletin of the Museum of Comparative Zoology*, 144: 27–61). More recently, *Echinerpeton* has been redescribed by Mann and Paterson (*Journal of Systematic Palaeontology*, 2020, 18: 529–39). The Romer expeditions to Nova Scotia, including the discovery of the skeleton-bearing tree stumps, is recounted by Sues et al. (*Atlantic Geology*, 2013, 49: 90–103).

Robert Reisz's career is profiled in a touching biographical article by Laurin and Sues (*Comptes Rendus Palevol*, 2013, 12: 393–404).

Emma Dunne's paper on the Carboniferous Rainforest Collapse, written with several colleagues, was published in *Proceedings of the Royal Society, Series B* (2018: 20172730). It follows up—in some cases updating, in other cases contrasting with—an earlier study by Sarda Sahney and colleagues (*Geology*, 2010, 38: 1079–82). Another fascinating paper on climate changes across the Carboniferous-Permian transition, and its impact on vertebrate evolution and distribution, was recently published by Jason Pardo and colleagues in *Nature Ecology & Evolution* (2019, 3: 200–206). Extinctions and diversification in the plant fossil record—including the finding that there have been only two mass extinction events—is covered by Cascales-Miñana and Cleal (*Terra Nova*, 2014, 26: 195–200). For more information on the Carboniferous and Permian ice caps, and why they formed, consult the paper by Georg Feulner (*Proceedings of the National Academy of Sciences [USA]*, 2017, 114: 11333–37), and references therein.

The pelycosaurs are what paleontologists call a "grade" of species. They do not form a "clade," which is defined as a group that includes a common ancestor and all of its descendants. Instead, a grade is a series of species on the line toward a clade—an ancestral stock. So when I talk about pelycosaurs I am talking about a succession of species on the line toward the more advanced therapsid clade (which includes mammals). I generally do not like to talk about grades, or give them names, but in this case, it is convenient because the pelycosaurs are generally similar in anatomy and biology and are the stock that the therapsids evolved from. In effect, the therapsid clade evolved from a single common ancestor that *was* part of the pelycosaur grade. There is a huge literature on *Dimetrodon* and other pelycosaurs, including papers by many of the leaders of nineteenth-century and early-twentieth-century paleontology: Cope, Case, Matthew, Olson, Sternberg, Romer, Vaughn, and so on. This is expertly summarized by Tom Kemp in his book *The Origin and Evolution of Mammals*, which I used extensively.

If you still doubt that *Dimetrodon* is more closely related to us than to dinosaurs, check out the clear and well-written essay by Ken Angielczyk—one of the world's leading experts on early synapsid evolution (*Evolution: Education and Outreach*, 2009, 2: 257–71). This essay is also an excellent introduction to "tree thinking": how paleontologists construct and talk about family trees. The grade vs. clade issue should be a lot clearer after reading Ken's essay, as should the step-by-step sequence of changes across the mammal stem lineage, from pelycosaurs to therapsids to mammals.

The collapse of pelycosaurs around the Early-Middle Permian boundary is part of an extinction event called Olson's Extinction, named for the paleontologist (E. C. Olson) who first noted it (*Geological Society of America Special Papers*, 1982, 190: 501–12). Olson, a fellow alum (although many decades earlier) of the University of Chicago's geology program, was a prolific researcher on Permian synapsids, and published landmark papers such as his 1962 monograph comparing North American and Russian species (*Transactions of the American Philosophical Society*, 52: 1–224). There has been debate about whether Olson's Extinction was a real event or a mirage, caused by a biased and unevenly sampled fossil record (a view suggested by Benson and Upchurch, *Geology*, 2013, 41: 43–46). Recently, another young, statistically minded paleontologist of Emma Dunne's generation—Neil Brocklehurst—led a team that addressed this debate using big databases and statistical analyses, which found that the extinction was a real event (*Proceedings of the Royal Society, Series B*, 2017, 284: 20170231).

A key concept I try to articulate in this chapter is that the characteristics that today make mammals unique (compared to other tetrapods, like birds, lizards, and amphibians) did not all evolve at once but were acquired piecemeal over millions of years of evolution, along the mammal "stem lineage": the series of synapsid groups on the line to mammals, including pelycosaurs, therapsids, and cynodonts (note that therapsids and cynodonts are both clades; so mammals are technically part of each group!). As usual, Tom Kemp does the best job of summarizing a long and deep literature on the subject. His book *The Origin and Evolution of Mammals*, particularly chapters 3 and 4, is required reading for anyone interested in the subject, as are two essay-style reviews (*Journal of Evolutionary Biology*, 2006, 19: 1231–47; *Acta Zoologica*, 2007, 88: 3–22) and his chapter in the 2012 book *The Forerunners of Mammals*. Bruce Rubidge and Chris Sidor have also written an influential review (*Annual Review of Ecology and Systematics*, 2001, 32: 449–80), and more recently Ken Angielczyk and Christian Kammerer have done a marvelous job of summarizing where the evidence stands now (in their chapter in the *Handbook of Zoology: Mammalian Evolution, Diversity and Systematics*, DeGruyter, 2018). For a more technical look at this issue, consult Sidor and Hopson (*Paleobiology*, 1998, 24: 254–73).

The Karoo Basin of South Africa is the world's premier place for preserving fossils of Permian therapsids. An accessible review of the basin, its rocks, and its fossils is given by Roger Smith and colleagues in their chapter in the 2012 book *The Forerunners of Mammals* (Indiana University Press). The story of Andrew Geddes Bain's first therapsid discoveries and Richard Owen's early work on these "mammal-like reptiles" (forgive my use of the term) are covered by David Rains Wallace in his book *Beasts of Eden*. Owen's two most important works on therapsids, which I mention, were published in 1845 (*Transactions of the Geological Society of London*, 7: 59–84) and 1876 (*Descriptive and Illustrated Catalogue of the Fossil Reptilia of South Africa in the Collection of the British Museum*, Taylor & Francis, London). Edward Drinker Cope outlined the link between "reptilian" ancestors like the Karoo fossils, pelycosaurs, and mammals in 1884, as recounted by Henry Fairfield Osborn (*The American Naturalist*, 1898, 32: 309–34), and Angielczyk and Kammerer in their review chapter cited above.

Robert Broom, his life, and his research are touchingly recalled in an obituary by D. M. S. Watson (*Obituary Notices of Fellows of the Royal Society*, 1952, 8: 36–70) and a review essay by Bruce Rubidge, the grandson of Broom's most productive farmer-collector, Sidney Rubidge (*Transactions of the Royal Society of South Africa*, 2013, 68: 41–52). Broom's magnum opus, his 1910 monograph linking pelycosaurs and therapsids, was published in the *Bulletin of the American Museum of Natural History* (28: 197–234). At this point, I must acknowledge that, while his work set the foundation for mammal origins, Broom was a vexing figure. He claimed that spirits guided him to fossils, argued that spirits inside animals acted on their chromosomes to cause evolutionary change, and most troubling, espoused racist ideas and was involved in grave robbing (in addition to mammal origins, he also studied human origins). For a discussion of the legacy of racism in human origins research, including Broom's work, see Christa Kuljian's book *Darwin's Hunch: Science, Race and the Search for Human Origins* (Jacana Media, 2016).

Many of my descriptions of dicynodonts, dinocephalians, and gorgonopsians are inspired by Tom Kemp's meticulous prose in *The Origin and Evolution of Mammals*. My description of the common ancestor of therapsids is taken from Kemp's 2006 paper, cited above.

Christian Kammerer's revision of *Dicynodon* was published in 2011 (*Society of Vertebrate Paleontology Memoir*, 11: 1–158). This monograph also includes a historical review of *Dicynodon* research (it is here where the term "taxonomic dumping ground," which I use, was coined!), and a comprehensive genealogical analysis of dicynodonts, which Christian and colleagues updated in 2013 (*PLoS ONE*, 8: e64203), and continue to update, with the most recent version (as of this writing) published in 2021 (Kammerer and Ordoñez, *Journal of South American Earth Sciences*, 108: 103171). Two other essential works on dicynodonts were published by G. M. King, including a review article in the *Handbuch der Paläoherpetologie*, Gustav Fischer Verlag, 1988) and a book (*The Dicynodonts: A Study in Palaeobiology*, Chapman & Hall, 1990).

Dinocephalian head-butting was proposed by Barghusen (*Paleobiology*, 1975, 1: 295–311) and has recently been explored in more detail through synchrotron scanning of the internal skull anatomy of *Moschops*, which shows that its brain and other neural structures were encased in extra thick bone to protect from impact (Benoit et al., *PeerJ*, 2017, 5: e3496). Information on the giant *Anteosaurus* was gleaned from Boonstra (*Annals of the South African Museum*, 1954, 42: 108–48) and Van Valkenburgh and Jenkins (*Paleontological Society Papers*, 2002, 8: 267–88).

There is debate about the jaw mechanics of gorgonopsians. The wide-gape hypothesis is supported by the work of Tom Kemp (*Philosophical Transactions of the Royal Society of London, Series B*, 1969, 256: 1–83) and L. P. Tatarinov (*Russian Journal of Herpetology*, 2000, 7: 29–40), whereas a dissenting view was articulated by Michel Laurin (*Journal of Vertebrate Paleontology*, 1998, 18: 765–76). The brain anatomy and sensory systems of gorgonopsians were recently described by Ricardo Araújo and colleagues, using computed tomography (CT) scan data (*PeerJ*, 2017, 5: e3119).

For an excellent review on the origin of higher metabolisms and finer temperature control in therapsids, hypotheses for why this happened, and a thorough review of the diverse literature on this subject, check out the paper by James Hopson published in 2012 (*Fieldiana*, 5: 126–48).

Anusuya Chinsamy-Turan described her methods of making and studying bone thin sections in her 2005 book *The Microstructure of Dinosaur Bone* (Johns Hopkins University Press). She wrote or coauthored several chapters in the 2012 book *The Forerunners of Mammals* (which she edited), on the bone texture, growth, and metabolism of mammal antecedents. Other key papers on this subject are her coauthored study with Sanghamitra Ray and Jennifer Botha (*Journal of Vertebrate Paleontology*, 2004, 24: 634–48), and studies by Huttenlocker and Botha-Brink (*PeerJ*, 2014, 2: e325), Olivier et al. (*Biological Journal of the Linnean Society*, 2017, 121: 409–19), and Rey et al. (2017, *eLife*, 6: e28589). Details of Anusuya's life and career were taken from an interview with her, published online: https://scibraai.co.za/anusuya-chinsamy-turan-breathing-life-bones-extinct-animals/.

The evolution of more upright locomotion in therapsids is discussed by Blob (*Paleobiology*, 2001, 27: 14–38), and in the dicynodont works of King cited above. A recent study by another brilliant PhD student, Jacqueline Lungmus, and her advisor, Ken Angielczyk, showed how therapsids developed a greater variety of forelimb shapes and motions, allowing them to ecologically diversify (*Proceedings of the National Academy of Sciences*, 2019, 116: 6903–07). Liam Drew's book *I, Mammal* has a fantastic chapter on the origin of hair, which has a detailed description of the sensory, display, and waterproofing hypothesis, and of how hair became co-opted for physiological reasons. Permian coprolites with hairlike structures have been described by Bajdek et al. (*Lethaia*, 2016, 49: 455–77) and Smith and Botha-Brink (*Palaeogeography, Palaeoclimatology, Palaeoecology*, 2011, 312: 40–53). The blood vessels and nerves in the skull that innervate the hair are reviewed by Benoit et al. (*Scientific Reports*, 2016, 6: 25604). Currently, it seems as if the facial bones of early therapsids have equivocal evidence for hair, but there is no doubt that later therapsids like cynodonts and close relatives had evolved whiskers and hair.

CHAPTER 2: MAKING A MAMMAL

The story of the burrowing *Thrinaxodon*, waiting out the dry season and setting out to eat and mate when the rains came, is based on the fossil and rock record of the Karoo Basin across the Permian-Triassic boundary. My primary sources were papers by Smith and Botha-Brink (*Palaeogeography, Palaeoclimatology, Palaeoecology*, 2014, 396: 40–53) and Botha et al. (*Palaeogeography, Palaeoclimatology, Palaeoecology*, 2020, 540: 109467), along with work by Peter Ward and colleagues (*Science*, 2000, 289: 1740–43; *Science*, 2005, 307: 709–14).

The best pop-science book on mass extinctions is *The Ends of the World*, by Peter Brannen (Ecco, 2017). I think Peter is one of the finest science writers working today, and his earth science writing is on par with my favorite all-time geologizing author, John McPhee. There are two excellent pop-science books on the end-Permian extinction, one by my former master's advisor Michael Benton (*When Life Nearly Died*, Thames & Hudson, 2003) and the other by Douglas Erwin (*Extinction: How Life on Earth Nearly Ended 250 Million Years Ago*, Princeton University Press, 2006). Zhong-Qiang Chen and Mike Benton wrote an accessible review of the extinction and subsequent recovery for *Nature Geoscience* (2012, 5: 375–83). Updated information on the timing and nature of the volcanic eruptions that caused the extinction was published by Seth Burgess and colleagues (*Proceedings of the National Academy of Sciences USA*, 2014, 11: 3316–21; *Science Advances*, 2015, 1: e1500470). An important paper on climate change and warming during the extinction was published by Joachimski and colleagues (*Geology*, 2012, 40: 195–98); this is my source for the figure of 9–14 degrees Fahrenheit (5–8 degrees Celsius) warming.

The ecological collapse of the Karoo ecosystems, and the prolonged recovery, has been studied by Peter Roopnarine, Ken Angielczyk, and colleagues, using ecological food web modeling (*Proceedings of the Royal Society, Series B*, 2007, 274: 2077–86; *Science*, 2015, 350: 90–93; *Earth-Science Reviews*, 2019,

189: 244–63). Adam Huttenlocker's work on the Lilliput effect was published in *PLoS ONE* (2014, 9: e87553), and Adam was part of a team led by Jennifer Botha-Brink that addressed the wider question of cynodont survival at the end-Permian, which proposed the hypothesis that growing young and breeding fast were key (*Scientific Reports*, 2016, 6: 24053). Other important work on body size evolution in early synapsids was published by Roland Sookias and colleagues (*Proceedings of the Royal Society, Series B*, 2012, 279: 2180–87; *Biology Letters*, 2012, 8: 674–77). Chris Sidor and colleagues published an important study on the distribution of species across Pangea during the Permian-Triassic transition (*Proceedings of the National Academy of Sciences, USA*, 2013, 110: 8129–33).

There is a wealth of literature on *Thrinaxodon*, the hero cynodont of the chapter. Tom Kemp, as usual, has a scrupulous summary in his book *The Origin and Evolution of Mammals*. One of the key descriptive papers is by Richard Estes (*Bulletin of the Museum of Comparative Zoology, Harvard University*, 1961, 125: 165–80), and an important study of its teeth was published by A.W. "Fuzz" Crompton (*Annals of the South African Museum*, 1963, 46: 479–521). Robert Broom himself described the cranial anatomy of *Thrinaxodon* in a 1938 paper (*Annals of the Transvaal Museum*, 19: 263–69), by cutting up a skull into eighteen slices from front to back. He grumbled that, while he wanted many more, thinner sections, a "medical practitioner has to be content with simpler and less perfect technique," which is ironic because contemporary paleontologists use CAT scanners to make digital x-ray slices of fossil skulls, and often need to beg medical doctors and hospitals for access to scanners!

Important burrows with *Thrinaxodon* skeletons inside were described by Damiani et al. (*Proceedings of the Royal Society, Series B*, 2003, 270: 1747–51) and Fernandez et al. (*PLoS ONE*, 2013, 8: e64978); the latter paper describes the remarkable fossil of a *Thrinaxodon* and an amphibian crowded together into a single burrow. The posture of *Thrinaxodon* and other cynodonts was expertly studied by Farish Jenkins, and two of his most important works are his 1971 monograph *The Postcranial Skeleton of African Cynodonts* (*Peabody Museum of Natural History Bulletin*, 36: 1–216) and a review paper published in *Evolution* (1970, 24: 230–52). Further information on cynodont posture was gleaned from Richard Blob's 2001 paper, cited above. Bone histology and growth of *Thrinaxodon* was described by Jennifer Botha and Anusuya Chinsamy (*Palaeontology*, 2005, 48: 385–94). The teeth, jaws, and jaw-closing muscles of *Thrinaxodon*—including how they changed during growth—have been the focus of publications by Sandra Jasinoski, Fernando Abdala, and Vincent Fernandez (*Journal of Vertebrate Paleontology*, 2013, 33: 1408–31; *The Anatomical Record*, 2015, 298: 1440–64), and Jasinoski and Abdala described social aggregations and parental care in a paper in *PeerJ* (2017, 5: e2875). The Antarctic fossils of *Thrinaxodon* were described by James Kitching (the son of the road-building Karoo fossil collector Croonie Kitching from chapter 1) and colleagues (*Science*, 1972, 175: 524–27; *American Museum Novitates*, 1977, 2611: 1–30).

I thank Christian Kammerer and his Twitter feed for drawing my attention to the remarkable life of Walter Kühne. Kühne's 1956 monograph of *Oligokyphus* was published by the British Museum of Natural History and is freely available online (https://www.biodiversitylibrary.org/item/206348#page/5/mode/1up). Some details of Kühne's life and incarceration are outlined in the monograph, but I sourced further information from Zofia Kielan-Jaworowska's book *In Pursuit of Early Mammals* (Indiana University Press, 2012), Alfred Romer's review of Kühne's monograph in the *Quarterly Review of Biology*, and Rex Parrington's paper on British Triassic mammals, which is a great general resource on the many cave discoveries such as *Morganucodon*, *Kuehneotherium*, and *Eozostrodon* (*Philosophical Transactions of the Royal Society, Series B*, 261: 231–72). The dismissive quote from the British Museum curator is paraphrased from Romer's review.

The phylogenetic relationships of cynodonts have been the subject of extensive analysis, reanalysis, and debate over the last several decades. Tritylodontids like *Oligokyphus* are recognized as some of the closest relatives of mammals, along with two groups called the tritheledontids and brasilodontids. These are all advanced groups of cynodonts that were blossoming around the same time, in the Late Triassic. My conception of cynodont genealogy is based on a recent study by Marcello Ruta and colleagues, on which I was one of the peer reviewers (*Proceedings of the Royal Society, Series B*, 2013, 280: 20131865). Another important recent work was published by Jun Liu and Paul Olsen (*Journal of Mammalian Evolution*, 2010, 17: 151–76).

Changes on the cynodont lineage—linking *Thrinaxodon* to *Oligokyphus* and mammals—are expertly covered in Tom Kemp's book. Fuzz Crompton and Farish Jenkins wrote an influential review on the subject in the very first volume of *Annual Review of Earth and Planetary Sciences* (1973, 1: 131–55). Changes in the vertebral column in cynodonts have been scrutinized by Katrina Jones—another winner of the Romer Prize, in 2014, the year after Adam Huttenlocker—and colleagues (*Science*, 2018, 361: 1249–52; *Nature Communications*, 2019, 10: 5071). The papers by Farish Jenkins and Richard Blob, cited above, discuss the postural and locomotor changes in more detail, particularly the development of full mammalian-style upright walking through a semi-sprawling transitional phase.

When it comes to the story of dinosaur origins and early evolution, I will not so humbly point readers in the direction of my book *The Rise and Fall of the Dinosaurs* (William Morrow, 2018), and a review paper I wrote on the subject, with my colleagues Sterling Nesbitt, Randy Irmis, Richard Butler, Mike Benton, and Mark Norell (*Earth-Science Reviews*, 2010, 101: 68–100). In my book, I also provide a more detailed description of Pangea and its climate and outline the most important sources.

Several authors have opined about a "nocturnal bottleneck" in early mammal evolution, with some pinpointing this phase to the origin of mammals in the Triassic, and others using it to refer to the small, night-living mammals that may have preferentially survived the dinosaur extinction. An illuminating study by Ken Angielczyk and Lars Schmitz showed—using eye measurements—that nocturnal behaviors probably first evolved early in the synapsid line, and various pelycosaurs, therapsids, and cynodonts lived in the night (*Proceedings of the Royal Society, Series B*, 2014, 281: 20141642). Other papers to consider are those by Margaret Hall and colleagues (*Proceedings of the Royal Society, Series B*, 2012, 279: 4962–68), Jiaqi Wu and colleagues (*Current Biology*, 2017, 27: 3025–33), and Roi Maor and colleagues (*Nature Ecology & Evolution*, 2017, 1: 1889–95). My comment about mammals going "all in" for scent and touch was inspired by an interview with my PhD advisor, Mark Norell, published by, of all things, Marvel Comics in 2019.

The way I discuss warm-bloodedness (endothermy) relied extensively on Tom Kemp's *The Origin and Evolution of Mammals*, and the discussion section in Katrina Jones and colleagues' *Nature Communications* paper on mammal spine evolution, cited above. The figure that warm-blooded mammals can run eight times faster than lizards was cited in the latter paper, based on work by Kemp (*Zoological Journal of the Linnean Society*, 2006, 147: 473–88) and Bennett and Ruben (*Science*, 1979, 206: 649–54). The increasing prevalence of fibrolamellar bone in cynodonts is covered in two very readable chapters in *The Forerunners of Mammals*, one by Jennifer Botha-Brink and colleagues (ch. 9) and the other by Jørn Hurum and Anusuya Chinsamy (ch. 10). The decreasing size of bone cells, and by extension red blood cells, was noted by Adam Huttenlocker and Colleen Farmer (*Current Biology*, 2017, 27: 48–54). Kévin Rey and team published their oxygen isotope work in *eLife* (2017, 6: e28589). Carrier's constraint was christened by Richard Cowen, named for the scientist who first articulated it: David Carrier, in *Paleobiology* (1987, 13: 326–

41). A nice paper on the evolution of nasal turbinates was written by Crompton and colleagues (*Journal of Vertebrate Paleontology*, 2017, e1269116). There is some uncertainty about exactly when respiratory turbinates covered in blood vessels first evolved. It seems like some nonmammal cynodonts may have had them, although they may have been made of cartilage rather than bone, but the oldest unequivocal bony evidence is from early mammals. It is very difficult to determine the size, shape, and blood vessel coverage of these delicate structures from fossils.

The evolution of the three-part mammalian jaw muscle system was comprehensively reviewed by Lautenschlager et al. (*Biological Reviews*, 2017, 92: 1910–40), with copious references to the historical literature on the subject. The same team followed with an important paper in *Nature* (2018, 561: 533–37) which used engineering software to test jaw function in a series of fossil species, leading them to argue that miniaturization was the primary driver for the evolution of the new dentary-squamosal jaw joint. As mammal ancestors got smaller, there was a sweet spot in which small size lowered stress and strain disproportionately to the loss of absolute bite force that came with a smaller jaw. Chris Sidor authored a study in *Evolution* (2001, 55: 1419–42) showing how the upper skull of mammal ancestors simplified (= fewer and more fused bones).

The definition of mammals that I use throughout the book—any descendant of the first cynodont to develop a robust dentary-squamosal jaw joint—is prevalent in the historical literature. It is more or less the definition used by Kielan-Jaworowska, Cifelli, and Luo in their magisterial overview of early mammals, *Mammals from the Age of Dinosaurs* (Columbia University Press, 2004). (Technically, they define mammals as "a clade defined by the shared common ancestor of *Sinoconodon*, morganucodontans, docodontans, Monotremata, Marsupialia, and Placentalia, plus any extinct taxa that are shown to be nested within this clade," which is basically equivalent to the group on the family tree that developed the dentary-squamosal joint). This group—what I call "mammals"—is referred to as Mammaliaformes by those researchers who prefer a "crown group" definition for mammals, which limits the name "mammals" to the group on the family tree including the modern mammals (monotremes, marsupials, placentals) and all descendants of their most recent common ancestor. Timothy Rowe's 1988 *Journal of Vertebrate Paleontology* (8: 241–64) paper first set out the crown-based definition of Mammalia and the new name Mammaliaformes for the larger group with the dentary-squamosal joint. And that's all I'll say about classification—an exercise in semantics more than science—before pleading again for my colleagues to forgive me in not using the crown group definition here.

There is a large literature on mammalian chewing, and once again, Tom Kemp's *The Origin and Evolution of Mammals* and Kielan-Jaworowska et al.'s *Mammals from the Age of Dinosaurs* were invaluable sources and are necessary reading for anyone interested in the subject. Kai Jäger and colleagues recently published a key paper on *Morganucodon* chewing and dental occlusion (*Journal of Vertebrate Paleontology*, 2019, 39: e1635135). Bhart-Anjan Bhullar and Armita Manafzadeh and team published an intriguing study in *Nature* (2019, 566: 528–32) that used x-ray analysis of actual living marsupials chewing food to argue that the rolling action of the lower jaw—which evolved around the same time as the dentary-squamosal jaw joint—was central to the chewing motions. David Grossnickle wrote a reply to this article (*Nature*, 2020, 582: E6–E8), which mostly addressed other aspects of the paper. My discussions with Dave greatly helped me understand the evolution of mammalian chewing. I should also mention here that Anjan Bhullar used the term "internal furnace" when describing endothermy in an interview, so I borrowed that phrase in this chapter.

There is a huge literature on *Morganucodon*, the exemplar for early mammals. It was first described and named by Kühne in 1949 (*Proceedings of the Zoological Society of London*, 119: 345–50), and more complete fossils were later described by Kenneth Kermack, Frances Mussett, and Harold Rigney in two *Zoological Journal of the Linnean Society* papers (1973, 53: 86–175; 1981, 71: 1–158). Rigney described the Chinese *Morganucodon* skull, the source of so much trouble, in *Nature* (1963, 197: 1122–23). His autobiography, *Four Years in a Red Hell*, was published by Henry Renery, Chicago. Diphyodonty—two generational tooth replacement—in *Morganucodon* was described by Rex Parrington (*Philosophical Transactions of the Royal Society, Series B*, 261: 231–72), and the postcranial skeleton was described by Jenkins and Parrington (*Philosophical Transactions of the Royal Society, Series B*, 1976, 273: 387–431).

Tim Rowe has published two important studies on the brain evolution of early mammals, including the development of larger olfactory bulbs and the neocortex. The first in *Science* (1996, 273: 651–54) and the second coauthored with Ted Macrini and Zhe-Xi Luo in *Science* (2011, 332: 955–57). These present CT scans of *Thrinaxodon*, *Morganucodon*, and other key species on the cynodont-mammal line.

Relevant literature on other, non-*Morganucodon* early mammals include Parrington's description of *Eozostrodon* (*Annals and Magazine of Natural History*, 1941, 11: 140–44), Diane Kermack's description of *Kuehneotherium* (*Journal of the Linnean Society [Zoology]*, 1968, 47: 407–23), Crompton and Jenkins's description of Ione Rudner's *Megazostrodon* (*Biological Reviews*, 1968,

43: 427–58), and Luo et al.'s description of *Hadrocodium* (*Science*, 2001, 292: 1535–40). Pam Gill and her team published their study of diet in *Morganucodon* and *Kuehneotherium*—using not only tooth wear, but also engineering models of the jaws—in *Nature* (2014, 512: 303–5). A related paper, also supporting feeding differences between these species, was written by Conith et al. (*Journal of the Royal Society Interface*, 2016, 13: 20160713).

Farish Jenkins is a legend in my field, and although I didn't know him personally, I will always remember being in attendance when he won the Romer-Simpson medal of the Society of Vertebrate Paleontology in 2009, while he was battling cancer. He passed away three years later, in 2012, reportedly telling friends that he was at peace because, as a paleontologist, he was familiar with extinction. I gleaned this quote, along with other biographical details, from obituaries published in the *New York Times*, the *Economist*, the *Boston Globe*, and *Nature* (written by Neil Shubin)—a sign of his prominence. The famous illustration of *Megazostrodon* was published by Jenkins and Parrington in their 1976 paper, cited above. In their acknowledgments, they credit the final artistic reconstructions to Laszlo Meszoly, an artist at Harvard's Museum of Comparative Zoology who was profiled in a *Harvard Gazette* article in 2003.

The idea that diversification of major new mammal groups often begins in the small-bodied insect-eating niche was presented by Dave Grossnickle and colleagues in a wonderfully written 2019 review paper, published in *Trends in Ecology and Evolution* (2019, 34: 936–49).

CHAPTER 3: MAMMALS AND DINOSAURS

The fascinating life of William Buckland is recounted in many books on the early history of paleontology, including my favorite: Deborah Cadbury's *The Dinosaur Hunters* (Fourth Estate, 2000), which was also published internationally under the title *Terrible Lizard*. Buckland's role in the study of early mammals was recounted in Wallace's *Beasts of Eden*, and I also gleaned additional details from a biography in the University of Oxford's online "Learning More" series, and an article in the *Guardian* about Buckland's gastronomic proclivities ("*The Man Who Ate Everything*," February 2008). Buckland described *Megalosaurus* and the tiny mammal jaws in a written version of his Geological Society address (*Transactions of the Geological Society of London*, 1824, 2: 390–96). Several decades later, Richard Owen published a landmark review of Mesozoic mammals known at the time (Monograph of the Fossil Mammalia of the Mesozoic Formations, *Monographs of the Palaeontographical Society*, 1871).

The stereotype of Mesozoic mammals as small, dull generalists was articulated in the two most important reviews of the early twentieth century, both written by the eminent mammal expert and evolutionary biologist George Gaylord Simpson (*A Catalogue of the Mesozoic Mammalia in the Geological Department of the British Museum*, Oxford University Press, 1928; American Mesozoic Mammalia, *Memoirs of the Peabody Museum*, 1929, 3: 1–235).

For more information on the end-Triassic extinction, I point readers to the discussion in my book *The Rise and Fall of the Dinosaurs*, and the references cited therein. The book *Triassic Life on Land: The Great Transition* (Columbia University Press, 2010) by Nicholas Fraser and Hans-Dieter Sues is a great summary of the Triassic world, its inhabitants, its physical geography, and the extinction. The lava erupting at the end of the Triassic created a huge amount of basaltic rock that covers part of four continents today, called the Central Atlantic Magmatic Province (or CAMP), which has been well described by Marzoli and colleagues (*Science*, 1999, 284: 616–18). The timing of the CAMP eruptions has been studied by Blackburn and colleagues (*Science*, 2013, 340: 941–45), which shows the eruptions took place in four large pulses over 600,000 years. Work by Jessica Whiteside, Paul Olsen, and colleagues shows that the extinctions on land and in the sea happened at the same time at the end of the Triassic, and that the first hints of extinction are synchronous with the first lava flows in Morocco (*Proceedings of the National Academy of Sciences USA*, 2010, 107: 6721–25). Changes across the Triassic-Jurassic boundary in atmospheric carbon dioxide, global temperature, and plant communities have been studied by, among others, McElwain et al. (*Science*, 1999, 285: 1386–90; *Paleobiology*, 2007, 33: 547–73) and Belcher et al. (*Nature Geoscience*, 2010, 3: 426–29).

The new image of Jurassic-Cretaceous mammals as diverse, dynamic, and exciting was first articulated to a big readership by Zhe-Xi Luo's review in *Nature* (2007, 450: 1011–19), which summarized the first decade of discoveries from Liaoning. In 2014, Meng Jin wrote an updated review of the Chinese fossils, further demonstrating their unexpected diversity (*National Science Review*, 1: 521–42). Roger Close and colleagues applied a variety of statistical methods to family trees of mammals, demonstrating that they underwent rapid rates of evolution in the Middle Jurassic (*Current Biology*, 2015, 25: 2137–42). One of these methods—a way of calculating rates of skeletal evolution—I helped develop with my colleagues Graeme Lloyd and Steve Wang (*Evolution*, 2012, 66: 330–48). Close et al. hypothesized that the breakup of Pangea may have caused these increased rates of evolution, and the overall explosive diversification of mammals in the Middle Jurassic.

The first fossil mammal reported from Liaoning, called *Zhangheotherium*, was described by Zhe-Xi Luo and his colleagues Yaoming Hu and Yuanqing

Wang et al. in 1997 (*Nature*, 390: 137–42). Two years later Luo, Ji Qiang, and Ji Shu-an described *Jeholodens*, also in *Nature* (1999, 398: 326–30). The dinosaur-eating *Repenomamus* was described in 2005 by Meng Jin and his team, led by first author Yaoming Hu (*Nature*, 433: 149–52).

Docodonts and haramiyidans are reviewed in books such as Kemp's *The Origin and Evolution of Mammals* and Kielan-Jaworowska et al.'s *Mammals from the Age of Dinosaurs*, but these are by now quite outdated, because of the rush of new discoveries from China. It's fun to read passages in these books— written in the early- to mid-2000s—bemoaning both groups as enduring mysteries, represented almost exclusively by fragmentary fossils. How things have changed! More information on the species mentioned in the text can be found in the papers describing them: *Microdocodon* (Zhou et al., *Science*, 2019, 365: 276–79), *Agilodocodon* (Meng et al., *Science*, 2015, 347: 764–68), *Docofossor* (Luo et al., *Science*, 2015, 347: 760–64), *Castorocauda* (Ji et al., *Science*, 2006, 311: 1123–27), *Vilevolodon* (Luo et al., *Nature*, 2017, 548: 326–29), *Maiopatagium* (Meng et al., *Nature*, 2017, 548: 291–96) and *Arboroharamiya* (Zheng et al., 203, *Nature*, 500: 199–202; Han et al., *Nature*, 2017, 551: 451–56). There was also an intriguing, late-surviving haramiyidan recently described from the Cretaceous of North America: *Cifelliodon*, named by Adam Huttenlocker and colleagues, in honor of the eminent fossil mammal expert Rich Cifelli (*Nature*, 2018, 558: 108–12).

As a brief pause, I note that there is currently huge debate about the placement of haramiyidans on the mammal family tree. There are two camps. One, led by Zhe-Xi Luo, argues that they are primitive mammals, placed on the stem of the family tree not too far from *Morganucodon*. The other camp, headed by Meng Jin, argues for a much more derived placement, within the crown group of mammals (the group that includes all the modern species and all descendants of their most recent common ancestor), as the sister group to the multituberculates, a group of plant-eating mammals that were diverse in the Cretaceous. I don't have a firm opinion either way. It may seem like a purely academic debate, but it does have one broad implication: because haramiyidans first appear in the Triassic, if they are crown mammals that means that the modern-type mammals go back further in time, to about 208 million years ago. If they are earlier stem mammals, then the crown group most likely first appeared in the Early Jurassic, around 178 million years ago.

We've described our dinosaur discoveries from Skye in a series of papers (Brusatte and Clark, *Scottish Journal of Geology*, 2015, 51: 157–64; Brusatte et al., *Scottish Journal of Geology*, 2016, 52: 1–9; dePolo et al., *Scottish Journal of Geology*, 2018, 54: 1–12; Young et al., *Scottish Journal of Geology*, 2019, 55:

7–19; dePolo et al., *PLoS ONE*, 2020, 15[3], e0229640). My work on Skye has been conducted with a great team of colleagues and students: Tom Challands, Mark Wilkinson, Dugald Ross, Paige dePolo, Davide Foffa, Neil Clark, and many others. Neil Clark has written several important papers on Skye dinosaurs, and Dugie Ross has discovered many of the most important fossils.

Hugh Miller's book *The Cruise of the Betsey* was published in 1858 in Edinburgh and can be found online here: https://minorvictorianwriters.org.uk /miller/b_betsey.htm. A fascinating biography of Miller was published by my colleague at the National Museum of Scotland, the paleontologist Michael Taylor (*Hugh Miller: Stonemason, Geologist, Writer*, National Museum of Scotland, 2007). Waldman and Savage published their description of *Borealestes* in 1972 (*Journal of the Geological Society*, 128: 119–25). Elsa's description of the *Borealestes* skull was published in 2021 (Panciroli et al., *Zoological Journal of the Linnean Society*, zla144), and she published a separate paper on the jaws and teeth (*Journal of Vertebrate Paleontology*, 2019, 39: e1621884) and another on the petrosal bone that encases the cochlea (*Papers in Palaeontology*, 2018, 5: 139–56). She's also written other papers on Skye mammals, which turn out to be quite diverse. There's *Stereognathus*, a tritylodontid: a member of the group of not-quite-mammals that also includes *Oligokyphus* and *Kayentatherium*; it was described initially by Waldman and Savage in their 1972 paper, and then redescribed by Elsa, myself, and colleagues in a 2017 paper (*Journal of Vertebrate Paleontology*, e1351448). There's *Wareolestes*, a primitive *Morganucodon*-type mammal (Panciroli et al., *Papers in Palaeontology*, 2017, 3: 373–86), and *Palaeoxonodon*, a more derived mammal closely related to the therians (placentals and marsupials) (Panciroli et al., *Acta Paleontologica Polonica*, 2018, 63: 197–206).

Liam Drew has an excellent discussion of the origin of lactation in his book *I, Mammal*. While writing this section, I also relied extensively on Olav Oftedal's fascinating review of mammary glands and lactation, published in 2002 (*Journal of Mammary Gland Biology and Neoplasia*, 7: 225–52). Eva Hoffman and Tim Rowe described their sublime *Kayentatherium* fossil family in *Nature* (2018, 561: 104–8). Zhou et al.'s paper on *Microdocodon*, cited above, is the best source for information on hyoid and throat musculature evolution.

There is a vast literature on mammalian middle ear bones, and I point readers to four papers to begin. First, Zhe-Xi Luo wrote a fantastic review of ear evolution in 2011 (*Annual Review of Ecology, Evolution, and Systematics*, 42: 355–80), which covers the anatomy of the ear, the homologies of the ear bones, the evolutionary sequence between many-jaw-boned cynodonts and mammals, and the genetic, developmental, and embryonic data that help us understand how the ear evolved. Second, Neal Anthwal and colleagues presented a review that blends a historical recounting of important work on the mammalian ear and the anatomical, genetic, and embryological evidence for ear evolution (*Journal of Anatomy*, 2012, 222: 147–60). Third, Wolfgang Maier and Irina Ruf wrote a historical piece explaining how researchers—dating back to the sixteenth century—studied the mammalian ear bones and came to understand their origins and evolutionary trajectory (*Journal of Anatomy*, 2015, 228: 270–83). Finally, the landmark paper of Edgar Allin—which laid out the evolutionary sequence of mammalian ear evolution—is well worth a read (*Journal of Morphology*, 1975, 147: 403–38).

Other important papers to consider, which expand on species I mention in the text, are Meng et al.'s description of *Liaoconodon*, the mammal with a transitional middle ear (*Nature*, 2011, 472: 181–85); Mao et al.'s description of *Origolestes*, the mammal with a middle ear detached from the bony strip, technically called the Meckelian element (*Science*, 2019, 367: 305–8; I note here that Zhe-Xi Luo has proposed an alternative explanation that the "detachment" of the cartilage is actually a fracture in the fossil and not genuine); Wang et al.'s description of the multituberculate *Jeholbaatar*, which revealed that the ear joint reflects the shape of the ancestral chewing motions when it was a jaw joint (*Nature*, 2019, 576: 102–5); Rich et al.'s description of the ears and jaws of early monotremes, which revealed that they evolved their detached middle ears independently of marsupials and placentals (*Science*, 2005, 307: 910–14); and Han et al.'s description of *Arboroharamiya allinhopsoni*, the haramiyidan with ear ossicles separated from the jaw (*Nature*, 2017, 551: 451–56). Some of these papers can be quite opaque to a nonspecialist (or to a paleontologist originally trained to study dinosaurs), so this reader was grateful to the commentary pieces by Anne Weil—a leading expert on multituberculate mammals—that accompanied some of these *Nature* papers.

I note that several months after drafting the chapter in the text, my former PhD student Sarah Shelley, my colleague and mentor John Wible, and their colleagues published an important paper on a haramiyidan ear, with broader implications for understanding the multiple ear bone detachments in mammalian history (Wang et al., *Nature*, 2021, 590: 279–83). They redefined some terms with historical baggage, and I follow their terms here. Most importantly, they use the term "detached middle ear" to refer to middle ear bones completely lacking any bony or cartilaginous attachment to the jaw; this is what many previous workers called a Definitive Mammalian Middle Ear (or DMME). Finally: the Wang et al. (2021) paper presented a new hypothesis to explain the differently shaped middle jaw joints of monotremes and the marsupial plus placental group: instead of being differently shaped because they reflect different chewing motions of the ancestral jaw bones (as proposed by the Wang et al. [2019] paper cited above, and as I outline in the text of chapter 3), they propose that the overlapping-type joint of monotremes is an evolutionary precursor to the more intricate interlocking joint of ours. This is an active debate!

Two recent studies, by the same team of embryologists and paleontologists, show how straightforward it is to sever the Meckel's cartilage from the ear bones in living mammals. One, by Anthwal et al. (*Nature Ecology & Evolution*, 2017, 1: 0093) focuses on chondroclasts in mice (a placental mammal), and the other, by Urban et al. (*Proceedings of the Royal Society, Series B*, 2017, 284: 20162416) focuses on cell death in possums (a marsupial mammal). Although I didn't cover it in the main text, there is another fascinating aspect of the jaw-ear story: some of the same genes that are expressed in the jaws of reptiles are expressed in the ears of mammals (for instance, the gene *Bapx1*), more definitive proof that jaw bones became ear bones. This work was published in 2004 by Abigail Tucker and team, including my Edinburgh colleague, the legendary geneticist Bob Hill (*Development*, 131: 1235–45).

CHAPTER 4: THE MAMMALIAN REVOLUTION

Zofia Kielan-Jaworowska tells her own life story in her book *In Pursuit of Early Mammals* (Indiana University Press, 2012), which also provides succinct overviews of the origin of mammals, the cynodont-mammal transition, and Mesozoic mammal groups. Other aspects of Zofia's biography come from my discussion with her on that summer afternoon in 2010, as recorded in my field notes. She also wrote a firsthand account of the first few Polish-Mongolian expeditions in her 1969 book *Hunting for Dinosaurs* (MIT Press). The Polish-Mongolian team's discoveries were described in a vast series of papers, which are extensively cited in Zofia's 2012 book. Many of these appeared in *Palaeontologia Polonica*—a quick perusal of this journal's back catalog online or in a library will reveal a wealth of information. Among these are her critical 1970 and 1974 papers on multituberculates.

Roy Chapman Andrews's life, and his Central Asiatic Expeditions, are the subject of Charles Gallenkamp's book *Dragon Hunter* (Viking, 2001). The American Museum of Natural History-Mongolian Academy of Sciences expeditions of the early 1990s are chronicled by Mike Novacek in his addictively readable book *Dinosaurs of the Flaming Cliffs* (Anchor Books, 1996), one of my favorite reads as a high school fossil enthusiast. The Ukhaa Tolgod locality—the site of so many multituberculate fossil discoveries—was initially described by Dashzeveg, Novacek, Norell, and colleagues (*Nature*, 1995, 374: 446–49).

Information on the geology of the site, and the detailed forensic evidence that fossils were formed in flood-collapsed sand dunes, can be found in papers by Loope et al. (*Geology*, 1998, 26: 27–30) and Dingus et al. (*American Museum Novitates*, 2008, 3616: 1–40).

Details of the Jurassic-Cretaceous transition are covered in my book *The Rise and Fall of the Dinosaurs*, with relevant references cited. The most useful general description of the climate and environmental changes is a review paper by Jon Tennant and colleagues (*Biological Reviews*, 2017, 92: 776–814). Jon, a young paleontologist and outspoken advocate of open science and open access publishing, tragically died in a motorbike accident in spring 2020, while I was writing this chapter.

Good general resources on multituberculates can be found in Zofia's 2012 book (cited above), Tom Kemp's *The Origin and Evolution of Mammals*, and the magnificent encyclopedia that Zofia edited with Zhe-Xi Luo and Richard Cifelli (*Mammals from the Age of Dinosaurs*, also cited above). The number I cite—that multituberculates constitute 70 percent of Gobi mammal faunas—comes from a paper by Chinsamy and Hurum on the bone microstructure and growth of early mammals (*Acta Palaeontologica Polonica*, 2006, 51: 325–38).

Important work on multituberculate feeding was published by Philip Gingerich, who presented evidence for the backward chewing stroke in a 1977 chapter in the book *Patterns of Evolution* (Elsevier); Zofia and her colleague Peter Gambaryan, who described the cranial musculature (*Acta Palaeontologica Polonica*, 1995, 40: 45–108); and David Krause (*Paleobiology*, 1982, 8: 265–313). Regarding multituberculate locomotion, Farish Jenkins and Krause described the reversible ankle and tree-climbing abilities (*Science*, 1983, 220: 712–15; *Bulletin of the Museum of Comparative Zoology*, 1983, 150: 199–246); my undergraduate advisor Paul Sereno and Malcolm McKenna described more advanced, fast-moving capabilities (*Nature*, 1995, 377: 144–47); and Zofia and Gambaryan published other important work (*Fossils and Strata*, 1996, 36; *Acta Palaeontologica Polonica*, 1997, 42: 13–44).

The Jurassic multituberculate *Rugosodon*—currently the oldest well preserved fossil of the group—was described by Chong-Xi Yuan, Luo, and their team (*Science*, 2013, 341: 779–83). Greg Wilson's study of multituberculate dental evolution was published in *Nature* (2012, 483: 457–60), and another study by David Grossnickle and David Polly found similar patterns of increasing tooth diversity with a different data set (*Proceedings of the Royal Society, Series B*, 2013, 280: 20132110). Our team—Zoltán Csiki-Sava, Mátyás Vremir, Meng Jin, Mark Norell, and myself—described *Litovoi* in 2018 (*Proceedings of the National Academy of Sciences [USA]*, 115: 4857–62), a paper that also reviews the Romanian island-living kogaionids more broadly. In 2021, Luke Weaver—a PhD student at the time—led a team describing the discovery of a social group of multituberculates in Montana (*Nature Ecology & Evolution*, 5: 32–37), work that in part earned him the Romer Prize of the Society of Vertebrate Paleontology.

A good source of general information on angiosperm origins and evolution is the book *Early Flowers and Angiosperm Evolution*, by Friis, Crane, and Pedersen (Cambridge University Press, 2011). The oldest good angiosperm vegetation fossils—called *Archaefructus*—were described from Liaoning by Sun et al. (*Science*, 2002, 296: 899–904). Important references that I consulted on early angiosperms, and why they eventually proved so adaptable, included papers by: Wing and Boucher (*Annual Review of Earth and Planetary Sciences*, 1998, 26: 379–421), Boyce et al. (*Proceedings of the Royal Society, Series B*, 2009, 276: 1771–76), Feild et al. (*Proceedings of the National Academy of Sciences [USA]*, 2011, 108: 8363–66), Coiffard et al. (*Proceedings of the National Academy of Sciences [USA]*, 2012, 109: 20955–59), deBoer et al. (*Nature Communications*, 2012, 3: 1221), Chaboureau et al. (*Proceedings of the National Academy of Sciences [USA]*, 2014, 111: 14066–70). One of these paleobotanists—Kevin Boyce—was an undergraduate instructor of mine at the University of Chicago. A few years after I took his class, he won a MacArthur Genius Grant!

The term "Cretaceous Terrestrial Revolution" was coined by some of my closest friends and colleagues in the field: Graeme Lloyd, Marcello Ruta, Mike Benton (my three MScR supervisors at the University of Bristol) and their colleagues in a 2008 paper on dinosaur evolution (*Proceedings of the Royal Society, Series B*, 275: 2483–90). Discussions with Mike Benton provided additional information on the Revolution, particularly the evolution of insects.

There is a vast literature on the anatomy, function, and evolution of the tribosphenic molar of therians—a complex topic that, by necessity, I had to greatly abridge in the main text so as not to dull everyone to boredom with pages after pages of molar cusp descriptions (as I tried to do in my first draft, before the red pens of my editor and wife set me straight). Two classic studies are Bryan Patterson's 1956 paper "Early Cretaceous mammals and the evolution of mammalian molar teeth," published in *Fieldiana* (13, 1–105), and Fuzz Crompton's 1971 paper "The Origin of the Tribosphenic Molar," published in *Zoological Journal of the Linnean Society* (50, supplement 1: 65–87). More recently, Brian Davis published a key paper on the origin and function of the tribosphenic molar, which also lays out the different wear patterns in the "tribosphenic-like" molars of the southern australosphenidans (*Journal of Mammalian Evolution*, 2011, 18: 227–44), and Julia Schultz and Thomas

Martin used 3-D models to describe how tribosphenic molars chew, in detail (*Naturwissenschaften*, 2014, 101: 771–871). Although I don't go into detail in the main text, the tribosphenic molars—with their complex interlocking shearing and grinding surfaces—would have required very precise chewing motions to function properly. There is currently debate about the jaw mechanics of the early tribosphenic therians, and it seems like they had either enhanced rotational movements (Bhullar et al., *Nature*, 2019, 566: 528–32) or enhanced yaw (side-to-side pivoting) motions (Grossnickle, *Scientific Reports*, 2017, 7: 45094), or possibly both. The description of the currently oldest known tribosphenic therian—*Juramaia*—was published by Zhe-Xi Luo and colleagues in 2011 (Nature, 476: 442–45).

David Grossnickle has published several important studies on how the evolution of the tribosphenic molar affected therian evolution, and mammalian evolution more generally. These include his review paper (written with Stephanie Smith and Greg Wilson) arguing that mammalian innovation often originates in the small-bodied insectivore niche (*Trends in Ecology and Evolution*, 2019, 34: 936–49); his paper with David Polly on mammal tooth and jaw shape evolution over time (*Proceedings of the Royal Society, Series B*, 2013, 280: 20132110); and his study with Elis Newham on the diversification of tribosphenic therians during and after the Cretaceous Terrestrial Revolution (*Proceedings of the Royal Society, Series B*, 2016, 283: 20160256; note that they argue most of this diversification was after the "Revolution" and not necessarily during it). I first met Dave when he joined our New Mexico fieldwork crew in 2013, on the invitation of the multituberculate expert Anne Weil. Dave went on to do his PhD with Zhe-Xi Luo in Chicago and has quickly become a leading expert on the evolution of mammals during the Jurassic and Cretaceous. He's also one of the funniest and most subversive people in the field (in a good way).

The genetic underpinning of therian tribosphenic tooth versatility has been studied by many developmental biologists and paleontologists. Key papers are by Jernvall et al. (*Proceedings of the National Academy of Sciences [USA]*, 2000. 97: 14444–48), Kavanagh et al. (*Nature*, 2007, 432: 211–14), Salazar-Ciudad et al. (*Nature*, 2010, 464: 583–86), and Harjunmaa et al. (*Nature*, 2014, 512: 44–48).

The effects of tribosphenic molar evolution (and many other mammal innovations) on community structure and ecology was covered in a recent paper by Meng Chen, Caroline Strömberg, and Greg Wilson, in *Proceedings of the National Academy of Sciences [USA]* (2019, 116: 9931–40).

Many Cretaceous eutherians and metatherians have been described in recent years, ranging from gorgeous Liaoning skeletons to fragmentary—but very important—teeth from North America (many of which have been studied by Richard Cifelli, Brian Davis, and colleagues). The problematic *Sinodelphys* from Liaoning was described as the oldest metatherian by Luo and colleagues (*Science*, 2003, 302: 1934–1940), but recently reinterpreted as a basal eutherian by Shundong Bi and his team, in their description of a new Liaoning eutherian called *Ambolestes* (*Nature*, 2018, 558: 390–95). A key resource on early metatherian evolution is a review paper that I contributed to, led by Tom Williamson and also including Greg Wilson as an author (*ZooKeys*, 2014, 465: 1–76). This is a follow-up to a genealogical analysis of Cretaceous-Paleogene metatherians that Tom and I published with a larger team (*Journal of Systematic Palaeontology*, 2012, 10: 625–51). The books *Mammals from the Age of Dinosaurs* and *In Pursuit of Early Mammals* (cited above) have great summaries of Gobi eutherians and metatherians, with references to all the important historical literature. More recently, the American Museum team published several important new *Deltatheridium* specimens, affirming their link to metatherians (Rougier et al., *Nature*, 1998, 396: 459–63). Another interesting recent paper focuses on the North American Late Cretaceous metatherian *Didelphodon*, also a fierce hunter in the small-size niche (Wilson et al., *Nature Communications*, 2017, 7: 13734).

The first encounters between humans and monotremes (platypuses and echidnas) in Australia is unrecorded by history, and the Aboriginal peoples had many thousands of years of experience with these peculiar animals. But the first European dealings with these animals has been covered in many retellings. I base my story on Brian Hall's 1999 paper on the platypus (*Bio-Science*, 49: 211–18) and Liam Drew's compelling storytelling in his book *I, Mammal*. I gleaned information on John Hunter from many online sources, including a nice Wikipedia biography (yes, even scientists click on Wikipedia sometimes, especially for an entry point to subjects outside of our immediate expertise). A more thorough reference is a 2009 biography of Hunter written by Robert Barnes (*An Unlikely Leader: The Life and Times of Captain John Hunter*, Sydney University Press).

Key references on the fossil monotremes and monotreme-line australosphenidans include papers on *Obdurodon* (Woodburne and Tedford, *American Museum Novitates*, 1975, 2588: 1–11; Archer et al., *Australian Zoologist*, 1978, 20: 9–27; Archer et al., *Platypus and Echidnas*, 1992, Royal Zoological Society of New South Wales; Musser and Archer, *Philosophical Transactions of the Royal Society of London, Series B*, 1998, 353: 1063–79); *Steropodon* (Archer et al., *Nature*, 1985, 318: 363–66; Rowe et al., *Proceedings of the National Academy of Sciences [USA]*, 2008, 105: 1238–42); *Ausktribosphenos* (Rich et al., *Science*,

1997, 278: 1438–42); *Ambondro* (Flynn et al., *Nature*, 1999, 401: 57–60); *Asfaltomylos* (Rauhut et al., *Nature*, 2002, 416: 165–68). There's another important one that I don't mention in the main text since: *Teinolophos*, named by the Rich and Vickers-Rich team in 1999 (*Records of the Queen Victoria Museum*, 106: 1–34), and more recently described in detail (*Alcheringa*, 2016, 40: 475–501).

Zhe-Xi Luo, Zofia Kielan-Jaworowska, and Richard Cifelli published their genealogical analysis, finding separate groups of northern tribosphenic therians and southern false-tribosphenic australosphenidans, in *Nature* (2001, 409: 53–57). I acknowledge here that not all mammal workers accept this phylogeny, and there have been arguments from the Rich and Vickers-Rich team that some of the southern species are closely related to therians and have true tribosphenic teeth. The nuances of this debate are outside of the scope of this book, but I feel the preponderance of evidence is on the side of separate northern and southern lineages. The independent origin of tribosphenic-like molars is supported by some astounding fossils that I do not talk about in the main text: the so-called pseudotribosphenic mammals, like *Shuotherium* (Chow and Rich, *Australian Mammalogy*, 1982, 5: 127–42) and *Pseudotribos* (Luo et al., Nature, 2007, 450: 93–97), which have backward tribosphenic-like teeth, with the talonid basin in front of the trigonid crests. These strange mammals—which might cluster with the southern australosphenidans on the family tree—are strong proof that different groups of mammals were evolving tribosphenic-like teeth independently, possibly many times. This is a classic case of convergent evolution: when similar ecological or other selective pressures lead to the independent evolution of similar-looking anatomical structures (like teeth) in distantly related groups.

The skull of *Vintana* was described by Krause and colleagues in 2014 in *Nature* (515: 512–17), and then later monographed in exceptional detail in a series of papers published as a *Memoir of the Society of Vertebrate Paleontology* in 2014, with contributions from several authors. Later, Krause's team described *Adalatherium* in a short report (*Nature*, 2020: 581, 421-27) and a series of papers in another *Memoir of the Society of Vertebrate Paleontology* in 2020. Broader information on gonwanatherians can be found in *Mammals from the Age of Dinosaurs* and *In Pursuit of Early Mammals*. The Argentine dryolestoid *Cronopio* was described by Rougier and team (*Nature*, 2011, 479: 98–102).

I also note here that in the final paragraph of this chapter, I state that there were different groups of mammals eating different foodstuffs at the end of the Cretaceous (e.g., insectivorous eutherians, herbivorous multituberculates). This was true—but it's not the case that all eutherians were insectivorous, for example. Most of the groups mentioned exhibited at least some dietary diversity.

CHAPTER 5: DINOSAURS DIE, MAMMALS SURVIVE

We described *Kimbetopsalis*, the new multituberculate species Carissa Raymond discovered, in 2016 (Williamson et al., *Zoological Journal of the Linnean Society*, 177: 183–208). The quotes I used in this chapter were taken from my field notes and recordings, a press release put out by the University of Nebraska, and Carissa and Tom's National Public Radio interview. Further information on the New Mexico fossils is given later in this section.

There is a huge literature on the end-Cretaceous extinction. I describe what the asteroid impact may have been like for dinosaurs and mammals in North America in *The Rise and Fall of the Dinosaurs* and also cite many pertinent references therein. The hypothesis that an asteroid caused the extinction was first proposed by the father-and-son team of Luis and Walter Alvarez and their colleagues (*Science*, 1980, 208: 1095–1108), and also independently by Dutch geologist Jan Smit around the same time. Walter Alvarez wrote a fantastic pop-science book, *T. rex and the Crater of Doom* (Princeton University Press, 1997), that tells the story of how he found the chemical fingerprint of iridium in end-Cretaceous rocks that pointed to an asteroid, and how over the next decade evidence for his theory continued to accumulate, until the Chicxulub Crater was discovered in Mexico, definitively proving that an asteroid (or comet) hit the earth circa 66 million years ago. Walter's book has references to all the key literature up until the time it was written.

There does remain some debate about whether the asteroid caused the extinction of the non-bird dinosaurs and other animals at the end of the Cretaceous. Critics of the asteroid theory instead implicate the Deccan volcanic eruptions in India—megavolcanoes on the scale of the eruptions that caused extinctions at the end of the Permian and Triassic. When it comes to dinosaurs, I led a team of paleontologists that reviewed all the evidence and firmly concluded that the asteroid was main culprit (*Biological Reviews*, 2015, 90: 628–42), a view that I also articulated in a *Scientific American* article (Dec. 2015, 313: 54–59). Recent work by Pincelli Hull and colleagues—a large team of authors including my fellow Edinburgh professor Dick Kroon and fieldwork compatriots Dan Peppe and Jessica Whiteside—forcefully argues that the asteroid was responsible for the entire extinction, and that the Indian volcanism played little if any role (*Science*, 2020, 367: 266–72), and a similar argument was made by Ale Chiarenza and colleagues based on climate and ecological modeling (*Proceedings of the National Academy of Sciences [USA]*, 2020, 117: 17084–93). Although I doubt this will be the last word on the subject, for me, there is little doubt that if no asteroid impact had occurred, there would have been no extinction, although it may be the case that the volcanism made the extinction worse, or prolonged the recovery. Another recent study, published as I was writing this chapter, found that the angle the asteroid impacted the earth made it even deadlier (Collins et al., *Nature Communications*, 2020, 11: 1480).

Bill Clemens was a warm and kind gentleman, who I knew from many pleasant chats at Society of Vertebrate Paleontology meetings. He was also generous with his time and expertise in communicating with my PhD student Sarah Shelley, when she was writing her thesis. Bill has published numerous papers on the mammals of the Hell Creek and Fort Union formations of Montana. Among the most important is a chapter in the 2002 edited volume *The Hell Creek Formation and the Cretaceous-Tertiary Boundary in the Northern Great Plains* (*Geological Society of America Special Paper*, 361: 217–45). Bill and Joseph Hartman wrote a review of the history of fossil collecting in the Hell Creek Formation, which was part of the 2014 edited volume *Through the End of the Cretaceous in the Type Locality of the Hell Creek Formation in Montana and Adjacent Areas* (*Geological Society of America Special Paper*, 503: 1–87). Early in his career, Bill wrote a three-part magnum opus on the mammals of the Late Cretaceous Lance Formation of Wyoming, which is roughly equivalent in age to the Hell Creek Formation; these were published in the *University of California Publications in Geological Sciences* (1964, 1966, and 1973). The details of how Bill strangely came into the orbit of the Unabomber were gleaned from a discussion with Anne Weil, whose admiration for her PhD advisor runs deep. Anne also told me another story of a weird brush with violence: one of the ranches in Montana on which Anne and Bill collected fossils was owned by the casino magnate Ted Binion, who was sadly murdered in a high-profile incident in 1998.

Greg Wilson Mantilla and his students and colleagues have written many important papers on mammal evolution across the end-Cretaceous extinction in Montana, which reveals what died and what survived, and why. These include sole-authored papers in *Journal of Mammalian Evolution* (2005, 12: 53–75), *Paleobiology* (2013, 39: 429–69), and in the 2014 edited volume cited above (*Geological Society of America Special Paper*, 503: 365–92). Greg and colleagues described the stunning new fossils of *Didelphodon* in 2016 (*Nature Communications*, 7:13734). Most recently, Greg and Bill were part of a team led by Greg's PhD student Stephanie Smith—one of the leading young mammal paleontologists today—that in 2018 described the Z-Line Quarry site and other Paleocene-aged localities from soon after the extinction—these are the best glimpses at the mammal communities that lived after the asteroid impact (*Geological Society of America Bulletin*, 130: 2000–2014). In unrelated studies,

Longrich and colleagues looked at Cretaceous and Paleocene mammals across all of western North America (*Journal of Evolutionary Biology*, 2016, 29: 1495–512), and Pires and colleagues looked at rates of extinction in multituberculates, metatherians, and eutherians in North America (*Biology Letters*, 2018, 14: 20180458).

There is more detailed discussion of why dinosaurs died at the end-Cretaceous in *The Rise and Fall of the Dinosaurs*. I credit my colleague Greg Erickson with using the term "dead man's hand"—borrowed from the sad tale of Wild Bill Hickok (who was actually born very close to my hometown of Ottawa, Illinois)—to describe the bad luck of the dinosaurs when the asteroid hit.

The historical story of Cope, the Wheeler Survey, Baldwin, and other discoveries in the San Juan Basin was based on discussion with Tom Williamson and Sarah Shelley, and a series of engrossing papers by the eminent mammal expert and writer George Gaylord Simpson, who did his own fieldwork in New Mexico, although mostly focused on younger, Eocene-aged mammals. These include works published in 1948 (*American Journal of Science*, 246: 257–82), 1951 (*Proceedings of the Academy of Natural Sciences of Philadelphia*, 103: 1–21), 1959 (*American Museum Novitates*, 57: 1–22), and 1981 (a chapter in *Advances in San Juan Basin Paleontology*, University of New Mexico Press). I discuss the "Bone Wars" rivalry of Cope and Marsh in *The Rise and Fall of the Dinosaurs*. There is a wealth of further information in John Foster's excellent book *Jurassic West: The Dinosaurs of the Morrison Formation and Their World* (Indiana University Press, 2007).

Cope published so many papers on the mammals from the "Puerco marls" that it is impossible to cite them all here. Tom Williamson's magisterial study of the Nacimiento Formation mammals, the subject of his PhD and published in 1996 (*New Mexico Museum of Natural History and Science Bulletin*, 8: 1–141), has references to all the important historical literature. Key papers are Cope's 1875 Wheeler Survey report (*Annual Report of the Chief of Engineers* for 1875, pp. 61–97) that first noted the "Puerco marls," several short papers he published in *The American Naturalist* during the 1880s, and his lengthy (that's putting it mildly) tome—often referred to as Cope's Bible—published in 1884 (*The Vertebrata of the Tertiary Formations of the West*. Book 1. Report of the U.S. Geological Survey of the Territories [Hayden Survey], pp. 1–1009). Another critical study of the New Mexico mammals was published by W. D. Matthew in 1937 (*Transactions of the American Philosophical Society*, 30: 1–510).

The best general source of information on Paleocene mammals like condylarths, taeniodonts, and pantodonts is the book *The Beginning of the Age of Mammals*, written by the eminent mammal paleontologist Kenneth Rose

(Johns Hopkins University Press, 2016). Sarah Shelley completed her PhD thesis at the University of Edinburgh in 2017 and published a large chunk of it—a monographic description of the condylarth *Periptychus*, a close relative of *Ectoconus*, also from New Mexico—in 2018 (*PLoS ONE*, 13[7]: e0200132). Other chapters will be published soon! Taeniodonts were comprehensively studied by Robert Schoch (*Bulletin of the Peabody Museum of Natural History, Yale University*, 1986, 42: 1–307), and Tom Williamson and I described fossils of *Wortmania* from New Mexico in 2013 (*PLoS ONE*, 8[9]: e75886). Pantodonts were comprehensively studied by Elwyn Simons (*Transactions of the American Philosophical Society*, 1960, 50: 1–99). I have current PhD students studying condylarths (Sofia Holpin, Hans Püschel), taeniodonts (Zoi Kynigopoulou), and pantodonts (Paige dePolo), so keep an eye on them and their publications!

Mammal placentas are fascinating and complex, and worthy of much more space than I was able to give them in this chapter. The best place to start learning about how they work and how they evolved is Liam Drew's book *I, Mammal*. In his book *Some Assembly Required*, Neil Shubin has a fascinating discussion of how placentas co-opted virus DNA to prevent themselves from being expelled by the mother, and how the mother's uterine cells developed ways to allow the placenta to literally invade the walls of the uterus. Interesting papers on placenta evolution include those by Chavan et al. (*Placenta*, 2016, 40: 40–51), Wildman et al. (*Proceedings of the National Academy of Sciences [USA]*, 2006, 103: 3203–08), and Roberts et al. (*Reproduction*, 2016, 152: R179–R189). Zofia Kielan-Jaworowska first reported the probable presence of epipubis bones in Cretaceous eutherians, based on her team's discovery of the pelvises of the eutherians *Barunlestes* and *Zalambdalestes* with articular notches for an epibuis bone (*Nature*, 1975, 255: 698–99). Actual epipubis bones were later found in the skeletons of the Gobi eutherians *Ukhaatherium* and a species that is most likely *Zalambdalestes*, by Novacek et al. (*Nature*, 1997, 389: 483–86). My description of the egg as a "care package" and the placenta as a multitasking organ were inspired by quotes from researcher Kelsey Coolahan, from a January 2020 interview on the Pulse radio program.

My research group has published several recent papers on the brains and senses of Paleocene mammals. This work was led by Ornella Bertrand, who was a postdoc in my lab (*Journal of Anatomy*, 2020, 236: 21–49); Joe Cameron, who was a master's student in my lab (*The Anatomical Record*, 2019, 302: 306–24); and James Napoli, who was an undergraduate student in my lab (*Journal of Mammalian Evolution*, 2018, 25: 179–95). Just as this book was going to press, Ornella's master study on early placental brain evolution was accepted for publication in *Science*! Among the most important work on mammal brain evolution is the pioneering research of Harry Jerison, most notably his 1973 book *Evolution of the Brain and Intelligence* (Academic Press).

There have been several important studies on body size evolution in Paleocene mammals, showing that they rapidly exploded in mass after the end-Cretaceous extinction. Two of the most influential are papers by the illustrious paleo-statistician John Alroy (*Systematic Biology*, 1999, 48: 107–18) and Graham Slater, an expert on using statistical models to study evolutionary trends (*Methods in Ecology and Evolution*, 2013, 4: 734–44). A number of young paleontologists have recently studied how various aspects of mammal biology changed across the Cretaceous-Paleocene extinction boundary: David Grossnickle and Elis Newham on molar shapes, a proxy for diet (*Proceedings of the Royal Society, Series B*, 2016, 283: 20160256), Gemma Benevento on jaw shape, another proxy for diet (*Proceedings of the Royal Society, Series B*, 2019, 286: 20190347), and Thomas Halliday on skeletal characteristics, a proxy for overall anatomy (*Biological Journal of the Linnean Society*, 2016, 118: 152–68).

CHAPTER 6: MAMMALS MODERNIZE

All the animals that I mention in my story are known from the Messel fossil site, and many of the details (such as the single fetus in the mare's womb, the anatomical features of the animals, and what they ate) are informed from actual fossils. In writing this story, I relied on the best source of information on Messel, the book *Messel: An Ancient Greenhouse Ecosystem*, edited by Krister Smith, Stephan Schaal, and Jörg Habersetzer (Senckenberg Museum, Frankfurt, 2018), which has chapters on all the mammal groups, plus the other animals (birds, crocodiles, turtles, etc.), the plants, the environment, and details on how the lake was formed from a volcanic eruption and how gases most likely killed the many fossilized animals. Other important sources were UNESCO's website (https://whc.unesco.org/en/list/720/); Gerhard Storch's article in *Scientific American* (1992, 266[2]: 64–69); and Ken Rose's short review of Messel mammals (*Palaeobiodiversity and Palaeoenvironments*, 2012, 92: 631–47).

There are several excellent papers on our hero, the mare *Eurohippus*. It was named by Jenz Lorenz Franzen, one of the deans of Messel research, in 2006 (*Senckenbergiana Lethaea*, 86: 97–102). For many years, until Franzen's paper, it was thought that *Eurohippus* was synonymous with another Messel horse, *Propalaeotherium*. Franzen and colleagues described a gorgeous skeleton of *Eurohippus* with a fetus inside in a series of papers (*PLoS ONE*, 2015, 10[10]: e0137985; *Palaeobiodiversity and Palaeoenvironments*, 2017, 97: 807–32). Franzen and his colleagues—including Phil Gingerich, who we meet later in the chapter—described the primate *Darwinius* to international fanfare in 2009 (*PLoS ONE*, 4[5]: e5723).

Mammal phylogenetics—the construction of their family trees—has a long and convoluted history. For general overviews of the current state of knowledge, and how we got there through many decades of debate, I recommend the relevant sections in Liam Drew's book *I, Mammal*, a general overview of mammal genealogy by Nicole Foley and colleagues (*Philosophical Transactions of the Royal Society, Series B*, 2016, 371: 20150140), and a more detailed summary of mammal relationships written by Robert Asher—an expert on early mammal evolution and a fantastic writer—published in *Handbook of Zoology: Mammalian Evolution, Diversity and Systematics* (DeGruyter, 2018). George Gaylord Simpson's famous 1945 family tree was published in the *Bulletin of the American Museum of Natural History* (85: 1–350), and Michael Novacek's later tree was published in *Nature* (1992, 356: 121–25). The stories about Simpson's life were gleaned from David Rains Wallace's *Beasts of Eden*, and you can read more about Simpson in a biography written by Léo Laporte (*George Gaylord Simpson: Paleontologist and Evolutionist*, Columbia University Press, 2000).

Over the past twenty-five years, there have been numerous DNA-based family trees of mammals, some surveying mammals as a whole, others focusing in on the detailed species-level relationships of individual groups like primates or rodents. The key early papers that established the DNA tree of mammals—and recognized the four main clusters of Afrotheria, Xenarthra, Euarchontoglires, and Laurasiatheria—were published by Mark Springer and his colleagues, including Ole Madsen, Michael Stanhope, William Murphy, Stephen O'Brien, Emma Teeling, and many others. The most important include: Springer et al. (*Nature*, 1997, 388: 61–64), Stanhope et al. (*Proceedings of the National Academy of Sciences, USA*, 1998, 95: 9967–72), Madsen et al. (*Nature*, 2001, 409: 610–14), Murphy et al. (*Nature*, 2001, 409: 614–18), and Murphy et al. (*Science*, 2001, 294: 2348–51). More recent workers have started to combine DNA and anatomical features to build "total evidence" trees, most prominently the genealogy of placentals published by Maureen O'Leary and her colleagues on the NSF-funded Mammal Tree of Life project (*Science*, 2013, 339: 662–67). In my description of the unexpected groupings in the DNA tree, I say that Afrotheria was "a most unusual union that nobody had ever predicted from anatomy"—which is true, although it should be noted that Edward Cope himself, in the late 1800s, used anatomical features to argue that golden moles (now known to be members of the Afrotheria group with elephants and tenrecs) were very different than European moles.

Anatomy-only trees are becoming less common; one was recently published by Thomas Halliday and colleagues, which attempted to untangle the relationships of the Paleocene "archaic" placentals (*Biological Reviews*, 2017, 92: 521–50). My European Research Council–funded team is working independently, to try to use anatomy and DNA to clarify these relationships further—particularly how the "archaic" species slot in with the modern species. We have so far published preliminary results as abstracts, but the research is still ongoing as I write this. A special thanks here to my co-PIs John Wible and Tom Williamson, who have also received funding for our big project through the National Science Foundation. Stay tuned for our publications!

There is a vast literature using the DNA clock to predict the origin times of placental mammals as a whole, and constituent subgroups. It is looking more and more likely that Placentalia itself originated back in the Cretaceous, during the time of dinosaurs, and that some subgroups did, too, but that the most explosive phase of their evolution occurred after the asteroid impact, in the Paleocene. However, this is all based on the DNA clock, as nobody, anywhere, has yet to find a convincing fossil of a Cretaceous placental. It might be that they were rare or localized to particular parts of the world at the time, or that they were common but we just have a hard time recognizing them as placentals . . . or that the DNA clock is wrong. For good overviews of this debate, consult the review papers of Archibald and Deutschmann (*Journal of Mammalian Evolution*, 2001, 8: 107–24) and Goswami (*EvoDevo*, 2012, 3:18).

The Paleocene-Eocene Thermal Maximum (PETM)—the global warming spurt circa 56 million years ago—has been the subject of intense study, from geologists, climatologists, biologists, and many other scientists. The best general summary of the PETM, its causes, and its duration is a review paper written by Francesca McInerney and Scott Wing (*Annual Review of Earth and Planetary Sciences*, 2011, 39: 489–516). It cites all the important work on the PETM up until 2011. There has been more recent geological and climatological work, which has convincingly (in my view) identified the North Atlantic volcanoes and their rock-baking magmas as the culprit (Gutjahr et al., *Nature*, 2017, 548: 573–77; Jones et al., 2019, *Nature Communications*, 10: 5547). These new studies essentially corroborate the hypotheses of Svensen et al. (*Nature*, 2004, 429: 542–45) and Storey et al. (*Science*, 2007, 316: 587–89), who noted that the PETM happened at the same time that volcanoes tore open the North Atlantic. For a more popular and poetic look at the PETM, the science writer extraordinaire Peter Brannen penned an evocative article for the *Atlantic* (August 2018).

The PETM had innumerable effects on the environment. Globally, the obscenely high arctic land temperatures were identified by Weijers et al. (*Earth and Planetary Science Letters*, 2007, 261: 230–38) and Eberle et al. (*Earth and Planetary Science Letters*, 2010, 296: 481–86); midlatitude temperatures were measured by Naafs et al. (*Nature Geoscience*, 2018, 11: 766–71); and the boiling tropical temperatures were studied by Aze et al. (*Geology*, 2014, 42: 739–42). Locally in the Bighorn Basin of Wyoming, Kraus and Riggins described evidence for transient drying (*Palaeogeography, Palaeoclimatology, Palaeoecology*, 2007, 245: 444–61); Ross Secord and colleagues described in detail the temperature increase (*Nature*, 2010, 467: 955–58); and Scott Wing and colleagues described floral changes (*Science*, 2005, 310: 993–96).

The Bighorn Basin mammal record, and how it responded to the PETM, has been the subject of Philip Gingerich's life work, and the work of many of his students. For a brief look at Gingerich's background, including his upbringing as a Mennonite in Iowa, see Tom Mueller's article in the August 2010 issue of *National Geographic*. In 2006 Gingerich published an accessible overview of the issue (*Trends in Ecology and Evolution*, 21: 246–53), which followed from two more technical papers that detailed how mammal diversity and body size changed during the PETM: one that was led by his graduate student William Clyde (Clyde and Gingerich, *Geology*, 1998, 26: 1011–14) and one that he wrote (*Geological Society of America Special Papers*, 2003, 369: 463–78). Gingerich also edited two important volumes on Bighorn Basin geology and paleontology (*University of Michigan Papers on Paleontology*, 1980, 24; *University of Michigan Papers on Paleontology*, 2001, 33).

Key papers on local Bighorn Basin exposures that record the PETM and mammal response were published by Gingerich (*University of Michigan Papers on Paleontology*, 1989, 28), Gingerich and his Belgian colleague Thierry Smith (*Contributions from the Museum of Paleontology*, The University of Michigan, 2006, 31: 245–303) and Kenneth Rose and colleagues (*University of Michigan Papers on Paleontology*, 2012, 24). Ross Secord presented his exceptional work on PETM horse dwarfing in *Science* (2012, 335: 959–62), and later work by Abigail D'Ambrosia and team found that mammals dwarfed in a similar way during later global warming events (*Science Advances*, 2017, 3: e1601430). Other important papers are those on the forest environments of the Eocene Bighorn Basin (Secord et al., *Paleobiology*, 2008, 34: 282–300) and Amy Chew's longer-term study of the Eocene mammals of the area (*Paleobiology*, 2009, 35: 13–31).

The migration of the PETM Trinity—primates, artiodactyls, and perissodactyls—during the temperature spike is clear in the fossil record: these animals show up all across the northern continents, essentially simultaneously. The primates have been studied by Thierry Smith and colleagues

(*Proceedings of the National Academy of Sciences, USA*, 2006, 103: 11223–27) and Chris Beard (*Proceedings of the National Academy of Sciences, USA*, 2008, 105: 3815–18). General dispersal between Asia and other continents has been assessed by Bowen et al. (*Science*, 2002, 295: 2062–65) and Bai et al. (*Communications Biology*, 2018, 1: 115), and European faunas are discussed by Smith et al. (*PLoS ONE*, 2014, 9[1]: e86229).

For more information on the actual members of the PETM Trinity that spread around the world, I recommend the following sources.

Primates: One of the most enjoyable and accessible guides to early primate evolution is a book by the eminent paleontologist Chris Beard, a recipient of a MacArthur Genius Grant (*The Hunt for the Dawn Monkey*, University of California Press, 2004). The PETM primate *Teilhardina* has been described by Ni et al. (*Nature*, 2004, 427: 65–68), Rose et al. (*American Journal of Physical Anthropology*, 2011, 146: 281–305), and Morse et al. (*Journal of Human Evolution*, 2019, 128: 103–31). Other interesting primates from around the same time are *Cantius* (Gingerich, *Nature*, 1986, 319: 319–21) and *Archicebus* (Ni et al., *Nature*, 2013, 498: 60–64).

Artiodactyls: The pioneering artiodactyl *Diacodexis* was named by Cope and described by Ken Rose, one of the world's most respected experts on early mammal anatomy and evolution (*Science*, 1982, 216: 621–23). Further descriptions of its postcranial anatomy were written by Thewissen and Hussain (*Anatomia, Histologia, Embryologia*, 1990, 19: 37–48), and Maëva Orliac and colleagues have used CT scans to describe its brain (*Proceedings of the Royal Society, Series B*, 2012, 279: 3670–77) and inner ear (*Journal of Anatomy*, 2012, 221: 417–26).

Perissodactyls: In many older publications, *Sifrhippus* is referred to as *Hyracotherium*—a well-known genus of early horse that became a wastebasket grouping many distinct species, which was clarified by David Froehlich, who created the name *Sifrhippus* (*Zoological Journal of the Linnean Society*, 2002, 134: 141–256). The anatomy of *Sifrhippus*—under the name *Hyracotherium*—was described by Wood et al. (*Journal of Mammalian Evolution*, 2011, 18: 1–32). Ken Rose and colleagues proposed a provocative theory that perissodactyls may have originated in India, and then spread across Asia when the two continents collided in the Eocene (*Nature Communications*, 2011, 5: 5570; *Society of Vertebrate Paleontology Memoir*, 2020, 20: 1–147). The idea of India as a "Noah's Ark" is intriguing, but what is not clear is how the ancestors of perissodactyls would have reached the then-island of India, in the Cretaceous or Paleocene.

One of the best sources of information on Eocene primates, perissodactyls, artiodactyls, and other mammals generally is Don Prothero's *Princeton Field Guide to Prehistoric Mammals* (Princeton University Press, 2017), which I used extensively. Also useful for the hoofed species is a book that Don wrote with Robert Schoch: *Horns, Tusks, and Flippers* (The Johns Hopkins University Press, 2002). Both these books have copious information on the bizarre brontotheres and chalicotheres. The single best technical resource on bronotheres is Matthew Mihlbachler's magisterial monograph, published in the *Bulletin of the American Museum of Natural History* (2008, 311: 1–475), which cleans up over a century of sloppy descriptive and taxonomic work on the group and presents an updated classification.

The historian Adrienne Mayor tells the tale of the Thunder Beasts and other Native American fossil discoveries in her 2005 book *Fossil Legends of the First Americans* (Princeton University Press) and a 2007 article (*Geological Society of London, Special Publications*, 273: 245–61).

Did rodents really outcompete multituberculates, and drive them to extinction? Or was it more of an opportunistic replacement? This was assessed using a clever biomechanical approach by the young paleontologist Neil Adams, while a master's student, and published in 2019 (*Royal Society Open Science*, 6: 181536). The verdict: maybe. Rodents actually have higher stresses on their skull bones when they bite compared to multituberculates, but they are able to optimize their bite forces, meaning the two groups didn't chew in exactly the same way, and it's not really clear if one or the other was "superior."

The best source of information—and real facts—about Charles Darwin's *Beagle* journey come from Darwin himself, in his books (both of which I quote): *The Voyage of the Beagle* (1839) and *The Origin of Species* (1859). Darwin's mammal discoveries were reviewed by the paleontologist Juan Fernicola and colleagues (*Revista de la Asociación Geológica Argentina*, 2009, 64: 147–59) and were celebrated for a more popular audience in David Quammen's *National Geographic* article (February 2009). Fernicola's paper also recounts stories of South American Indigenous peoples coming into contact with large fossil bones.

Darwin's South American Ungulates were recently reviewed by Darin Croft and colleagues (*Annual Review of Earth and Planetary Sciences*, 2020, 48: 11.1–11.32) and are also covered in both of Don Prothero's books cited above. Information on Richard Owen and the Ameghinos was gleaned from Wallace's *Beasts of Eden*. The "paternity test" studies linking them to perissodactyls are based on proteins (Welker et al., *Nature*, 2015, 522: 81–84; Buckley, *Proceedings of the Royal Society, Series B*, 2015, 282: 20142671) and DNA (Westbury et al., *Nature Communications*, 2017, 8: 15951). It should be noted that only two subgroups of Darwin's Ungulates—notably the litopterns (the *Macrauchenia* group) and notoungulates (the *Toxodon* group)—have been subjected to the paternity test, so it is not yet clear if the other subgroups are linked to perissodactyls too. The dispersal of Darwin's Ungulates from South America to Antarctica is examined by Reguero et al. (*Global and Planetary Change*, 2014, 123: 400–413).

The insular South American mammal fauna is expertly profiled by Darin Croft in his book *Horned Armadillos and Rafting Monkeys* (Indiana University Press, 2016). George Gaylord Simpson's views were featured in his book *Splendid Isolation* (Yale University Press, 1980). Important papers on sparassodonts include those by: Argot (*Zoological Journal of the Linnean Society*, 2004, 140: 487–521), Forasiepi (*Monografías del Museo Argentino de Ciencias Naturales*, 2009, 6: 1–174), Goswami et al. (*Proceedings of the Royal Society, Series B*, 2011, 278: 1831–39), Prevosti et al. (*Journal of Mammalian Evolution*, 2013, 20: 3–21), Croft et al. (*Proceedings of the Royal Society, Series B*, 2017, 285: 20172012), Muizon et al. (*Geodiversitas*, 2018, 40: 363–459), and Janis et al. (*PeerJ*, 2020, 8:e9346). The latter paper uses biomechanical analysis to argue that the "marsupial saber tooth" *Thylacosmilus* used its canines differently than true saber-toothed cats, more as belly-opening tools than throat slashers. Sparassodont bite marks on Darwin's Ungulate bones were described by Tomassini et al. (*Journal of South American Earth Sciences*, 2017, 73: 33–41).

The improbable but true story of primates and rodents rafting from Africa to South America is covered in several important papers. Mariano Bond and colleagues described the oldest New World monkey, *Perupithecus*, from the Eocene of South America (*Nature*, 2015, 520: 538–41). Stunningly, Seiffert and colleagues recently described a second lineage of South America primates, also nested within an African group, which may have rafted westward independent of the New World monkeys (*Science*, 2020, 368: 194–97). The rafting caviomorphs are discussed by Antoine et al. (*Proceedings of the Royal Society, Series B*, 2012, 279: 1319–26). The massive cow-size rodent *Josephoartigasia*—imagine that, a guinea pig–looking thing as big as some cars!—is examined by Rinderknecht and Blanco (*Proceedings of the Royal Society, Series B*, 2008, 275: 923–28) and Millien (*Proceedings of the Royal Society, Series B*, 2008, 275: 1953–55). Before rodents and primates went from Africa to South America, they needed to travel from Asia (or Europe) to Africa, which is discussed by Sallam et al. (*Proceedings of the National Academy of Sciences, USA*, 2009, 106: 16722–27), Jaeger et al. (*Nature*, 2010, 467: 1095–98), and Chris Beard in his book, cited above.

CHAPTER 7: EXTREME MAMMALS

The chapter title "Extreme Mammals" was inspired by an exhibit of the same name, originally shown at the American Museum of Natural History and curated by one of my PhD committee members and mentors, the widely admired mammal expert John Flynn.

This chapter highlights elephants, bats, and whales. There are many sources of further reading for each group, expanded on below. The most engaging general resource that covers the evolution of all three groups is Don Prothero's *Princeton Field Guide to Prehistoric Mammals* (Princeton University Press, 2017). For information on the evolutionary history and relationships of each group, Robert Asher's chapter in the *Handbook of Zoology: Mammalian Evolution, Diversity and Systematics* (cited above) is an excellent summary.

Elephants: Emmanuel Gheerbrant and his colleagues have published several papers on their transitional sequence of elephant fossils, showing how they supersized over time. These include works on *Eritherium* (*Proceedings of the National Academy of Sciences, USA*, 2009, 106: 10717–21), *Phosphatherium* (*Nature*, 1996, 383: 68–70), and *Daouitherium* (*Acta Palaeontologica Polonica*, 2002, 47: 493–506). Gheerbrant has also described fossils of primitive afrotherians, on the ancestral lineage toward elephants and the other modern species, including *Ocepia* (*PLoS ONE*, 2014, 9: e89739). and *Abdounodus* (*PLoS ONE*, 2016, 11: e0157556), and he has done important work on the origin of the embrithopods, the extinct group that includes the bizarre huge-horned *Arsinotherium* (*Current Biology*, 2018, 28: 2167–73). Regarding other extinct afrotherians, three papers nicely exhibit the strange body types and large body sizes of prehistoric hyraxes (Schwartz et al., *Journal of Mammalogy*, 1995, 76: 1088–99; Rasmussen and Simmons, *Journal of Vertebrate Paleontology*, 2000, 20: 167–76; Tabuce, *Palaeovertebrata*, 2016, 40: e1–12). One final point on afrotherians, for clarity: while they seem to be African endemics, they would have had an ancestor that came from elsewhere in the latest Cretaceous or early Paleocene, and so it is possible that these earliest afrotheres were actually present—maybe even widespread—on other continents before they became restricted to Africa.

My discussion of elephant body sizes over time—including the size estimates for individual species—come from an important paper by Asier Larramendi (*Acta Palaeontologica Polonica*, 2016, 61: 537–74). This paper reviews the evidence for the massive body size estimates for *Palaeoloxodon*, which is admittedly based on extrapolating from fragmentary fossils. This paper also discusses the sizes of Eocene-Oligocene rhinos like *Paraceratherium*, and how these animals compared to the largest elephants. From my reading of this and other literature, I don't think we can be confident (yet) about whether it was elephants like *Palaeoloxodon* or rhinos like *Paraceratherium* that held the "largest land mammal ever" title, but it doesn't matter that much: these animals got to be roughly the same size, and that size was monstrous.

My discussion of mammal body size evolution over time, and how it reached a peak around the Eocene-Oligocene boundary, was informed by two key papers from the same general research group, led by Felisa Smith (*Science*, 2010, 330: 1216–19) and Juha Saarinen (*Proceedings of the Royal Society of London Series B*, 2014, 281: 20132049), plus something of a rebuttal to the first paper by Roland Sookias and colleagues (*Biology Letters*, 2012, 8: 674–77).

For more information on how dinosaurs grew to enormous sizes, and which features of their anatomy enabled it, please see my book *The Rise and Fall of the Dinosaurs* and many references cited therein, including important work by Martin Sander and colleagues (*Biological Reviews*, 2011, 86: 117–55). For information on elephant brains, and how they got bigger over time, consult the study of Julien Benoit and team (*Scientific Reports*, 2019, 9: 9323).

Bats: When it comes to bats, Nancy Simmons and her colleagues described the oldest and most primitive fossil bat *Onychonycteris* on the cover of *Nature* (2008, 451: 818–21). They later published details on its throat and ear (*Nature*, 2010, 466: E8–E9, in response to the Vesekla et al. paper cited below), and on its wings and flying style, in a paper led by Lucila Amador (*Biology Letters*, 2019, 15: 20180857). Nancy was junior author on an engrossing review of bat origins, led by her late colleague Gregg Gunnell (*Journal of Mammalian Evolution*, 2005, 12: 209–46), and Nancy and Jonathan Geisler published a landmark monograph on bat genealogy in the *Bulletin of the American Museum of Natural History* (1998, 235: 1–182).

Other important Eocene bat fossils include *Icaronycteris* from the western United States (Jepsen, *Science*, 1966, 1333–39), *Australonycteris* from Australia (Hand et al., *Journal of Vertebrate Paleontology*, 1994, 14: 375–81), *Tanzanycteris* from Tanzania (Gunnell et al., *Palaeontologica Electronica*, 2003, 5[3]:1–10), an early Eocene specimen from Algeria (Ravel et al., *Naturwissenschaften*, 2011, 98: 397–405), several species from India (Smith et al., *Naturwissenschaften*, 2007, 94: 1003–09), and specimens from Portugal (Tabuce et al., *Journal of Vertebrate Paleontology*, 2009, 29: 627–30). The Messel bats from Germany are described in the book *Messel: An Ancient Greenhouse Ecosystem* (cited above), and in two papers by Jörg Habersetzer and colleagues (*Naturwissenschaften*, 1992, 79: 462–66; *Historical Biology*, 1994, 8: 235–60).

There is a rich literature on bat flight. One of the best resources on how bats fly, and how this is enabled by their skeletons and wing shapes, is a landmark monograph by Norberg and Rayner (*Philosophical Transactions of the Royal Society Series B*, 1987, 316: 335–427). The 100 mph speed I mention was recorded by McCracken and colleagues (*Royal Society Open Science*, 2016, 3: 160398). Karen Sears and colleagues published an important study on how bat wings develop in an embryo, and what this means for how they evolved (*Proceedings of the National Academy of Sciences, USA*, 2006, 103: 6581–86).

There is also a rich literature on bat echolocation. Accessible review papers include those by Arita and Fenton (*Trends in Ecology and Evolution*, 1997, 12: 53–58), Speakman (*Mammal Review*, 2001, 31: 111–30), and Jones and Teeling (*Trends in Ecology and Evolution*, 2006, 21: 149–56). Mike Novacek showed how cochlea size is related to echolocation (*Nature*, 1985, 315: 140–41), and Nina Veselka and team showed how the connection between throat and ear bones is related to echolocation (*Nature*, 2010, 463: 939–42). Other authors have focused on how echolocation evolved in bats, informed by the family tree of modern bats and the distribution of different types of echolocation among the modern species; chief among these are papers by Emma Teeling and colleagues (*Nature*, 2000, 403: 188–92) and Mark Springer and team (*Proceedings of the National Academy of Sciences, USA*, 2001, 98: 6241–46). Teeling—who like Nancy Simmons is universally regarded as a leading expert on bats—led a milestone study on the genealogical relationships of today's bats, using the DNA paternity test (*Science*, 2005, 307: 508–84).

I learned much about vampire bats, particularly about their hunting styles and how their brains are attuned to breathing rhythms, in papers by Gröger and Weigrebe (*BMC Biology*, 2006, 4: 18) and Schmidt et al. (*Journal of Comparative Physiology A*, 1991, 168: 45–51). The factoid about how much cow blood a vampire colony eats in a year comes from National Geographic (https://www .nationalgeographic.com/animals/mammals/c/common-vampire-bat/).

Whales: There is an enormous, broad, and deep literature on whales, covering how they evolved from walking ancestors and how these most extreme of mammals move, feed, reproduce, cognate, and communicate today. For a fast-paced, first-person account of the past, present, and future of whales, I cannot recommend enough Nick Pyenson's pop-science book *Spying on Whales* (2018, Viking). Nick is a curator at the Smithsonian, and has seemingly done everything in the field of whaleontology, from excavating and describing fossil whales to dissecting modern-day whales and tagging living whales to study their migration and diving patterns. Carl Zimmer, the peerless science writer, wrote an earlier pop-science book focusing, in part, on how whales went from walking to swimming, called *At the Water's Edge* (Simon & Schuster, 1999).

More recently, Hans Thewissen wrote a semitechnical, semipersonal account of whale evolution and his fossil discoveries, called *The Walking Whales* (University of California Press, 2019), and Annalisa Berta wrote a more general semitechnical book on all marine mammals, called *Return to the Sea* (University of California Press, 2012). The paleontologists Felix Marx, Olivier Lambert, and Mark Uhen teamed up to write a fantastic semitechnical review of whale evolutionary history, *Cetacean Paleobiology* (Wiley-Blackwell, 2016).

Before further diving into whales, a digression. If you want a quirky read, check out *The Stones of the Pyramids*, by Dietrich and Rosemary Klemm, which outlines what source rocks the Giza Pyramids, and other Egyptian monuments, were constructed from (De Gruyter, 2010).

When it comes to whale evolution—and the tale of how walkers turned into swimmers—there are several general review papers that tell the story well. Chief among these are essays by Hans Thewissen and E. M. Williams (*Annual Review of Ecology, Evolution, and Systematics*, 2002, 33: 73–90); Hans Thewissen, Lisa Noelle Cooper, and colleagues (*Evolution: Education and Outreach*, 2009, 2: 272–88); Sunil Bajpai and colleagues (*Journal of Biosciences*, 2009, 34: 673–86); Mark Uhen (*Annual Review of Earth and Planetary Sciences*, 2010, 38: 189–219); John Gatesy and colleagues (*Molecular Phylogenetics and Evolution*, 2013, 66: 479–506); and Nick Pyenson (*Current Biology*, 2017, 27: R558–R564). These reviews also cover the DNA and other molecular evidence linking whales to artiodactyls, and the history of this research, with citations to the key primary literature. Although I wish I could claim credit for it, the phrase "Bambi turning into Moby Dick" was inspired by the headline of Ian Sample's *Guardian* article on the discovery of *Indohyus* (from which I also gathered details of the discovery of the *Indohyus* fossils).

Two papers published nearly simultaneously announced the discovery of the signature artiodactyl double-pulley astragalus in Eocene whales: one by Thewissen and colleagues (*Nature*, 2001, 413: 277–81) and one by Philip Gingerich and colleagues (*Science*, 2001, 293: 2239–42).

I want to be very clear about a potential pitfall of my narrative style. When I talk about transitional whales, I might give the impression that a single *Indohyus* individual fled into the water, and this one individual was the ancestor of whales. Similarly, I might give the impression that *Indohyus* evolved into *Pakicetus*, which evolved into *Ambulocetus*, which evolved into *Rodhocetus*, which evolved into today's whales. These things are not strictly true. To the first point, there would have been a population of *Indohyus* (and/or closely related species) living on the Indian island, which began to experiment with living in the water. To the second point, the fossils I mention in the text form successive branches on the family tree, on the line to whales. These species were not strictly ancestral to each other, but are the fossils that paleontologists have happened to find so far, which are links in a much bigger chain that would have included many other species that we have yet to find. Also, the particular species I highlight are members of larger groups: *Indohyus* is part of Raoellidae, *Pakicetus* part of Pakicetidae, *Ambulocetus* part of Ambulocetidae, *Rodhocetus* part of Protocetidae, and *Basilosaurus* part of Basilosauridae. It is these *groups*—plus another one called Remingtonocetidae that I don't discuss in the text—that form the series of ancestral steps on the line to whales. The species I discuss in the text are the best exemplars of these groups: they are known from the best fossils and have been the subject of most intensive study, so they are the easiest ones to profile. Thus, the fossils I mention *represent* the progressive stages that the deerlike ancestors of whales morphed through as they became ever-better swimmers. Each one is a clue that reveals part of the story; it is their step-by-step arrangement on the family tree, on the line to whales, that provides the directionality to the story, even though these specific species would not have formed a strict ancestor-descendant chain. Who knows what other links in the chain are still out there to be found?

Here are the most essential sources for the chain of transitional species on the line to whales:

Indohyus: A. Ranga Rao described the first, scrappy fossils of this mammal in 1971 (*Journal of the Geological Society of India*, 12: 124–34). Hans Thewissen and colleagues later described new fossil material that revealed a link to whales (*Nature*, 2007, 250: 1190–94), and these fossils were then described in more comprehensive detail by Lisa Noelle Cooper, Thewissen, and colleagues in 2012 (*Historical Biology*, 24: 279–310). Thewissen's team included Indian colleagues Sunil Bajpai and B. N. Tiwari.

Pakicetus: Phil Gingerich and colleagues first described *Pakicetus* in *Science* (1983, 220: 403–6). S. I. Madar described the skeleton of *Pakicetus* and other pakicetids in more detail (*Journal of Paleontology*, 2007, 81: 176–200).

Ambulocetus: Hans Thewissen and colleagues first described *Ambulocetus* in *Science* (1994, 263: 210–12), and then later published a comprehensive description of its skeleton (*Courier Forsch.-Inst. Senckenberg*, 1996, 191: 1–86). In 2016, Konami Ando and Shin-ichi Fujiwara published an important study arguing, based on postcranial anatomy, that *Ambulocetus* was a strong swimmer and poor walker, and probably spent most of its time in the water (*Journal of Anatomy*, 229: 768–77). Sunil Bajpai and Gingerich described another important ambulocetid: *Himalayacetus*, which at approximately 52.5 million years old is the oldest whale currently known in the fossil record, meaning that the land-to-water transition had taken place by that time (*Proceedings of the National Academy of Sciences USA*, 1998, 95: 15464–68).

Protocetids: Phil Gingerich and colleagues described *Rodhocetus* in 1994 (*Nature*, 368: 844–47). The Belgian whale expert Olivier Lambert—with whom for a few years I coedited the journal *Acta Palaeontologica Polonica*—described the Peru pakicetid *Peregocetus* in a fascinating 2019 paper in *Current Biology* (29: 1352–59), which discusses the issue of early whale distribution and migration more generally. He was joined on the paper by several coauthors from Peru, Italy, and France—a global team studying the first global whales. There has been considerable work on the hearing abilities of whales in the pakicetid-ambulocetid-protocetid part of the family tree, most notably papers by Thewissen and Hussain (*Nature*, 1993, 361: 444–45), Nummella et al. (*Nature*, 2004, 430: 776–78), and Mourlam and Orliac (*Current Biology*, 2017, 27: 1776–81).

Basilosaurus and the Wadi al-Hitan Whales: *Basilosaurus* has a colorful history, first discovered in the southern United States in the 1830s and given its name—which means "king lizard"—because it looked like a sea serpent. It was our recurring villain, Richard Owen, who first realized that the animal was an early whale and not a reptile, but by the rules of zoological nomenclature, the name *Basilosaurus* had to stick. This history is chronicled in David Rains Wallace's *Beasts of Eden* and Don Prothero and Robert Schoch's *Horns, Tusks, & Flippers* (both cited above). Gingerich led a team that described the legs and feet of Egyptian specimens of *Basilosaurus* in 1990 (*Science*, 249: 154–57). In 1992, Gingerich published a scrupulous monograph on the Wadi al-Hitan whales, and other Egyptian Eocene whales, meticulously documenting where individual specimens were found both geographically and in the stratigraphic sequence of the Eocene rocks (*University of Michigan Papers on Paleontology*, 30: 1–84). Manja Voss led a team of authors (including Gingerich and Egyptian colleagues) describing the stunning fossil of a *Dorudon* inside a *Basilosaurus* (*PLoS ONE*, 2019, 14: e0209021). Tom Mueller's article in the August 2010 issue of *National Geographic* is an evocative pop-science portrayal of the Wadi al-Hitan whales and Gingerich's work.

Important references on the early evolution of odontocetes include papers on the three species mentioned by name in the text: *Cotylocara* (Geisler et al., *Nature*, 2004, 508: 383–86), *Echovenator* (Churchill et al., *Current Biology*, 2016, 26: 2144–49), and *Livyatan* (Lambert et al., *Nature*, 2010, 466: 105–8). Furthermore, other important studies on aspects of odontocete biology include a study of the origin and early evolution of echolocation by the early career paleontologist Travis Park and colleagues (*Biology Letters*, 2016, 12: 20160060), and papers on the evolution of the enormous odontocete brain by Lori Marino and colleagues (*The Anatomical Record*, 2004, 281A: 1247–55; *PLoS Biology*, 2007, 5: e139).

Important references on the early evolution of mysticetes include papers on the two species mentioned by name in the text: *Mystacodon* (Lambert et al., *Current Biology*, 2017, 27: 1535–41) and *Llanocetus* (Mitchell, *Canadian Journal of Fisheries and Aquatic Sciences*, 1989, 46: 2219–35; Fordyce and Marx, *Current Biology*, 2018, 28: 1670–76), plus the key species *Maiabalaena*, which has neither teeth nor baleen on its jaws (Peredo et al., *Current Biology*, 2018, 28: 3992–4000). Furthermore, other important studies on aspects of mysticete biology include papers on the origin and early evolution of baleen (Peredo et al., *Frontiers in Marine Science*, 2017, 4: 67; Hocking et al., *Biology Letters*, 2017, 13: 20170348; see also a different hypothesis from Demere et al. [*Systematic Biology*, 2008, 57: 15–37] and Geisler et al. [*Current Biology*, 2017, 27: 2036–42]) and on their hearing abilities (Park et al., *Proceedings of the Royal Society, Series B*, 2017, 284: 20162528).

My main sources for the biological talents of blue whales and the evolution of mysticete gigantism include papers on blue whale weaning and calf sizes (Lockyer, *FAO Fisheries Series*, 1981, 3: 379–487. Mizroch et al., *Marine Fisheries Review*, 1984, 46: 15–19); feeding and engulfing krill (Goldbogen et al., *Journal of Experimental Biology*, 2011, 214: 131–46; Fossette et al., *Ecology and Evolution*, 2017, 7: 9085–97); and body size evolution (Slater et al., *Proceedings of the Royal Society, Series B*, 2017, 284: 20170546; Goldbogen et al., *Science*, 2019, 366: 1367–72). Nick Pyenson's discussion of mysticete size evolution in his book *Spying on Whales* and his 2017 *Current Biology* review paper (cited above) is clear, gripping, and fascinating writing.

CHAPTER 8: MAMMALS AND CHANGING CLIMATES

My fictionalized tale of the ashfall apocalypse on the American Savanna was based on the fossils preserved at Ashfall Fossil Beds (the various species, how their skeletons were positioned in the ash, and the maladies observable on their bones), the geology of the site (the different layers of ash, their thicknesses and properties, and what that implies about the sequence of events), and a fun discussion with two of my volcanologist colleagues in Edinburgh, Eliza Calder and Isla Simmons.

The best sources of information on the Ashfall site have been written by the scientist who discovered it, Mike Voorhies. Particularly informative are his 1985 article in the *Research Reports of the National Geographic Society* (19: 671–88), his paper with Joseph Thomasson on the grass fossils preserved within the mouths and rib cages of the rhinos (*Science*, 1979, 206: 331–33), a popular piece in the *University of Nebraska State Museum, Museum Notes* (1992, 81: 1–4), and a chapter he cowrote with S. T. Tucker and colleagues in the book *Geologic Field Trips along the Boundary between the Central Lowlands and Great Plains* (*Geological Society of America Field Guide*, 2014, 36). The Ashfall Fossil Beds website is also a font of information (https://ashfall.unl.edu/), as is an article by Terri Cook in *Earth Magazine* in 2017. The age of the Ashfall deposits, and the geological detective work tracing their source to a Yellowstone eruption in Idaho, are discussed by Smith et al. (*PLoS ONE*, 2018, 13: e0207103).

I used several references to flesh out the biology, behaviors, diets, and herd structure of the Ashfall rhinos. Chief among these are papers by Alfred Mead (*Paleobiology*, 2000, 26: 689–706), Matthew Mihlbachler (*Paleobiology*, 2003, 29: 412–28), Nicholas Famoso and Darren Pagnac (*Transactions of the Nebraska Academy of Sciences*, 2011, 32: 98–107), and my New Mexico field buddies Bian Wang and Ross Secord (*Palaeogeography, Palaeoclimatology, Palaeoecology*, 2020, 542: 109411). There is also a fascinating conference abstract by D. K. Beck on the pathologies of the Ashfall rhinos, caused by ash poisoning (*Geological Society of America Abstracts with Programs*, 1995, 27: 38).

There is copious literature on the hothouse-to-coolhouse shift at the Eocene-Oligocene boundary, what caused it, how severe it was in different parts of the world, and how the temperature change affected precipitation and other aspects of climate. The single best reference on how the Earth's climate has changed over the past 66 million years, postasteroid, was published in *Science* in 2020 by Thomas Westerhold and colleagues, including Edinburgh's Dick Kroon (369: 1383–87). This paper includes easy-to-follow plots of temperature over time, indicating major shifts and marking when the Earth was in hothouse, coolhouse, and icehouse phases (note that they divide what I call a hothouse into harsher hothouse and milder warmhouse phases). Other key references on the Eocene-Oligocene are papers by DeConto and Pollard (*Nature*, 2003, 421: 246–49), Cox et al. (*Nature*, 2005, 433: 53–57), Scher and Martin (*Science*, 2006, 312: 428–30), Zanazzi et al. (*Nature*, 2007, 445: 639–42), Liu et al. (*Science*, 2009, 323: 1187–90), Katz et al. (*Science*, 2011, 332: 1076–79), and Spray et al. (*Paleoceanography and Paleoclimatology*, 2019, 34: 1124–38).

There is a huge literature on the spread of grasslands during the Oligocene and Miocene, much of it by Caroline Strömberg and her colleagues. Two of the most useful and readable general reviews are Caroline's paper in *Annual Review of Earth and Planetary Sciences* (2011, 39: 517–44) and a paper she contributed to in *Science*, led by Erika Edwards (2010, 328: 587–91). Caroline's PhD thesis work on grassland and mammal coevolution in North America was published in several key papers (*Palaeogeography, Palaeoclimatology, Palaeoecology,* 2004, 207: 239–75; *Proceedings of the National Academy of Sciences [USA],* 2005, 102: 11980–84; *Paleobiology,* 2006, 32: 236–58), and with colleagues, she also studied grasslands in Turkey (*Palaeogeography, Palaeoclimatology, Palae-oecology,* 2007, 250: 18–49) and South America (*Nature Communications,* 2013, 4:1478), including one study on South America led by her PhD student Regan Dunn (*Science,* 2015, 347: 258–61). Caroline and her Indian collaborators, led by Vandana Prasad, described their Late Cretaceous grass phytoliths in *Science* (2005, 310: 1177–80). Some details of Caroline's career and early research were gleaned from a Society of Vertebrate Paleontology biography announcing her Romer Prize.

The effects of grassland evolution on mammals, and the development of tooth hypsodonty in response, has been studied for many years by the eminent paleontologist Christine Janis, with a team of colleagues. The best and most comprehensive review of the subject was written by Christine and John Damuth, and this paper is the source for the mechanical pencil analogy that I use, along with the statistics on how much grit modern grazers consume and how quickly their teeth wear down (*Biological Reviews,* 2011, 86: 733–58). This paper is, in many respects, a follow-up to a landmark review that Christine published with Mikael Fortelius in the same journal in 1988 (63: 197–230). Christine was also part of a team, led by Borja Figueirido, that broadly examined how mammal evolution related to climate over the last 66 million years (*Proceedings of the National Academy of Sciences [USA],* 2019, 116: 12698–03), and another team, led by Phillip Jardine, that assessed patterns of hypsodonty evolution in horses and other high-toothed mammals on the American Savanna (*Palaeogeography, Palaeoclimatology, Palaeoecology,* 2012, 365–66: 1–10). The evolution of hypsodonty in South American mammals was assessed by Rodrigues et al. (*Proceedings of the National Academy of Sciences [USA],* 2014, 114: 1069–74). The relationship between hypsodonty and tooth wear—and the remarkable finding that grazing-related tooth wear appeared in horses long before hypsodonty—was the subject of a scintillating 2011 paper in *Science* by Matthew Mihlbachler and colleagues (331: 1178–81). The relationship between grazing, hypsodonty, and tooth enamel complexity was illuminated by Nicholas Famoso and coworkers (*Journal of Mammalian Evolution,* 2016, 23: 43–47), and the evolution of cursorial running mammals on the savannas was addressed by David Levering and team (*Palaeogeography, Palaeoclimatology, Palaeoecology,* 2017, 466: 279–86).

George Gaylord Simpson laid out the story of the Great Transformation in his 1951 book *Horses: The Story of the Horse Family in the Modern World and through Sixty Million Years of History* (Oxford University Press). The modern-day dean of horse research in North America is Bruce MacFadden, of the University of Florida. He published his own detailed book on horse evolution, *Fossil Horses: Systematics, Paleobiology, and Evolution of the Family Equinae* (1992, Cambridge University Press), along with reams of papers, including a brief but influential review in *Science* in 2005 (307: 1728–20). Other key works include his 1984 monograph on Miocene and Pliocene horses (*Bulletin of the American Museum of Natural History*, 179: 1–196), a 1988 paper with Richard Hulbert on the genealogy of early horses and the explosive radiation of Miocene grazers (*Nature*, 336: 466–68), a study of the diets and ecologies of horses during the waning days of their glory in the latest Miocene and early Pliocene (*Science*, 1999, 283: 824–27), and enjoyable reviews on the evolution of grazing mammals (*Trends in Ecology and Evolution*, 1997, 12: 182–87; *Annual Review of Ecology and Systematics*, 2000, 31: 33–59).

It wasn't only grass-grazers that thrived in the Miocene! Christine Janis, John Damuth, and Jessica Theodor wrote a provocative series of papers showing that leaf-eating browsers were still diversifying, too—and in fact were more diverse than in similar environments today (*Proceedings of the National Academy of Sciences [USA]*, 2000, 97: 7899–904; *Palaeogeography, Palaeoclimatology, Palaeoecology*, 2004, 207: 371–98). When it comes to the predators eating all these grazers and browsers, my key sources were reviews by the renowned carnivorous mammal expert Blaire van Valkenburgh (*Annual Review of Earth and Planetary Sciences*, 1999, 27: 463–93; *Paleontological Society Papers*, 2002, 8: 267–88). An intriguing study by Figueirido and team showed that meat-eating mammals still were mostly ambush predators, or pounce-pursuit predators that could track their prey for a short distance, during the American Savanna times, and it was only very recently, during the Ice Age, that pure long-distance pursuit predators evolved (*Nature Communications*, 2015, 6: 7976). And we can't forget about the small mammals! Joshua Samuels and Samantha Hopkins beautifully outline their evolution on the grasslands in a 2017 paper (*Global and Planetary Change*, 149: 36–52).

The section on Riversleigh and the evolution of Australia's marsupial fauna was based on the literature and discussions with Mike Archer and Robin Beck. Mike and colleagues Sue Hand and Hank Godthelp wrote a book on Riversleigh in 1994 (*Riversleigh: The Story of Animals in Ancient Rainforests of Inland Australia, Reed Books*)—and they need to write another to update with all their recent discoveries! Other important general reviews include a chapter on the rise of Australian marsupials led by Karen Black, one of Mike's many PhD students, in the 2012 book *Earth and Life* (edited by John Talent, published by

Springer), and a chapter that Rob wrote for the *Handbook of Australasian Biogeography* in 2017 (329–66). Rob also led two important papers on the earliest Australian marsupials from the Eocene (*PLoS ONE*, 2008, 3: e1858; *Naturwissenschaften*, 2012, 99: 715–29). I also must mention Mike's riveting and fun series of articles for *Nature Australia*, which cover so many of the Riversleigh finds and the general story of marsupial evolution; he really needs to update these and combine them into a pop-science book.

Mike, Sue, Hank, Rob, and their many Riversleigh compatriots—Derrick Arena, Marie Attard, Tim Flannery, Julien Louys, Anna Gillespie, Kenny Travouillon, Steve Wroe, and so many others—have published dozens and dozens of research papers on the Riversleigh fossils. The ones that I consulted to write this chapter include papers on: the Riversleigh rain forest environment (Travouillon et al., *Palaeogeography, Palaeoclimatology, Palaeoecology*, 2009, 276: 24–37; Travouillon et al., *Geology*, 2012, 40[6]: e273); the age of the Riversleigh fauna and its division into four zones spanning the Oligocene and Miocene (Arena et al., *Lethaia*, 2016, 49: 43–60; Woodhead et al., *Gondwana Research*, 2016, 29: 153–67); the preservation of Riversleigh fossils in cave settings (Arena et al., *Sedimentary Geology*, 2014, 304: 28–43); the overall diversity of Riversleigh mammals (Archer et al., *Alcheringa*, 2006, 30:S1: 1–17); and the Burdekin Plum fossil from Riversleigh (Rozefelds et al., *Alcheringa*, 2015, 39: 24–39).

For information on specific Riversleigh mammals I mentioned in the text, please see the following references: the giant tree-climbing wombat-relative *Nimbadon* (Black et al., *Journal of Vertebrate Paleontology*, 2010, 30: 993–1011); the carnivorous thylacine *Nimbacinus* (Attard et al., *PLoS ONE*, 2014, 9[4]: e93088); the fierce marsupial-lions (Gillespie et al., *Journal of Systematic Palaeontology*, 2019, 17: 59–89); primitive kangaroos (Kear et al., *Journal of Paleontology*, 2007, 81: 1147–67; Black et al., *PLoS ONE*, 2014, 9[11]: e112705); the potentially meat-eating rat-kangaroo *Ekaltadeta* (Archer and Flannery, *Journal of Paleontology*, 1985, 59: 1331–49; Wroe et al., *Journal of Paleontology*, 1998, 72: 738–51); primitive koalas (Louys et al., *Journal of Vertebrate Paleontology*, 2009, 29: 981–92; Black et al., *Gondwana Research*, 2014, 25: 1186–201); the rain forest marsupial mole *Naraboryctes* (Archer et al., *Proceedings of the Royal Society, Series B*, 2011, 278: 1498–506; Beck et al., *Memoirs of Museum Victoria*, 2016, 74: 151–71); the hammer-toothed snail eater *Malleodectes* (Arena et al., *Proceedings of the Royal Society, Series B*, 2011, 278: 3529–33; Archer et al., *Scientific Reports*, 2016, 6: 26911); and "Thingodonta" itself, which goes by the scientific name *Yalkaparidon* (Archer et al., *Science*, 1988, 239: 1528–31; Beck, *Biological Journal of the Linnean Society*, 2009, 97: 1–17).

CHAPTER 9: ICE AGE MAMMALS

The story of the Stono slaves and their mammoth discovery is told by Adrienne Mayor, the leading historian of ancient and indigenous encounters with fossils, in her book *Fossil Legends of the First Americans* (Princeton University Press, 2005), and an article she wrote in 2014 for *Wonders & Marvels* magazine. She also provided further information during an email chat. Her book is excellent and covers a wealth of information on Native American fossil discoveries, and how they interpreted the giant bones they found—including the tale of Big Bone Lick.

Thomas Jefferson's mammoth obsession is well documented in many historical and scientific works. You can read Jefferson's 1797 address on *Megalonyx*, as it was published as a research paper (*Transactions of the American Philosophical Society*, 1799, 4: 246–60). As I'm writing this passage in mid-January 2021, I'm imagining how improbable it is that our outgoing vice president, Mike Pence, would publish a peer-reviewed scientific paper. Maybe Buffon was correct about American degeneracy? Other important sources on Jefferson include two articles by early-twentieth-century paleontologist Henry Fairfield Osborn (*Science*, 1929, 69: 410–13; *Science*, 1935, 82: 533–38). Osborn was an avowed white supremacist, so you'll find no mention of the Stono slaves in these works. I also gleaned information on Jefferson, Buffon, Lewis and Clark, and mammoths from several articles, including those by Richard Conniff in *Smithsonian*, Cara Giaimo in *Atlas Obscura*, Phil Edwards in *Vox*, Emily Petsko in *Mental Floss*, and Keith Thomson in *American Scientist*, along with the Monticello website, which has information on Jefferson's fossil collection. Finally: huge thanks to Ted Daeschler in Philadelphia, for showing me the *Megalonyx* bones as a grad student.

Much of my knowledge of Illinois's glacial topography comes from the class I took with Joe Jakupcak in high school, supplemented by countless conversations with him over the years. The Illinois State Geological Survey (ISGS) has a variety of resources on Ice Age Illinois, ranging from their website (https://isgs.illinois.edu/outreach/geology-resources) to their series of *Field Trip Guidebooks*. Of the latter, the most relevant to the story told here are guides 1986B (to my hometown of Ottawa), 1995C (to the Streator-Pontiac area, south of Ottawa), and 2002A (to the Hennepin area, east of Ottawa—a field trip I attended as a senior in high school, with Mr. Jakupcak). Other facts and figures in this section were taken from the 1942 ISGS Bulletin (number 66) on the *Geology and Mineral Resources of the Marseilles, Ottawa, and Streator Quadrangles*, and a United States Geological Survey report on the hydrogeology of LaSalle County (Scientific Investigations Report 2016-5154). The moraine south of Ottawa, by the way, is the Farm Ridge Moraine, sometimes called the Grand Ridge Moraine, after the agricultural village of Grand Ridge perched on top of it.

One of the best, most lavishly illustrated, and most digestible summaries of the Ice Age and its megafauna is Ross MacPhee's book *End of the Megafauna* (W.W. Norton & Company, 2019). Along with Peter Schouten's stunning illustrations of megafauna in their environments, the book includes maps of glacial coverage during the last glacial advance (called the Wisconsin Glacial Advance in North America; the glaciers reached as far south as central Illinois, but in earlier glacial periods the ice sheets went even farther south), maps of the Mammoth Steppe and other Ice Age biomes, and plots showing changes in temperature and ice volume over time.

Important resources on Ice Age climate include works by Zhang et al. on carbon dioxide levels over time (*Philosophical Transactions of the Royal Society, Series A*, 2013, 371: 20130096), Sarnthein et al. on how changes in Atlantic Ocean circulation fed the ice caps (*Climate of the Past*, 2009, 5: 269–83), Bailey et al. on the start of the polar ice cap's advance onto North America (*Quaternary Science Reviews*, 2013, 75: 181–94), and Spray et al. on the timing of the Northern Hemisphere ice cap formation (*Paleoceanography and Paleoclimatology*, 2019, 34: 1124–38). The latest research on the celestial cycles and how they controlled glacial pulses—which is a bit more complicated than the musical analogy I gave in the text—can be found in Bajo et al.'s 2020 *Science* paper (367: 1235–39). There are three celestial cycles, which are technically called the Milankovitch Cycles: eccentricity (shape of Earth's orbit around the sun), obliquity (the tilt of Earth's axis), and precession (the wobble of Earth's axis). Bajo et al. found that obliquity, in particular, was a major driver of the initiation and duration of glaciations, with contributions from the other cycles.

Haley O'Brien's paper on the African dome-headed wildebeest *Rusingoryx* was published in *Current Biology* (2016, 26: 503–6), and additional information can be found in a paper by Tyler Faith and colleagues (*Quaternary Research*, 2011, 75: 697–707). Christine Janis and colleagues wrote an intriguing paper on the hopping abilities—or lack thereof—in the giant Australian Ice Age kangaroos (*PLoS ONE*, 2014, 9[10]: e109888).

Mammoths are an endless source of fascination, and thus an endless source of literature. A good general overview can be found in Adrian Lister and Paul Bahn's book *Mammoths: Giants of the Ice Age* (University of California Press, 2007) and Lister's sole-authored, similar-named *Mammoths: Ice Age Giants* (Natural History Museum, 2014). Other readable accounts can be found in Jordi Agustí and Mauricio Antón's book *Mammoths, Sabertooths, and Hominids*

(Columbia University Press, 2002), Don Prothero's *Princeton Field Guide to Prehistoric Mammals* and Prothero and Schoch's *Horns, Tusks, and Flippers,* both cited above.

The mad Mammoth Hunters of Siberia have been profiled in Helen Pilcher's book *Bring Back the King* (Bloomsbury, 2016), her article in BBC *Science Focus,* Sabrina Weiss's article in *Wired,* and a Radio Free Europe exposé, in which the photographer Amos Chapple accompanied a team on the ivory trail. In case modern-day mammoth hunting seems romantic, just stop. It's dangerous, it pollutes the environment, and it is illegal. Plus, there is the argument—which resonates with me—that selling mammoth tusks keeps the ivory market afloat, promoting the poaching of our last remaining African and Indian elephants.

The complete woolly mammoth genome was published by Eleftheria Palkopoulou and colleagues in *Current Biology* in 2015 (25: 1395–1400). Later that year, Vincent Lynch and colleagues published a paper in *Cell Reports* describing two additional mammoth genomes, which they used to identify genetic changes related to its cold habitat and frosty lifestyle (12: 217–28). Then, as I was finishing this chapter, a sensational study by Tom van der Valk et al. reported mammoth DNA more than one million years old—a record for ancient DNA (*Nature,* 2021, 591: 265–69)! These genetic studies follow over a decade of work on mammoth DNA. Other key papers over the years, sequentially building on each other, are those by: Poinar et al. (*Science,* 2006, 311: 392–94), Krause et al. (*Nature,* 2006, 439: 724–27), Rogaev et al. (*PLoS Biology,* 2006, 4[3]: e73), Gilbert et al. (*Proceedings of the National Academy of Sciences [USA],* 2008, 105: 8327–32), Miller et al. (*Nature,* 2008, 456: 387–90), Debruyne et al. (*Current Biology,* 2008, 18: 1320–26), Campbell et al. (*Nature Genetics,* 2010, 42: 536–40; this paper describes the cold-adapted hemoglobin mutations), Rohland et al. (*PLoS Biology,* 2011, 8[12]: e1000564; this paper also describes mastodon DNA); Enk et al. (*Genome Biology,* 2011, 12: R51; this paper presents evidence for woolly mammoth and Columbian mammoth inbreeding).

Mammoth hair has been the subject of both descriptive and genetic work. In 2006, Römpler and colleagues identified a nuclear gene that indicated different hair colors (*Science,* 313: 62). Later, Claire Workman and her team surveyed forty-seven mammoths, sampled their DNA, and found that the genetic combination causing lighter hair was exceptionally rare (*Quaternary Science Reviews,* 2011, 30: 2304–08). Using a different tack in 2014 Silvana Tridico and colleagues microscopically examined over four hundred hairs from various mammoth mummies and described a kaleidoscope of color differences, including major color distinctions between the outer guard hairs and inner coat (*Quaternary Science Reviews,* 2014, 68–75).

Adrian Lister—mammoth expert extraordinaire at the Natural History Museum in London—has published a catalog of papers on woolly mammoth evolution and migrations. Two that were especially useful in writing this chapter were his coauthored 2005 review of mammoth evolution in Eurasia (*Quaternary International*, 126–128: 49–64), and his stirring 2015 *Science* paper on the multiple migrations of mammoths into North America (350: 805–9).

On the subject of mammoth herds and social lives, the Canadian tracksite discussed in the text was described by McNeill et al. (*Quaternary Science Reviews*, 2005, 24: 1253–59). On the subject of mammoth growth and nursing, Metcalfe et al. used isotopic studies of teeth and bone to show that mammoth mothers fed their babies milk for at least three years (*Palaeogeography, Palaeoclimatology, Palaeoecology*, 2010, 298: 257–70). Much of what we know about mammoth reproduction, childhood, and child-rearing comes from the spectacular one-month-old Siberian ice mummy, named Lyuba. The details of her discovery are told with gusto in Tom Mueller's May 2009 *National Geographic* cover article. Key research papers on Lyuba include those by van Geel et al. on her diet and stomach contents (*Quaternary Science Reviews*, 2011, 30: 3935–46), Fisher et al. on her death and preservation (*Quaternary International*, 2012, 255: 94–105), and Rountrey et al. on her development and season of birth (*Quaternary International*, 2012, 255: 106–205).

Smilodon and other saber-toothed tigers are, like mammoths, an endless subject of fascination and research. Three of the best general overviews are Alan Turner and Mauricio Antón's book *The Big Cats and Their Fossil Relatives* (Columbia University Press, 1997), Antón's book *Sabertooth* (Indiana University Press, 2013), and *Smilodon: The Iconic Sabertooth* (Johns Hopkins University Press, 2018), a volume of technical papers edited by Lars Werdelin, Gregory McDonald, and Christopher Shaw. A nice summary of the Rancho La Brea deposits in Los Angeles can be found in the collection of papers edited by John Harris, *La Brea and Beyond: The Paleontology of Asphalt-Preserved Biotas* (*Natural History Museum of Los Angeles County Science Series*, 2015, 42: 1–174). My statistics on *Smilodon* body size were taken from Christiansen and Harris (*Journal of Morphology*, 2005, 266: 369–84).

Although the focus of my narrative is *Smilodon* itself, it is part of a broader saber-toothed family, Machairodontinae. The relationships of this family to modern cats has been clarified by genetic studies of *Smilodon* and other machairodontines, in a series of papers including those by Janczewski et al. (*Proceedings of the National Academy of Sciences [USA]*, 1992, 89: 9769–73), Paijmans et al. (*Current Biology*, 2017, 27: 3330–36), and two papers by Ross Barnett and colleagues (*Current Biology*, 2005, 15: R589–R590; *Current Biology*, 2020,

30: 1–8). Barnett wrote a fascinating book on the evolution and extinction of the megafauna in Britain: *The Missing Lynx* (Bloomsbury, 2019). The evolution and distribution of *Smilodon* in South America, after its double crossing, is explored by Manzuetti et al. (*Quaternary Science Reviews*, 2018, 180: 57–62).

How did saber-toothed tigers use their canines to hunt and kill? This question has fascinated paleontologists for generations and spawned an expansive literature. First things first, the Texas cave with mammoth bones in a saber-toothed lair was described by Marean and Ehrhardt (*Journal of Human Evolution*, 1995, 29: 515–47). More generally, Blaire Van Valkenburgh led an intriguing paper arguing that large Ice Age predators, like *Smilodon*, would have been able to feed on the juveniles of the largest megafauna, like mammoths (*Proceedings of the National Academy of Sciences [USA]*, 2016, 113: 862–67). Isotopic evidence that *Smilodon* preferred forest-dwelling species, and dire wolves grassland species, was presented by Larisa DeSantis and colleagues (*Current Biology*, 2019, 29: 2488–95). DeSantis is a leader is using isotopic analysis to study the diets and habitats of fossil vertebrates, and I've long admired her work in blending paleontology and chemistry.

Two scintillating papers have used computer modeling to study the *Smilodon* bite: a landmark study by McHenry et al. using techniques employed by engineers (*Proceedings of the National Academy of Sciences [USA]*, 2007, 104: 16010–15) and a more recent study by Figueirido et al. (*Current Biology*, 2018, 28: 3260–66). The two works differ in some detail; for instance, the former study argues for an exceptionally weak bite and skull in *Smilodon*, and the latter for a stronger skull that could withstand greater stresses. Regardless, both agree that a saber puncture (technically a "canine shear bite") was the most likely way *Smilodon* killed. A paper by Julie Meachen-Samuels and Blaire Van Valkenburgh described evidence of particularly strong and robust forelimbs in *Smilodon*—evidence that the arms were used to wrestle prey before the saber bite delivered the coup de grâce (*PLoS ONE*, 2010, 5[7]: e11412).

The tough lives, pathologic bones, and broken teeth of *Smilodon* have been chronicled by Van Valkenburgh and Hertel (*Science*, 1993, 261: 456–59), Rothschild and Martin (in *The Other Saber-Tooths*, edited by Naples, Martin, and Babiarz, Johns Hopkins University Press, 2011), and Brown et al. (*Nature Ecology & Evolution*, 2017, 1: 0131). Chris Carbone and colleagues described evidence for *Smilodon* sociality in a 2009 paper (*Biology Letters*, 5: 81–85), which instigated a series of back-and-forth responses arguing whether the evidence was strong enough, or not. The information on *Smilodon* hyoids and roaring was gleaned from a *Scientific American* by John Pickrell, quoting in-progressing research by Christopher Shaw, presented at the 2018 Society of

Vertebrate Paleontology annual meeting and published in abstract form in the accompanying meeting volume.

The fossilized *Smilodon* family—mother and two offspring—was described by Ashley Reynolds, Kevin Seymour, and David Evans, who like me is nominally a dinosaur researcher, although dabbles in many things (*iScience*, 2021, 101916). The timing and pattern of *Smilodon* tooth growth were enumerated by Wysocki et al. (*PLoS ONE*, 2015, 10[7]: e0129847), and the robust skeletons of juvenile *Smilodon* were described by Long et al. (*PLoS ONE*, 2017, 12[9]: e0183175).

One of my greatest regrets in editing this book down to a digestible length was that I could not devote more attention to dire wolves—iconic and monstrous fossil dogs made famous by *Game of Thrones*. Yes, I can assure you, they were real. I briefly mention them in this chapter, as counterparts to saber-toothed tigers at La Brea. In fact, their bones outnumber *Smilodon* bones in the Los Angeles tar pits. Dire wolves were among the first successful pursuit predators, a subject touched on in the last chapter, and in papers by Blaire Van Valkenburgh and Borja Figueirido cited in the chapter 8 references above. As I was writing this chapter, a stunning new study on the genetics of dire wolves was published: it turns out they were an ancient group of homegrown North American wolves, not close relatives of today's North American gray wolves and coyotes whose ancestors colonized the continent more recently (Perri et al., *Nature*, 2021, 591: 87–91.

The dwarf mammoths of Wrangel Island—the last and strangest of the extinct megafauna—and their bizarre, bottlenecked genomes have been studied by Nyström et al. (*Proceedings of the Royal Society, Series B*, 2010, 277: 2331–37), Rogers and Slatkin (*PLoS Genetics*, 2017, 13[3]: e1006601), and Arppe et al. (*Quaternary Science Reviews*, 2019, 222: 105884).

CHAPTER 10: HUMAN MAMMALS

The mammoth hunting story opening this chapter is not pure fiction, but based on two discoveries of mammoth skeletons near Kenosha, Wisconsin, at the Hebior and Schaefer sites. These mammoth skeletons are marked by cuts and wedges made by stone tools, some of which were found adjacent to the bones. The sites, their ages and environments, and the evidence for mammoth-human interaction have been described by Overstreet and Kolb (*Geoarchaeology*, 2003, 18: 91–114) and Joyce (*Quaternary International*, 2006, 142–143: 44–57).

Details of Leigh Van Valen's remarkable life and scientific achievements are summarized in a touching obituary by some of his former students, published in the journal *Evolution* (Liow et al., 2011, doi:10.1111/j.1558-5646

.2011.01242.x). I gathered other information from my own reminisces of him in Chicago, a chat with Christian Kammerer, and obituaries published in the *New York Times* (by Douglas Martin 2010) and by the University of Chicago.

Van Valen and Robert Sloan described and named *Purgatorius* in their 1965 *Science* paper (150: 743–45), and Van Valen much later monographed it (and other plesiadapiforms) in detail in another of his self-published journals, *Evolutionary Monographs* (15: 1–79). It is in this latter paper that he outlines why he identified *Purgatorius* and other plesiadapiforms as early primates, based on the shapes of their cusps; his rationale is not that clear in his shorter 1965 paper, at least to those (like me) uninitiated in 1960s terms for tooth cusps and ridges. Other important works on *Purgatorius* itself include papers on jaw and dental material by Bill Clemens (*Science*, 1974, 184: 903–5; *Bulletin of Carnegie Museum of Natural History*, 2004, 36: 3–13); studies of what were the oldest-known specimens when I started writing this chapter, from Saskatchewan, written by Richard Fox and Craig Scott (*Journal of Paleontology*, 2011, 85: 537–48; *Canadian Journal of Earth Sciences*, 2016, 53: 343–54); a study on what became the oldest-known specimen when I revised this chapter, from Montana, written by Greg Wilson, Stephen Chester, Bill Clemens, and colleagues (*Royal Society Open Science*, 2021, 8: 210050); and Stephen's paper, with colleagues including Bill Clemens, describing ankle bones that show it was a capable climber (*Proceedings of the National Academy of Sciences [USA]*, 2015, 112: 1487–92).

Leigh's great insight was that *Purgatorius*—which lived so soon after the end-Cretaceous extinction—was an early plesiadapiform, and thus an early primate. Earlier paleontologists had recognized the link between later-living plesiadapiforms and primates, including James Gidley (*Proceedings of the US National Museum*, 1923, 63: 1–38) and our recurring mammal phylogeneticist, George Gaylord Simpson (*American Museum Novitates*, 1935, 817: 1–28; *United States National Museum Bulletin*, 1937, 169: 1–287; *Bulletin of the American Museum of Natural History*, 1940, LXXVII: 185–212).

Mary Silcox has published widely on plesiadapiforms and early primates. Her 2001 PhD, at Johns Hopkins University, included a large phylogenetic analysis that corroborated Van Valen's, Gidley's, and Simpson's earlier hypotheses that plesiadapiforms are primates. I do note here that some workers, including Christopher Beard and Xijun Ni, have argued that some plesiadapiforms might be more closely related to dermopterans (the "flying lemurs") than primates. Mary and Sergi López-Torres review these debates in their masterful and enjoyable review of primate origins and early evolution, published in 2017 in *Annual Review of Earth and Planetary Sciences* (45: 113–37). She also collaborated with Gregg Gunnell to pen a more detailed review of plesiadapiform taxonomy, anatomy, and evolution, published as a chapter in the 2008 book *Evolution of Tertiary Mammals of North America: Volume 2* (Cambridge University Press), and then a later review in 2017 (*Evolutionary Anthropology*, 26: 74–94). She published her work on plesiadapiform and early primate brain evolution in *Proceedings of the National Academy of Sciences (USA)* in 2009 (106: 10987–92). This was followed by another important study on plesiadapiform brains, by Maeva Orliac and colleagues (*Proceedings of the Royal Society, Series B*, 2014, 281: 20132792).

Other important plesiadapiform studies alluded to in the text are the descriptions of the *Torrejonia* skeleton, which was discovered by Tom Williamson and his sons Ryan and Taylor (Chester et al., *Royal Society Open Science*, 2017, 4: 170329); Jonathan Bloch and Doug Boyer's description of the Eocene-aged *Carpolestes* with long fingers and opposable toes (*Science*, 2002, 298: 1606–10); and Bloch's paper (with Mary, Doug Boyer, and Eric Sargis) on Paleocene plesiadapiform locomotion and phylogeny (*Proceedings of the National Academy of Sciences [USA]*, 2007, 104: 1159–64). While all bona fide plesiadapiforms are thus far Paleocene or younger, DNA-based phylogenies of primates imply a Cretaceous origin (see, for example, Springer et al., *PLoS ONE*, 2012, 7[11]: e49521).

The evolution, diversification, and dispersal of lemurs is fascinating. Gregg Gunnell and colleagues presented evidence for multiple dispersals to Madagascar (*Nature Communications*, 2018, 9: 3193), which Ali and Huber showed would have been possible on eastward-moving currents implied by ocean circulation models for that time (*Nature*, 2010, 463: 653–56).

Interesting sources on Oligocene primate evolution across the world include studies on Europe (Köhler and Moyà-Solà, *Proceedings of the National Academy of Sciences [USA]*, 1999, 96: 14664–67); Asia (Marivaux et al., *Science*, 2001, 294: 587–91; Marivaux et al., *Proceedings of the National Academy of Sciences [USA]*, 2005, 102: 8436–41; Ni et al., *Science*, 2016, 352: 673–77); Africa (Stevens et al. *Nature*, 2013, 497: 611–14); and the Middle East (Zalmout et al. *Nature*, 2010, 466: 360–64). The issue of New World monkeys and their lack of dispersal northward is discussed by Bloch et al. (*Nature*, 2016, 533: 243–46).

The early evolution of apes and close relatives has been reviewed by Williams et al. (*Proceedings of the National Academy of Sciences [USA]*, 2010, 107: 4797–4804). The divergence between chimps and humans, and its complex nature and timing, is discussed by Kumar et al. (*Proceedings of the National Academy of Sciences [USA]*, 2005, 102: 18842–47) and Patterson et al. (*Nature*, 2006, 441: 1103–08). The chimpanzee genome was fully sequenced in 2005 and is extremely similar to our genome (Mikkelsen et al., *Nature*, 431: 69–87).

The deeper origins of human-style bipedalism—or at least its first wobbly start—among apes is a subject of deep contention. I point readers in the direction of papers by Thorpe et al. (*Science*, 2007, 316: 1328–31) and Böhme et al. (*Nature*, 2019, 575: 489–93).

This is not a book about humans! It is a book about all mammals, including humans, which is why I give only a single chapter to us and our hominin kin. There is an *enormous* literature on early human evolution, so here I will keep this section to key books and papers that helped me shape my narrative.

To start, I will mention that there are several recent books on human evolution that are excellent, among them: *Fossil Men* by Kermit Pattison (William Morrow, 2020), which chronicles Tim White and Berhane Asfaw's work in Ethiopia, and from which I gleaned the story of Gadi and his *Ardipithecus* discovery; *Sediments of Time* by Meave Leakey (Houghton Mifflin Harcourt, 2020), an autobiography by the once-scion and now-matriarch of the great Leakey paleoanthropology dynasty; *The World Before Us* by Tom Higham (Viking, 2021), which is a meticulous outline of the timing of human origins and migrations and how we know these things based on DNA evidence and dating of rocks; *The Origin of Our Species* (Allen Lane, 2011) and *Lone Survivors* (Melia, 2012) by Chris Stringer of the Natural History Museum, a great popularizer of paleoanthropology; and *Almost Human* by Lee Berger and John Hawkes (National Geographic, 2017). *Nature* editor and fine writer Henry Gee provides a breezy and fun overview of human evolution in his book *A (Very) Short History of Life on Earth* (St. Martin's Press, 2021). For a slightly more iconoclastic take on the early evolution of apes and humans, check out Madelaine Böhme's *Ancient Bones* (Greystone Books, 2020). I also recommend anything written by Kate Wong of *Scientific American*, the leading journalistic sage on human origins, and one of my favorite editors.

The following is a breakdown of important references on early hominins mentioned in the text, their biology and evolution, and their world.

Ardipithecus: Kermit Pattison's book (cited above) is a magnificent work of journalism and explains the discovery and importance of *Ardipithecus* in detail. Tim White, Gen Suwa, and Berhane Asfaw named the species *ramidus*, based on Gadi's initial discovery of the teeth (which they credited as "found by Gada Hamed on Wednesday 29 December 1993") in *Nature* in 1994 (371: 306–12). They initially placed *ramidus* in the genus *Australopithecus*, but the next year reassigned it to the new genus *Ardipithecus*, in a short follow-up in *Nature* (375: 88). The *Ardipithecus* skeleton—which was found near Gadi's original teeth, but belongs to a different individual—was described in detail in a special issue of *Science*, published on October 2, 2009 (vol. 326).

Australopithecus: The discovery of the Lucy skeleton is told with gusto in Donald Johanson's books of the Lucy series, the first of which was published in 1981. Johanson and Tim White scientifically described the Lucy skeleton in a 1979 paper in *Science* (203: 321–30). The footprints mentioned in the text are the famous Laetoli trackways, discovered by the eminent Mary Leakey in the mid 1970s. The brain of *Australopithecus* was studied by Phillip Gunz and colleagues in 2020 (*Science Advances*, 6: eaaz4729). Other important papers on *Australopithecus*, its age and origination, and the many species assigned to it include those by Leakey et al. (*Nature*, 1995, 376: 565–71; *Nature*, 1998, 393: 62–66), Asfaw et al. (*Science*, 1999, 284: 629–35), White et al. (*Nature*, 2006, 440: 883–89), Berger et al. (*Science*, 2010, 328: 195–204), Haile-Selassie et al. (*Nature*, 2015, 521: 483–88; *Nature*, 2019, 573: 214–19).

Other early hominins before *Homo*: A summary of the many human species coexisting in the Pliocene are provided by Haile-Selassie et al. (*Proceedings of the National Academy of Sciences [USA]*, 2016, 113: 6364–71) and a commentary piece by Fred Spoor in *Nature* accompanying the Haile-Selassie et al. (2015) paper cited above; both articles contain useful timelines showing which human species were living when. The environments that these early hominins lived in—and the issue of shrinking forests and growing grasslands—has been examined by Cerling et al. (*Nature*, 2011, 476: 51–56). The hard-object-feeding hominins mentioned in the text are the "robust australopithecines," which are usually assigned to the genus *Paranthropus*. Meave Leakey and her team described the flat-faced *Kenyanthropus* in 2001 (*Nature*, 410: 433–40). The oldest stone tools are found in the vicinity of these hominins, although it is difficult to prove that they, and not another early human species, were the stonemasons (Harmand et al., *Nature*, 2015, 521: 310–15). The oldest tool cut marks on bone are slightly older and were described by McPherron et al. (*Nature*, 2010, 466: 857–860). I note here that distinguishing between human cut marks and animal bites can be difficult, leading to debate about the marks described by McPherron et al. and others (see, for example: Sahle et al., *Proceedings of the National Academy of Sciences [USA]*, 2017, 114: 13164–69). The origin of meat-eating, and how it was a game-changer for hominins, is elucidated by Zink and Lieberman (*Nature*, 2016, 531: 500–503). The diversity of African environments settled by early hominins is discussed by Mercader et al. (*Nature Communications*, 2021, 12:3).

Early *Homo*: The evolution of early *Homo* is reviewed by Antón et al. (*Science*, 2014, 345: 6192). Currently, the oldest-known fossils of our genus, *Homo*, are 2.8 million years old, come from Ethiopia, and were described by Brian Villmoore and colleagues (*Science*, 2015, 347: 1352–55). However, the oldest-known fossils often underestimate the actual date of origin of a species. My PhD student Hans Püschel led a study—which also includes his brother Thomas, a noted human evolution expert, and my postdoc Ornella Bertrand—that used statistical techniques to predict that *Homo* most likely diverged around 3.3 million years ago, and perhaps as long as 4.3 million years ago (*Nature Ecology & Evolution*, 2021, 5, 808–19). The environments of the earliest *Homo* fossils were outlined by Erin DiMaggio and team (*Science*, 2015, 347: 1355–59), and later by Zeresenay Alemseged and colleagues (*Nature Communications*, 2020, 11: 2480).

Homo erectus: The violent nature of early humans is the subject of an interesting paper by Gomez et al. (*Nature*, 2016, 538: 233–37), which uses phylogenetic methods to put humans into the context of animals broadly, demonstrating that we come from a particularly violent part of the family tree. For information on how humans began to use fire, consult Gowlett (*Philosophical Transactions of the Royal Society, Series B*, 2016, 371: 20150164). The beautiful stone tools of *Homo erectus* are those of the Acheulian type. Information on locomotion and sociality in *Homo erectus* is covered by Hatala et al. (*Scientific Reports*, 2016, 6: 28766). Evidence for the mixing of *Homo erectus* and *Australopithecus* (and *Paranthropus* too!) in southern Africa is presented by Herries et al. (*Science*, 2020, 368: eaaw7293). The oldest Asian *Homo* fossils were described by Zhu et al. (*Nature*, 2018, 559: 608–12), and the circa 750,000-year-old age of the Peking Man fossils from Beijing were determined with clarity by Shen et al. (*Nature*, 2009, 458: 198–200). The oldest *Homo* fossils from the Philippines were described by Ingicco et al. (*Nature*, 2018, 557: 233–37); *Homo luzonensis* was described by Détroit et al. (Nature, 2019, 568: 181–86); and *Homo floresiensis* was described by Brown et al. (*Nature*, 2004, 431: 1055–61) and in many subsequent papers, and its age was accurately determined by Sutikna et al. (*Nature*, 2016, 532: 366–69), and earlier *floresiensis*-like fossils were described from about 700,000 years ago on Flores by van den Bergh et al. (*Nature*, 2016, 534: 245–48). Best indications are that both Flores and Luzon were far enough offshore from the Southeastern Asian mainland, and separated by deep enough water, that they would have required a cross-water journey even during times of low sea level in the Ice Age. While it seems most plausible to me that early *Homo* constructed watercraft, it is possible that they passively rafted on mats of vegetation after storms, like those New World monkeys that crossed the Atlantic.

I note here that, based on current evidence, it seems that *Homo erectus* was the first hominin to leave Africa. However, our fossil record is poor, and new discoveries are coming fast. It may be that earlier hominins ventured out of Africa, and even deep into Asia. Who knows what the latest discoveries will be?

Homo sapiens: The currently oldest-known fossils of our species, *Homo sapiens*, come from Morocco and were described by Hublin et al. (*Nature*, 2017, 546: 289–92), and their age by Richter et al. (*Nature*, 2017, 546: 293–96). Our concept of *sapiens* origins is getting really complicated really fast, and older ideas about our species cleanly breaking from other *Homo* have been replaced by a pan-Africa network model, in which populations swapped genes and features until a modern-type *sapiens* body plan became fixed. It can be difficult to understand—for me, too, as I am used to thinking about anatomical features of fossils and not their genetic variation. For more information, please consult the excellent essays of Eleanor Scerri and colleagues (*Trends in Ecology & Evolution*, 2018, 33: 582–94; *Nature Ecology & Evolution*, 2019, 3: 1370–72), Chris Stringer (*Philosophical Transactions of the Royal Society, Series B*, 2016, 371: 20150237), two from Chris Stringer and Julia Galway-Witham in which they swapped first-authorship (*Nature*, 2017, 546: 212–14; *Science*, 2018, 360: 1296–98), a great review of *Homo* evolution over the last million years by Galway-Witham, Stringer, and James Cole (*Journal of Quaternary Science*, 2019, 34: 355–78), and a review of modern human origins published as I was writing this chapter (Bergström et al., 2021, *Nature* 590: 229–37).

Early *Homo sapiens*, and their close *Homo* relatives, migrated widely around Africa and the Mediterranean region (Middle East, Caucasus, parts of Europe). Timmermann and Friedrich explored how these migrations were likely driven by climate (*Nature*, 2016, 538: 92–95). The oldest reported European *Homo sapiens* fossils, from Greece, were described by Katerina Harvati and colleagues (*Nature*, 2019, 571: 500–504). As always, the relevance of these fossils comes down to the dating, as my colleague Huw Groucutt reminds me, and the quite old (ca. 210,000) dates for this Greek fossil need to be corroborated by other discoveries. It is undoubted, however, that by around 120,000 to 100,000 years ago some *Homo sapiens* were leaving Africa. Other important papers on early European *Homo sapiens*, and the close *Homo* relatives that were migrating around the same time, include those by Grun et al. (*Nature*, 2020, 580: 372–75) and Hublin et al. (*Nature*, 2020, 581: 299–302). Neanderthals, Denisovans, and *Homo sapiens* diverged from a common *Homo* ancestor, most likely between approximately 550,000 and 765,000 years ago (see: Prüfer et al., *Nature*, 2014, 505: 43–49; Meyer et al., *Nature*, 2016, 531: 504–7). This ancestor may have been a species like *Homo antecessor* or *Homo heidelbergensis*, or a very close relative; recently ancient proteins from *Homo antecessor*, *Homo erectus*, *Homo sapiens*, Neanderthals, and Denisovans have been compared, to build a family tree (Welker et al., *Nature*, 2020, 580: 235–38). Although this part of our family tree is extremely convoluted, what is clear is that the various *Homo* species were moving and interacting.

The big, globular brain of *Homo sapiens* appears to not only be a key part of our signature species body plan, but perhaps it helped bring about advances in our toolmaking and cognition. Brain evolution in *Homo sapiens* has been chronicled by Simon Neubauer and colleagues (*Science Advances*, 2018, 4: eaoo5961). Information on human cognitive evolution was gleaned from an essay by the noted anthropologist Richard Klein (*Evolutionary Anthropology*, 2000, 28: 179–88) and a review by McBrearty and Brooks (*Journal of Human Evolution*, 2000, 39: 453–563). My paleoanthropology colleagues Huw Groucutt, Bob Patalano, and Eleanor Scerri explained to me how once-popular ideas about a sudden "cognitive revolution" are now outdated (and based largely on the European archaeological record), and instead the African record shows that different groups of early *Homo sapiens* developed advances in technology and brainpower in a mosaic fashion over many tens of thousands of years, and these coalesced as *sapiens* populations expanded, migrated, and mixed. One prime example of an African record of symbolic and technological advances, from Kenya, was presented by Shipton et al. (*Nature Communications*, 2018, 9: 1832).

For those interested in when and how *Homo sapiens* populated North and South America after traversing the Bering Land Bridge, the recent review essay by Michael Waters is an excellent read (*Science*, 2019, 365: eeat5447). Traditionally, a date of around 15,000 years ago has been widely considered as the time *sapiens* crossed the land bridge, but there have been several tantalizing clues of older humans in the Americas, both fossils and artifacts. Two prime candidates pushing the arrival of humans earlier, between 20,000 and 30,000 years ago, were published in 2020 (Ardelean et al., *Nature*, 584: 87–92; Becerra-Valdivia and Higham, *Nature*, 584: 93–97). This active debate has important implications for the question of whether humans caused the extinction of the megafauna mammals, as much of that debate comes down to timing of human migration and settlement (see below). One of the latest papers on when humans first reached Australia—perhaps by 65,000 years ago—is a study from Clarkson et al. (*Nature*, 2017, 547: 306–10). A review of *Homo sapiens* migrations to Asia—including evidence for meanderings prior to the large wave "out of Africa" 50,000–60,000 years ago—was penned by Bae et al. (*Science*, 2017, 358: eaai9067).

Neanderthals: As I was writing, a fantastic book about Neanderthals was published by Rebecca Wragg Sykes, *Kindred* (Bloomsbury, 2020). It is a one-stop shop for everything you need to know about these close *Homo* cousins that we interbred with. Other references relevant to specifics mentioned in my text are papers on Neanderthal origins (Arsuaga et al., *Science*, 2014, 344: 1358–63), their cave constructions (Jaubert et al., *Nature*, 2016, 534: 111–14), and their cave art and use of pigments (Roebroeks et al., *Proceedings of the National Academy of Sciences [USA]*, 2012, 109: 1889–1984; Hoffmann et al., *Science*, 2018, 359: 912–15; Hoffmann et al., *Science Advances*, 2018, 4: eaar5255).

Denisovans: The very existence of these near-*sapiens* relatives was recognized in 2010, by David Reich, Svante Pääbo, and colleagues (*Nature*, 468: 1053–60). Reich, an eminent expert on the genetics of ancient *Homo* populations and how to extract this information from fossils and archaeological materials, wrote a book in 2018 on this subject (*Who We Are and How We Got Here*, Pantheon). Denny—the Denisovan and Neanderthal hybrid—was described in 2018 by Viviane Slon, Pääbo, and their team (*Nature*, 561: 113–16). The age of the Denisova Cave specimens was explained by Douka et al. (*Nature*, 2019, 565: 640–44). Other important papers on Denisovan DNA, their population structure, and how their genes have endured in Asian *Homo sapiens* populations today include studies by Meyer et al. (*Science*, 2012, 338: 222–26), Huerta-Sánchez et al. (*Nature*, 2014, 512: 194–97), Malaspinas et al. (*Nature*, 2016, 538: 207–14), Chen et al. (*Nature*, 2019, 569: 409–12), Massilani et al. (*Science*, 2020, 370: 579–83), and Zhang et al. (*Science*, 2020, 370: 584–87).

Landmark studies on the genetics of modern *Homo sapiens*, and how Neanderthal and Denisovan DNA remains in our genome, were published by the Simons Genome Diversity Project in 2016 (Mallick et al., *Nature*, 538: 201–6) and Pagani et al. (*Nature*, 2016, 538: 238–42). For a readable review of human migrations and interbreeding over time, and how these can be traced by ancient DNA analysis, check out the *Nature* essay by Rasmus Nielsen and colleagues (2017, 541: 302–10). When it comes to a readable and thought-provoking big history of *Homo sapiens*, I enjoyed Yuval Noah Harari's *Sapiens* (Vintage, 2015), although I can't vouch for how accurate and up-to-date its discussions of early human archaeology are, and I didn't use it as source material for this chapter.

The extinction of the megafauna is expertly and tactfully tackled by Ross MacPhee in his book *End of the Megafauna* (W.W. Norton & Company, 2019), which cites all the most important literature on the subject. Excellent and digestible reviews on the subject are papers by Anthony Barnosky and colleagues (*Science*, 2004, 306: 70–75) and Paul Koch and Barnosky (*Annual Review of Ecology, Evolution, and Systematics*, 2006, 37: 215–50).

Paul Martin presented his idea of blitzkrieg in a 1973 paper in *Science* (179: 969–74), and fully fleshed it out in his popular book *Twilight of the Mammoths* (University of California Press, 2005). Some paleontologists and ecologists have pushed back and have targeted climate change as the cause of the extinctions. This is well articulated in a 2013 essay by Stephen Wroe and colleagues (including our old friend Michael Archer from chapter 8) published in *Proceedings of the National Academy of Sciences (USA)* (110: 8777–81), and a paper published after I wrote this chapter (Stewart et al., *Nature Communications*, 2021, 12: 965). For a balanced and critical review of the subject, check out the essay by David Meltzer (*Proceedings of the National Academy of Sciences [USA]*, 2020, 117: 28555–63).

The most recent studies, from a global perspective, present strong evidence that humans were the overriding cause of the extinctions, which in some cases were exacerbated by climate changes during the last glacial-interglacial transition (Sandom et al., *Proceedings of the Royal Society, Series B*, 2014, 281: 20133254; Bartlett et al., *Ecography*, 2016, 39: 152–61; Araujo et al., *Quaternary International*, 2017, 431: 216–22). More focused studies on particular landmasses also have identified humans as the main factor in the extinction, including those in Australia and adjoining lands (Rule et al., *Science*, 2012, 335: 1483–86; Johnson et al., *Proceedings of the Royal Society, Series B*, 2016, 283: 20152399; Saltré et al., *Nature Communications*, 2016, 7: 10511) and South America (Barnosky et al., *Quaternary International*, 2010, 217: 10–29; Metcalf et al., *Science Advances*, 2016, 2: e1501682; Polis et al., *Science Advances*, 2019, 5: eaau4546). The scintillating study of how humans may have coalesced extirpations started by warming shifts in the northern Holarctic was published by Cooper et al. (*Science*, 2015, 349: 602–6).

On the subject of domestication, scientist and ace science popularizer Alice Roberts wrote a book, *Tamed* (Hutchinson, 2017), that profiles ten major domesticated species, including dogs, cows, and horses, and key agricultural cultivars. Important works on the domestication of dogs include papers by Ní Leathlobhair et al. (*Science*, 2018, 361: 81–85) and Perri et al. (*Proceedings of the National Academy of Sciences [USA]*, 2021, 118: e2010083118). The numbers I cite for the percentage of domesticated mammal biomass on Earth today come from Bar-On et al. (*Proceedings of the National Academy of Sciences [USA]*, 2018, 115: 6506–11).

On the subject of mammoth cloning, I recommend Beth Shapiro's book *How to Clone a Mammoth* (Princeton University Press, 2015) and Helen Pilcher's book *Bring Back the King* (Bloomsbury, 2016), along with the section on cloning in Ross MacPhee's book (cited above).

EPILOGUE: FUTURE MAMMALS

The numbers on mammal extinctions over the last approximately 125,000 years, and predicted extinctions in the future, come from a study by Tobias Andermann and colleagues (*Science Advances*, 2020, 6: eabb2313). Figures for rates of background and current mammal extinction come from a study by Gerardo Ceballos and team (*Science Advances*, 2015, 1: e1400253). The prediction that if all currently threatened mammals go extinct then half the diversity from 125,000 years ago will be gone comes from a study by Felisa Smith and colleagues (*Science*, 2018, 360: 310–13). This study also explores body size trends in mammalian extinctions and makes the prediction that future mammal communities with be more homogenized and overrun with rodents, and that the largest mammals of the future might be domestic cows. Recovery rates for mammals, if the extinctions were to cease, are discussed by Davis et al. (*Proceedings of the National Academy of Sciences [USA]*, 2018, 115: 11262–67), and predictions for what mammal communities of the future will look like (hint: overrun with small, fast-living, fast-breeding, insect-eating generalists, like rodents) are given by Cooke et al. (*Nature Communications*, 2019, 10: 2279). Mammal migration patterns with climate change are detailed in a paper by Silvia Pineda-Munoz and team (*Proceedings of the National Academy of Sciences [USA]*, 2021, 118(2): e1922859118).

Other useful and interesting studies on how human activities have impacted mammalian communities and ecosystems are papers by Faurby and Svenning (*Diversity and Distributions*, 2015, 21: 1155–66), Boivin et al. (*Proceedings of the National Academy of Sciences [USA]*, 2016, 113: 6388–96), Lyons et al. (*Nature*, 2016, 529: 80–83), Smith et al. (*Quaternary Science Reviews*, 2019, 211: 1–16), Tóth et al. (*Science*, 2019, 365: 1305–08), and Enquist et al. (*Nature Communications*, 2020, 11: 699).

There is a huge literature on climate and temperature change, and how humans are causing it. In general, I point interested readers in the direction of the United Nations Intergovernmental Panel on Climate Change reports, which can be accessed at https://www.ipcc.ch/. The projections for temperature rise over the next few centuries, and comparisons with Pliocene and Eocene climates, come from papers by Burke et al. (*Proceedings of the National Academy of Sciences [USA]*, 2018, 115: 13288–93) and Westerhold et al. (*Science*, 2020, 369: 1383–87).

The Sixth Extinction is the subject of Elizabeth Kolbert's Pulitzer Prize–winning book of the same title (Henry Holt and Company, 2014), and an excellent review by Anthony Barnosky and colleagues (*Nature*, 2011, 471: 51–57). It is also covered in detail by Peter Brannen in his book *The Ends of the World* (Ecco, 2017), from which I took the collapsing power grid analogy. You might note that I don't use the term *Anthropocene* in my narrative. This formal name has been proposed for the subdivision of the geological time scale during which humans have significantly impacted the planet. But when all is said and done, I don't think human activities will leave much of a mark in the rock record at all. It was one of Brannen's articles in the *Atlantic*, entitled "The Arrogance of the Anthropocene" (2019), that convinced me once and for all.

Finally, for accuracy's sake I must state that I do know why the Chicago Bears were thus named. It is because many early professional American football teams were named after baseball teams in the same city, so the Bears were named in reference to the Chicago Cubs. As a White Sox fan, who went to school on the South Side and comes from a long line of south suburban family, this pains me. For many years I thought our human species would go extinct before the Cubs won another World Series, but alas, 2016.

ACKNOWLEDGMENTS

WRITING THIS BOOK during a year of pandemic lockdown, with an infant at home and while having to shift all teaching and lab management and student supervision online, has been the most difficult thing I've done in my professional life. My deepest thanks to my wife, Anne, who made sure I had a block of time each day to devote to work, and for continued love and emotional support as we learned the ropes of parenthood, we juggled the stresses of work and family, and I felt despondent while separated from my parents, brothers, and wider family back home in Illinois. I am so lucky to have found such a supportive spouse and live in a country now where she was able to go on a real maternity leave—although a pandemic maternity leave was far from her ideal.

I am also grateful for all my Daddy time with Anthony, which not only inspired me to write, but gave me hope and optimism, and incredible insight into what it means to be a mammal: nursing, weaning, teething, and so on. When the (first) strict lockdown ended after the first few terrifying months of Covid, Anne's parents, Pete and Mary, were able to help us with childcare, and later Anne's sister Sarah did, too, and without them, I couldn't have finished this book, or kept my sanity. And to my postdocs, students, and colleagues: thank you for understanding that, with a book to write and a child to raise and a wife to love and support and a family back home to Facetime every day, I couldn't devote as much time as any of us wanted for supervisory meetings, lab meetings, and pastoral care. Finally—to the British National Health Service, the health care workers, the scientists developing vaccines, and the others keeping us safe and guiding a way out of the pandemic: THANKS!

My background is in dinosaurs, and then I got interested in mammals. That's almost entirely due to Tom Williamson, my friend and colleague in New Mexico, who started to convert me when I was a PhD student. Along with Tom, I've had great mammal mentors in John Wible, Zhe-Xi

Luo, Meng Jin, John Flynn, Michelle Spaulding, and Ross Secord. More recently, I've had a chance to mentor students, and I've learned so much from them. Sarah Shelley was the first person to take a risk and join my lab after I started in Edinburgh—how crazy of her to do a PhD with somebody who just finished their PhD, and do it on mammals, with a guy who was just starting to study mammals. I now have a fantastic, and very international and diverse, group of students and postdocs studying various aspects of mammal history. They know far more about mammals than I do, and they inspire me with their new ideas and insights, and they also (along with Sarah, Tom, and John Wible) read an entire draft of this book, pointed out errors, gave me suggestions for improvement, told me when I was being silly, and kept me motivated with kind words. So thank you to: Ornella Bertrand, Greg Funston, Paige dePolo, Sofia Holpin, Zoi Kynigopoulou, and Hans Püschel.

I've had so many other awesome students, who have studied things other than mammals, and thank them for keeping me on my toes and contributing to a caring environment in Edinburgh; I especially acknowledge my PhD students Davide Foffa, Natalia Jagielska, Michela Johnson, and Julia Schwab. My colleagues in Edinburgh have always been supportive, and a shout-out here to my faculty mentor and closest work colleague, Rachel Wood; my bosses Peter Mathieson, Simon Kelley, Bryne Ngwenya, and Mark Chapman; and my fellow paleontologists Dick Kroon, Sean McMahon, Sandy Hetherington, Tom Challands, and Mark Young.

Many colleagues answered my questions, provided information, consented to Zoom interviews, and read passages of the text: Mike Archer, Robin Beck, Eliza Calder, Stephen Chester, Dave Grossnickle, Huw Groucutt, Tom Holland, Adam Huttenlocker, Joe Jakupcak, Christian Kammerer, Adrienne Mayor, Robert Patalano, Michael Petraglia, Eleanor Scerri, Mary Silcox, Isla Simmons, Anne Weil, and more. Further thanks to students in the 2020 and 2021 third-year Palaeontology and Sedimentology course that I teach, who read drafts of the whale section and gave feedback as part of their coursework. But of course, any remaining errors are mine. I also thank many friends on Twitter for giving me suggestions for things I must cover in this book. While I couldn't end up including everything everyone asked for, I did my best!

My scientific research on mammals, at least thus far, has been minimal. I thank everyone who has welcomed me into the mammal community, collaborated with my students and me on projects, shared data, and taken part in joint fieldwork. While I will surely forget people, I especially thank, in addition to those mentioned above: Sophia Anderson, Joe Cameron, Stephen Chester, Ian Corfe, Zoltan Csiki-Sava, Nick Fraser, Luke Holbrook, Marina Jimenez, Tyler Lyson, James Napoli, Mike Novacek, Elsa Panciroli, Dan Peppe, Ken Rose, Helena Scullion, Thierry Smith, Calin Suteu, Radu Totoianu, Carl van Gent, Stig Walsh, Greg Wilson Mantilla, and more. Many of these same colleagues—and others—graciously helped me source images.

My deepest thanks, and godspeed, to three characters in this book who passed away while I was drafting and writing: my dear friend Mátyás Vremir in Romania (discoverer of *Litovoi*, and the one man I always trusted to be by my side in the field, no matter the situations we got ourselves into), my Chinese dinosaur pal Junchang Lü (who showed me that mystery mammal in Liaoning), and the great Cretaceous and Paleocene mammal expert Bill Clemens (always a gentleman to me and other young scholars).

My name is on the cover of this book, but it has been a team effort, and I know I have the best team in the science publishing industry. My agent, Jane von Mehren, made this book happen, and before that, the dinosaur book that made this book possible. She never ceases to amaze me, and her colleagues at Aevitas often help: Esmond Harmsworth, Chelsey Heller, Shenel Ekici-Moling, Erin Files, Nan Thornton, Allison Warren. I have the keenest, most encouraging, and savviest editor: Peter Hubbard, who saw a story in dinosaurs and then mammals, always has a helpful word, and irons out my worst tendencies. My British editor, Ravi Mirchandani, is equally awesome. And I am so lucky to have many others in the publishing industry behind me: most importantly Maureen Cole, Molly Gendell, and Kell Wilson. Massive gratitude to Todd Marshall and Sarah Shelley, two artists of the highest caliber whose illustrations do so much to enliven the prose in this book—after all, words alone can never convey the majesty of a mammoth or sabertooth. Thanks also to Colin Trevorrow, Kev Jenkins, Marc Gordon, Sandy Jarrell, Carrie-Rose Menocal, Erin Derham, Cathy Veisel, Mark Mannucci, Jon

Halperin, and Phillip Watson for keeping my creative energy flowing. Finally, thanks to so many editors and translators for bringing my words to an international audience, particularly Sylvain Collette, Thür-Bédert Prisca, May Yang, Lucas Giossi, and Elisa Montanucci.

Whatever writing ability I now possess—for better or worse—was forged in two places. First, the newsroom of the *Times* newspaper in my hometown of Ottawa, Illinois, where I worked summers and holidays as a teenager and college student. Writing so many diverse assignments on deadline taught me how to think fast and plow through loads of information to untangle the right threads, which is probably the handiest skill I've ever learned, and without which this particular book would have been impossible to write in lockdown. Lonny Cain, Mike Murphy, and Dave Wischnowsky: you were the best writing teachers I've ever had. And then, second, I wrote so many (too many) articles for amateur paleontology websites and magazines during my formative years as a teenage fossil obsessive. Thanks to Fred Bervoets, Lynne Clos, Allen Debus, and Mike Fredericks for not only dealing with me at that age, but in providing feedback, and a platform for my scribblings.

I must acknowledge the agencies that have funded my research: the European Research Council, the National Science Foundation, US Bureau of Land Management, the Leverhulme Trust, National Geographic, the Blavatnik Family Foundation, and the Royal Society. Thank you for the financial support to make my fieldwork and lab studies possible . . . and thanks to the taxpayers and benefactors who provide the actual money for these agencies to disperse to lucky scientists like me, and my students.

To my family back in Illinois, although as of this writing I haven't seen you in a very long time, I miss you terribly and thank you for all your support over the years. My parents, Jim and Roxanne; my brothers, Mike and Chris; Mike's wife, Stephenie, and their family: Lola, Luca, Giorgi. And to my best friends over the ocean, in particular Mr. Jakupcak, that one teacher who really did make it all click.

And finally, as I mention in the acknowledgments in my dinosaur book, I owe a huge debt of gratitude to all the unsung heroes, the folks who usually remain anonymous, but without whom our field would grind to extinction. This includes the fossil preparators, field technicians, undergraduate assistants, university secretaries and administrators, the patrons who visit museums and donate to universities, the science journalists and feature writers, the artists and photographers, the journal editors and peer-reviewers, the amateur collectors who do good and donate their fossils to museums, the folks who administer public lands and process our permits (particularly my friends at the Bureau of Land Management, Scottish Natural Heritage, and the Scottish government), the politicians and federal agencies who support science (and stand up to those who do not), all science teachers at all levels, and so many more.

INDEX

endothermy (warm-blooded animals),
  57–60, 63, 101, 403, 418*n*
end-Permian extinction event, 43–50,
  54–55, 63–64, 86, 174, 406, 415–16*n*
end-Triassic extinction, 46, 85–87,
  168, 406, 422*n*
entelodonts, 308
Eocene, *viii*, 179, 182, 188, 297–98, 301
  mammals, 179, 182, 184, 188, 191,
    196–205, 212, 218–25, *221*, 232,
    234, 236–37, 245, 246, 249, 295–96,
    305, 317–18, 440–41*n*
  Messel Fossil Site, 196–205, 211–12,
    296
  Paleocene-Eocene Thermal
    Maximum (PETM), 212–19, 223,
    232, 252, 270, 301, 372, 438–39*n*
  primates, 372–73
Eocene-Oligocene boundary, 254–55,
  278, 279, 281, 283, 311, 334, 450*n*
Eocene-Oligocene cooling, 298,
  301–2, 310–11, 378
Eoconodon, 186, *186*, 187, 190, 223–24
Eomanis, 202–3
Eozostrodon, 73
Epicyon, 294–95, 308
epipubis bones, 189–90, 435*n*
Equus, 310–11
Erickson, Greg, 434*n*
Eritherium, 250, 251, 252, 443n
Erwin, Douglas, 415*n*
Estes, Richard, 416*n*
Ethiopia, 374–78
Euarchontoglires, 209, 437*n*
Eurohippus, *202*, 202–3, 205, 217,
  436–37*n*
European Research Council, 211
eurypterids, 3
eutherians, 155, 163–65, 174–75,
  184–90, 429–30*n*, 431*n*, 435*n*

Evans, David, 459*n*
extinctions (extinction events), xix,
  xxii, 15, 469*n*. *See also* end-
  Cretaceous extinction event
  end-Permian extinction event, 43–50,
    54–55, 63–64, 86, 174, 415–16*n*
  end-Triassic extinction, 85–87, 422*n*
  megafauna and *Homo sapiens*, 393–97,
    467*n*
  Near Time Extinction, 393–96
  "Sixth Extinction," 407, 469–70*n*
extreme mammals, xx, 240–83, 443*n*.
  See also specific species

Falklands, 395
family tree, *ix*, xviii, 205–12, 437–38*n*.
  See also DNA
  convergent evolution, 206–9, 234,
    248, 268–69, 431*n*
Famoso, Nicholas, 450*n*
Farmer, Colleen, 418*n*
Farm Ridge Moraine, 455*n*
Fayum Oasis, 266
feathers, 83, 87, 91, 136, 258
ferns, 2, 40, 45, 136, 185, 198, 199
Feulner, Georg, 411*n*
fibrolamellar bone, 35–36, 59, 418*n*
Figueirido, Borja, 451*n*, 452*n*, 458*n*,
  459*n*
Florence fossils, 11
Flores Hobbit, 387, 393, 464*n*
Flynn, John, 151, 443*n*
Foffa, Davide, 424*n*
Foley, Nicole, 437*n*
Fort Beaufort, 24
Fortelius, Mikael, 451*n*
Fort Union Formation, 170, 173–77,
  433*n*
fossils (fossil records), 5–6. See also
  specific scientists and fossils